AF776811

# Sustainable Landscape Planning and Natural Resources Management

## IEREK Interdisciplinary Series for Sustainable Development


**Series Editor**

Mourad Amer, International Experts for Research Enrichment and Knowledge Exchange, Alexandria, Egypt

**Editorial Board**

Fabio Pollice, Monastero degli Olivetani, Università del Salento, Lecce, Italy

Amos Darko, Department of Building and Real Estate, Hong Kong Polytechnic University, Kowloon, Hong Kong

Muhamad Uznir Ujang, Department of Built Environment and Surveying, Universiti Teknologi Malaysia, Johor Bahru, Johor, Malaysia

Jesús Rodrigo-Comino, Department of Regional Geographic Analysis, Universitario de Cartuja, Granada, Spain

Mohamed El Kaftangui, Abu Dhabi, United Arab Emirates

Alessandra Battisti, PDTA, Sapienza University of Rome, Rome, Italy

Aiman Albatayneh, Amman, Jordan

Veysel Turan, Bingol University, Bingöl, Türkiye

Domenico M. Doronzo, Istituto Nazionale di Geofisica, Napoli, Italy

Alaa M. Morsy, Arab Academy for Science, Technology and Maritime Transport, Alexandria, Egypt

Moustafa Yehia, c/o ESGAR, Wien, Austria

Elisabetta Di Stefano, Dipartimento Scienze Umanistiche, University of Palermo, Palermo, Italy

Gasim Hayder Ahmed Salih, Department of Civil Engineering, Universiti Tenaga Nasional (UNITEN), Kajang, Selangor, Malaysia

Mina Michel, Galala University, Suez, Egypt

Vinita Vishwakarma, Galgotias University, Greater Noida, India

Ashraf Mortada, Ain Shams University, Cairo, Egypt

Alkan Mehmet, Yıldız Technical University, Istanbul, Türkiye

Mahesh Kumar Jat, Malaviya National Institute of Technology, Jaipur, India

Paola Gallo, Dipartimento DIDA, Università degli Studi di Firenze, Florence, Italy

M. M. El Aref, Cairo University, Giza, Egypt

Zakaria Hamimi, Geology Department, Faculty of Science, Benha University, Benha, Egypt

Ahmed Kalid Elewa, Helwan University, Cairo, Egypt

Ferdinando Trapani, University of Palermo, Palermo, Italy

Francesco Alberti, Department of Architecture (DIDA), University of Florence, Florence, Italy

Ibrahim Maarouf, Alexandria University, Alexandria, Egypt

Akram M. Soliman, Arab Academy for Science, Technology and Maritime Transport, Alexandria, Egypt

Sustainable Landscape Planning and Natural Resources Management (SLNR) is a peer-reviewed book series that presents groundbreaking research through the intersection of two main areas that of which are 'Sustainable Landscape Planning' and 'Natural Resources Management'. The series initially covers a wide range of academic disciplines in sustainability, including landscape ecology, green urbanism, urban regeneration, land restoration and renaturation, energy management, and nature conservation. Moreover, it also incorporates the latest developments and research in engineering topics that contribute to more sustainable cities, such as Geosciences, infrastructure, civil engineering, geology, mining, water, and natural resources conservation and management.

By showcasing methods for discovering, managing, and employing natural resources, the series aims to provide approaches that enable the responsible use of these resources. Ultimately, the SLNR series is committed to advancing the United Nations' Sustainable Development Goals for 2030.

The wide range of topics covered in this series are primarily related to sustainable development, involving durable infrastructure projects, civil engineering, urban regeneration, the built, sustainability, climate change mitigation, energy management, geosciences, geology, and the future of cities, all of which depend on the availability of natural resources and effective wealth management. The series also features conceptual and empirical contributions from various interrelated fields of earth science, with an emphasis on the era of new discovery. It focuses on providing practical solutions to ensure sustainable development and provides new case studies that offer concrete examples of how to manage natural resources effectively.

The series encompasses the best research papers from various international conferences organized by IEREK conferences, also accepts proposals from other international conferences, including interdisciplinary contributions. It exhibits innovative approaches and highlights how they can best support both economic and sustainable development through the responsible use of natural resources and global, local, and individual activities for the welfare of all societies on Earth.

The SLNR series is a valuable source for experts in government, international organizations, consultancies, industry, business sectors, and academics. Published in cooperation with IEREK, the Springer SLNR series displays important new research in various sectors, delivering valuable insights and practical solutions to the complex challenges of sustainable development for the readers.

Ashootosh Mandpe · Maulin P. Shah
Editors

# Wastewater Sludge Nexus

## A Comprehensive Exploration of Sludge Management and Applications

Springer

*Editors*
Ashootosh Mandpe
Department of Civil Engineering
Indian Institute of Technology Indore
Indore, Madhya Pradesh, India

Maulin P. Shah
Microbiology Lab
GIDC
Ankleshwar, Gujarat, India

ISSN 2948-1910 ISSN 2948-1929 (electronic)
Sustainable Landscape Planning and Natural Resources Management
ISBN 978-3-031-45757-9 ISBN 978-3-031-45758-6 (eBook)
https://doi.org/10.1007/978-3-031-45758-6

© The Editor(s) (if applicable) and The Author(s), under exclusive license to Springer Nature Switzerland AG 2025

This work is subject to copyright. All rights are solely and exclusively licensed by the Publisher, whether the whole or part of the material is concerned, specifically the rights of translation, reprinting, reuse of illustrations, recitation, broadcasting, reproduction on microfilms or in any other physical way, and transmission or information storage and retrieval, electronic adaptation, computer software, or by similar or dissimilar methodology now known or hereafter developed.
The use of general descriptive names, registered names, trademarks, service marks, etc. in this publication does not imply, even in the absence of a specific statement, that such names are exempt from the relevant protective laws and regulations and therefore free for general use.
The publisher, the authors and the editors are safe to assume that the advice and information in this book are believed to be true and accurate at the date of publication. Neither the publisher nor the authors or the editors give a warranty, expressed or implied, with respect to the material contained herein or for any errors or omissions that may have been made. The publisher remains neutral with regard to jurisdictional claims in published maps and institutional affiliations.

This Springer imprint is published by the registered company Springer Nature Switzerland AG
The registered company address is: Gewerbestrasse 11, 6330 Cham, Switzerland

If disposing of this product, please recycle the paper.

# Contents

**Principles of Wastewater Treatment and Sludge Generation** .................... 1
Debabrata Panda, Abhishek Kumar, and Krunal M. Gangawane

**Optimizing Sludge Dewatering Techniques** .................................... 13
Abhishek Kumar, Hara Mohan Jena, and Krunal M. Gangawane

**Innovative Approaches to Sludge Treatment and Dewatering** ................... 25
Mansi Jyani, Abhimanyu Bishnoi, and Urmila Gupta Phutela

**Managing Sludge from Textile and Leather Industries** ......................... 45
Ankita Sharma, Ankita Kumari, and Sudesh Kumar

**Textile and Leather Industry Residues: Managing Complex Sludge Challenges** ............................................................. 59
Shelley Oberoi

**Effective Sludge Management in the Petroleum and Chemical Sector** ............ 69
Abhishek Kumar, Debabrata Panda, and Krunal M. Gangawane

**Streamlining Sludge Management in Petroleum and Chemical Industries** ........ 79
Sunita Saklani and Nitesh Kumar

**Sludge Management in Petroleum and Chemical Industry** ...................... 95
Debjani Bhakta, Swapnadipa Patra, and Subhasis Roy

**Clay-Polymer Nanocomposites for Advanced Wastewater Sludge Management** ............................................................ 109
Baiju John

**Toxicity and Ecotoxicology of Sludge** ......................................... 123
Prasann Kumar

**Assessing Ecological and Human Health Risks Associated with Sludge** .......... 143
Vipin Nautiyal, Tanushree Prasad, and Sweta Kumari

**Understanding the Environmental and Health Implications of Sludge** ........... 157
Kundan Samal, Satya Ranjan Samal, and Rajesh Roshan Dash

**Unveiling Environmental and Health Risks Linked to Sludge** ................... 169
Nitesh Kumar, Stuti Rawat, Neha Thakur, Pratibha, Ruchika Devi, Arti Jamwal Sharma, and Sanjeev Kumar

**Sludge Management Strategies for Developing Nations** ......................... 187
Humira Assad and Ashish Kumar

**Design, Estimation and Implementation Strategy of Wastewater Treatment Plant—A Case Study at Centurion University of Technology and Management, Odisha** ............................................... 197
Prafulla Kumar Panda, Rajib Kumar Majhi, and Laxmidhar Behera

# Principles of Wastewater Treatment and Sludge Generation

Debabrata Panda, Abhishek Kumar, and Krunal M. Gangawane

**Abstract**

Sewage sludge is derived from the process of treating wastewater at wastewater management units. The sludge comprises two fundamental arrangements, namely primary and secondary sludge, which can be alternatively referred as actuated sludge in the context. Municipal sewage sludge (MSS) sometimes referred to as sewage is a residual substance that may exist in solid, semi-solid, or liquid form, characterised by its muddy consistency. The substance in question can be described as a semi-solid or liquid composition with a murky consistency. Sewage is a composite of both residential and industrial waste materials, mostly consisting of water more than 99%. Sludge refers to a partially solidified mixture, often known as slurry, which may be generated as sewage sludge via the many procedures involved in wastewater treatment. The sludge is comprised of several detrimental substances, polychlorinated biphenyls, organochlorine insecticides, chlorine by-products that have been absorbed and extracted, polycyclic aromatic hydrocarbons, phenols and its by-products, phthalate, and other similar compounds. Sewage organisation refers to the systematic removal of pollutants from wastewater resources, with a primary focus on eliminating impurities present in domestic sewage. The treatment of wastewater involves a combination of physical, chemical, and biological methods to eliminate the pollutants and provide preserved wastewater that is ecologically safe. The treatment process is comprised of three distinct steps, namely pre-treatment, main therapy, and secondary treatment. During the pre-treatment phase, the process of screening is used to eliminate sizable solids and grit. During the first stage of treatment, the water is allowed to undergo a period of sedimentation, whereby solid particles settle at the bottom, while oil and grease substances separate and accumulate on the water's surface. During the secondary treatment phase, the sludge undergoes further treatment processes inside sludge digesters.

**Keywords**

Wastewater treatment · Sludge treatment · Sewage sludge · Municipal sewage sludge · Sedimentation

## 1 Introduction

The process of wastewater treatment (WWT) involves the exclusion of various contaminants from wastewater, which encompasses several physical, chemical, and biological mechanisms before settled back to the natural atmosphere (Dentel and Dursun 2009; Bharathiraja et al. 2014). Sludge is a residual substance generated as a result of the progressions intricate in the treatment of water and wastewater. Typically, sewage treatment generates a residual substance known as sewage sludge, which is a semi-solid waste or slurry. In prior to its appropriate disposal or land application, this sewage sludge needs additional treatment procedures (Demirbas et al. 2016).

Sewage sludge is derived from the process of treating wastewater at sewage management facilities. The sludge is comprised of two fundamental types, namely primary sludge and secondary sludge, which is alternatively referred as the activated sludge development (Demirbaş 2003). The term

D. Panda (✉) · A. Kumar · K. M. Gangawane
Department of Chemical Engineering, National Institute of Technology Rourkela, Rourkela, Odisha, India
e-mail: debabratapanda89@gmail.com

K. M. Gangawane
Department of Chemical Engineering, Indian Institute of Technology Jodhpur, Jodhpur, Rajasthan, India

© The Author(s), under exclusive license to Springer Nature Switzerland AG 2025

A. Mandpe and M. P. Shah (eds.), *Wastewater Sludge Nexus*, Sustainable Landscape Planning and Natural Resources Management, https://doi.org/10.1007/978-3-031-45758-6_1

"wastewater bio solids" is occasionally used to describe the sludge resulting from biological treatment processes. Wastewater sludge comprises a diverse array of biological- and mineral-based constituents. The production of MSS has experienced a proportional rise in conjunction with the rapid process of industrialisation. The composition of sludge primarily consists of water, which is extracted from liquid wastewaters that include a lower concentration of solid substances (Demirbas et al. 2017b). The principal sludge comprises non-dissolved solids that have precipitated during the preliminary pre-treatment process with the clarifiers. The tributary or auxiliary sludge obtained through the process of separation in the secondary cleaners comprises the sewage sludge that has undergone purification in the secondary treatment bioreactors (Demirbas et al. 2017a).

The organisation of sludge derived from waste management units is multifaceted and expensive endeavour that, if executed inadequately, has the potential to undermine the anticipated environmental and sanitary benefits associated with treatment units. The significance of this practice was recognised by Agenda 21, a document that addressed the topic of ecologically sound supervision of solid waste and issues pertaining to sewage (Abad et al. 2005). It outlined the guidelines for their implementation: minimising waste production, maximising opportunities for reuse and recycling, and promoting environmentally sound methods for management and removal. The growing expectations of civilisation and ecological organisations for enhanced environmental quality standards have been evident in the actions taken by both public and commercial sanitation service administrators (Barwal and Chaudhary 2014). Given the prevailing poor catalogues of wastewater management in several emerging nations, it is reasonable to anticipate a future rise in wastewater management units. Therefore, it is anticipated that there would be a corresponding rise of the magnitude of sludge generation. In licencing processes of these countries, certain environmental agencies now mandate the inclusion of a technical specification pertaining to the ultimate disposal of sludge. These factors indicate that the management of solids is becoming an increasingly significant issue in numerous nations, with a projected escalation in the coming years due to the implementation of additional wastewater treatment facilities. The term "sludge" is commonly employed to refer to the solid residues generated during the progression of wastewater management. Within the realm of biological treatment operations, a portion of the organic waste undergoes absorption and transformation into microbial biomass, commonly referred to as biological or secondary sludge (Li and Fang 2007; Demirbas et al. 2016). The substance in question mostly consists of biological solids, so it is commonly referred to as a bio solid. The application of this terminology necessitates the compatibility of the chemical and biological properties of the sludge with its potential use, such as in agricultural practices. The word "bio solids" is employed to underscore its advantageous qualities, hence enhancing the significance of its productive applications as opposed to its mere non-productive disposal through landfills or cremation. The appropriate ultimate disposal of biosolids is crucial determinant of efficacy of the hygiene system. However, this particular endeavour can be disregarded in numerous emerging nations (Demirbaş 2003; Weiland 2010). While designing of WWT plants, the aspect of sludge organisation is often overlooked, resulting in the execution of this intricate task without adequate pre-planning by plant operators, sometimes in response to urgent circumstances. As a consequence, suboptimal methods of ultimate disposal have been used, resulting in a significant diminishment of the advantages achieved by the sewerage systems. The management of sludge, despite its relatively small proportion of 1–2% in treated wastewater volume, is a very intricate process that incurs significant expenditures typically accounting for 20–60% of overall operational expenses of treatment facility. In addition to its significant economic significance, the ultimate disposal of sludge is a multifaceted undertaking, often occurring beyond the confines of the treatment facility (Szabó et al. 2009; Shahbaz et al. 2021).

Sewage is comprised of a predominantly aqueous solution, consisting of over 99% water, which includes the combination of both domestic and industrialised waste materials. It also generated from various types of facilities including domestic, established institutes, viable, and trade entities. Sewage management refers to the systematic procedure employed to eliminate impurities from wastewater, with a primary focus on the treatment of domestic sewage. The treatment process encompasses several physical, chemical, and biological mechanisms aimed at eliminating these pollutants and generating treated wastewater that is environmentally benign (Kim et al. 2003; Demirbas et al. 2017b).

In terms of nomenclature, the distinction between sewage and sludge lies in their respective definitions. Sludge refers to a broad category encompassing solid substances that have been separated from a liquid suspension. Sludge refers to a partially semi-solid slurry that can be generated either as manure through the management of wastewater, or precipitated suspension acquired from unadventurous treatment and other industrial procedures (Abad et al. 2005). The magnitude of sludge production must be directed to the sludge processing units, can be quantified in terms of mass (grams of total solids per day, on a dry basis) and volume (cubic metres of sludge per day, on a wet basis). The approach for mass and volume computations is outlined in Sect. 2. In this analysis, a simplified methodology is employed to quantify sludge output, taking into account both per capita and

**Table 1** Features and magnitudes of sludge production in different wastewater management units

| Wastewater treatment system | Characteristics of the sludge production | | | |
|---|---|---|---|---|
| | kgSS/kgCOD applied | Dry solids content (%) | Mass of sludge (gSS/inhabitant) | Volume of sludge (L/inhabitant) |
| Primary treatment (conventional) | 0.35–0.45 | 2–6 | 35–45 | 0.6–2.2 |
| Primary treatment (septic tanks) | 0.20–0.30 | 3–6 | 20–30 | 0.3–1.0 |
| Facultative pond | 0.12–0.32 | 5–15 | 12–32 | 0.1–0.25 |
| Anaerobic pond | 0.26–0.55 | 5–7 | 26–55 | 0.15–0.45 |
| Complete mix aerated (sedimentated pond) | 0.11–0.13 | 5–8 | 11–13 | 0.15–0.25 |
| Septic tank + anaerobic filter | 0.27–0.39 | 1.4–5.4 | 27–39 | 0.5–2.8 |
| Conventional activated sludge | 0.60–0.80 | 1–2 | 60–80 | 3.1–8.2 |
| Activated sludge | 0.50–0.55 | 0.8–1.2 | 40–45 | 3.3–5.6 |
| High rate trickling filter | 0.55–0.75 | 1.5–4.0 | 55–75 | 1.4–5.2 |
| Submerged aerated biofilter | 0.60–0.80 | 1–2 | 60–80 | 3.1–8.2 |
| UASB + aerobic post treatment | 0.20–0.32 | 3–4 | 20–32 | 0.5–1.1 |

chemical oxygen demand (COD) factors. In the context of biological wastewater treatment, a portion of the chemical oxygen demand (COD) that is eliminated undergoes conversion into biomass, hence constituting the biological sludge. Table 1 illustrates the amount of SS that are discarded per unit of chemical oxygen demand (COD), taken into account for the customary efficiencies of COD removal in various wastewater management procedures, for the purpose of simplification. According to the data presented in Table 1, it can be observed that stabilisation ponds generate a relatively lesser volume of sludge compared to typical activated sludge systems, which produce the biggest amount of sludge that needs to be treated. The rationale behind this phenomenon is that the sludge generated in the ponds has a prolonged storage period, during which it undergoes digestion, resulting in the conversion of its constituents into water and gases (Andreoli et al. 2007). Additionally, this process leads to the thickening of the sludge, resulting in a significant reduction in its overall volume. In contrast, the typical activated sludge technique does not facilitate sludge digestion in the ventilation cistern due to insufficient time.

## 2 Sludge Characteristics at Each Treatment Stage

In contemporary discourse, the terminology "sewage treatment plant" has frequently been substituted with the more commonly used word "wastewater treatment plant." The process of sewage treatment can be categorised into three fundamental stages: preliminary treatment (also known as pre-treatment), primary treatment, and secondary treatment (Kartal et al. 2010).

### 2.1 Pre-treatment

Figure 1 depicts a rudimentary schematic representation of the process involved in sewage treatment. During the pre-treatment stage, the process of screening is employed to eliminate coarse solids, which are defined as solids with a diameter exceeding 2 cm, as well as grit, which refers to heavy solids. The exclusion of coarse debris from biosolids is evident. The initial stage of treatment may involve the utilisation of a porous sandstorms or porous grits, where the velocity inlet sewage is regulated to facilitate the sedimentation. The removal of the abrasive materials is necessary due to their potential to cause harm to pumps and other equipment (Logan and Regan 2006). In the process of wastewater treatment, the effluent undergoes filtration through a bar screen in order to eliminate sizable debris within the wastewater flow. The solid materials are gathered and subsequently either deposited in a landfill or subjected to incineration. Different sizes of bar screens or mesh screens can be employed to enhance the efficiency of solids removal (Logan and Regan 2006).

### 2.2 Primary Treatment

The removal of grit (small, solid particles) occurs in two different stages during primary treatment: pre-aeration and sedimentation. The removal of stable and variable resources is followed by the potential discharge or secondary treatment of the residual liquid. The floating debris may occasionally yield grease and oil that can be utilised for the purposes of saponification or biodiesel generation. The aeration of the wastewater is achieved by introducing air into the system

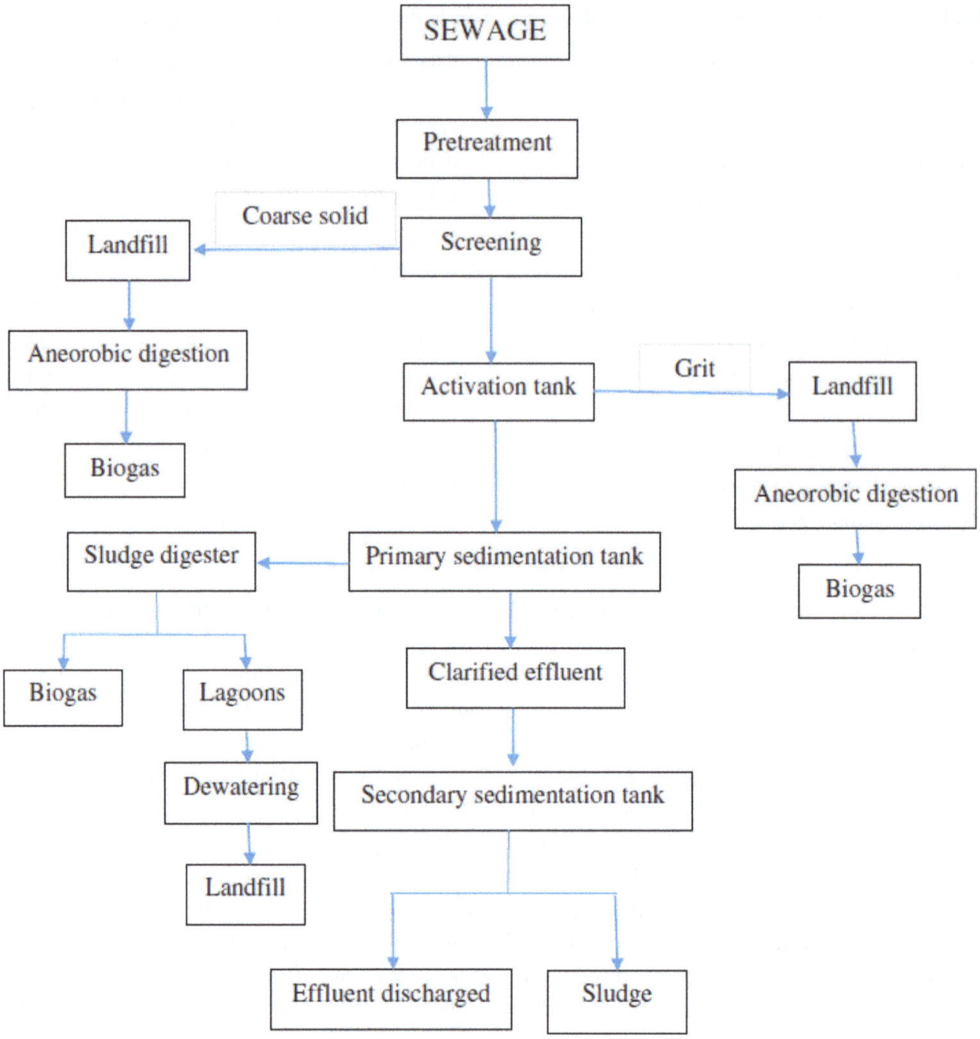


**Fig. 1** Flow diagram of sewage treatment

through perforated pipes positioned in close proximity to the tank floor.

The process of aeration reduces the density of the wastewater, resulting in the settling of grit particles. The strategic positioning of the air jets within the tanks induces a swirling motion of the water, so impeding the settling of suspended materials. The presence of oxygen in the air facilitates the availability of dissolved oxygen for bacterial utilisation at a later stage in the process. However, the duration of the wastewater's exposure to these tanks is insufficient to allow for significant bacterial activity to take place. The solid particles are extracted from the lower regions of the tanks through the utilisation of scrapers, while the scum is effectively eliminated through the application of water jets. The particulate matter and solid substances are transported to a shared gathering location for amalgamation to create sludge and subsequently dispatched for secondary treatment. The tanks utilised for the purpose of sedimentation, wherein sludge is allowed to settle, while grease and oils undergo flotation and are subsequently removed from the surface through skimming. The solid materials are removed from the bottom surface, while the residue is eliminated by the use of water jets. The amalgamation of these two chemicals results in the formation of sludge. The principal method of wastewater treatment entails the utilisation of gravity sedimentation to separate the filtered wastewater into settled solids. The resulting by-product of this procedure is a highly concentrated mixture, referred to as primary sludge, which undergoes additional treatment in order to be transformed into a substance known as biosolid.

## 2.3 Secondary Treatment

During the initial phase of sedimentation, sewages are sent through sizable tanks, pre-settling basins, sedimentation rocks, clarifiers. Tanks are employed for the purpose of sedimentation, and settle down the sludge, while oils ascend at the top are subsequently removed through skimming. The solid substances are removed from the lower surface, while the residue is eliminated by the use of water spray.

The principal method of wastewater management entails the utilisation of gravity sedimentation to separate the filtered wastewater into settled solids. A portion of the leftover stream is directed through preliminary operation, resulting in its suspension. The resulting by-product of this procedure is a highly concentrated mixture, referred to as primary sludge, which undergoes additional treatment in order to transform into a substance known as biosolid.

## 2.4 Tertiary Treatment

In instances where a superior standard of waste is necessary, such as for direct release into a drinking water supply, further processing methods are employed. The solid residue obtained during tertiary treatment primarily comprises chemicals that were introduced into the final recovered raw material prior to disposal, and as such, it is not considered as part of the biosolids.

The predominant components utilised in the manufacturing of municipal biosolids are a combination of primer and secondary effectors. Effective management throughout the entirety of the treatment process enables plant managers to exercise control over solid constituents, nutrients, and various other elements included in biosolids. The primary objective of tertiary treatment is to offer a conclusive step of treatment that enhances the quality of effluent prior to its release into the receiving environment, such as seas, rivers, lakes, wetlands, and groundwater. Multiple tertiary treatment processes may be employed at treatment plants. Disinfection is typically implemented as the concluding stage in a series of processes.

## 2.5 Final Treatment

Micropollutants, which encompass pharmaceuticals, household chemical constituents, chemicals utilised in small-scale businesses or industries, environmental persistent pharmaceutical pollutants (EPPP), and pesticides, may persist in water even after undergoing conventional treatment processes (including primary, secondary, and tertiary treatment), thereby contributing to water pollution. Micropollutants are currently removed through a fourth stage of treatment in sewage treatment processes. However, due to the high costs associated with these procedures, their routine implementation has not yet been widely adopted.

Sludge undergoes treatment through a variety of techniques that can be employed in diverse combinations. Typically, sewage sludge undergoes treatment through a series of processes, which may include lime stabilisation, thickening, dewatering, drying, anaerobic digestion, or composting. The solid components included in the muck possess valuable nutrients for plant growth as fertiliser. The elimination of the necessity for filtered sewage sludge is not yet feasible by any known process. Toxic compounds discharged into the sewer system as a result of industrial activities become contaminated by various sediments originating from industrial or commercial zones. Elevated levels of these compounds have the potential to enhance the agricultural viability of the mud, necessitating further disposal methods such as incineration or landfilling.

The disposal of sludge, which is the primary by-product generated by wastewater treatment plants, poses a significant environmental challenge within the context of wastewater treatment procedures. Prior to disposal, sludge must undergo a requisite level of treatment. The appropriate course of treatment is contingent upon the suggested mode of disposal (Harhangi et al. 2012). There exist three primary methods for the ultimate disposal of wastewater sludge and its components. According to (Ødegaard et al. 2002), sludge and its many components have the potential to be put on land, sea, or air. The activated sludge technique is a commonly employed method for the management of sewage and industrial wastewater. It involves the utilisation of a biological aggregate consisting of air, bacteria, and protozoa. The process of activated sludge involves the transformation of organic molecules within microorganisms into $CO_2$, water, and various inorganic chemicals (Demirbas et al. 2016).

The composting process entails the aerobic decomposition of organic materials. The primary aim of sludge composting is to achieve biological stabilisation of putrescible organic matter, eliminate pathogenic microorganisms, and decrease the overall waste volume (Bharathiraja et al. 2014). The process of composting involves the biological decomposition of organic material, leading to a decrease of approximately 20–30% in volatile solids. During the process of composting, aerobic bacteria facilitate the conversion of a significant portion of the organic matter into carbon dioxide, resulting in the production of a rather stable material that is free from unpleasant odours (Karvelas et al. 2003). The resultant final artefact exhibits stability and can be utilised in soil amendment in agricultural contexts. According to (Demirbas 2016), aerobic composting is a more prevalent method compared to anaerobic composting. One notable benefit of compost lies in its efficacy as a highly effective fertiliser; nonetheless, its utilisation remains relatively limited.

The process of anaerobic digestion entails the decline of organic particulate in sludge by biological activity, specifically in the absence of oxygen (Barwal and Chaudhary 2014). The process of anaerobic digestion encompasses two concurrent stages to decomposition of sludge. The initial step involves the hydrolysis of organic molecules with high molecular weight, followed by the transformation of organic acids by microorganism. The subsequent phase involves the conversion of organic acids into biogas by similar microorganisms (Demirbas 2016). The process of aerobic digestion for sludge management has similarities to the treatment method known as activated sludge. The primary purpose of aerobic digestion is to achieve the stabilisation of slurries by means of extended aeration, resulting in a decrease in biological oxygen demand (BOD) and the removal of volatile solids. The process of prolonged oxidation of cellular material in open tanks is a significant aspect of the degradation of biodegradable materials by microorganisms. During this temporal period, the quantity of biological material undergoes a reduction of around 50% from its initial quantity (Demirbas 2016).

The presence of elevated levels of sulphur compounds in sludge might give rise to operational challenges in the implementation of anaerobic digestion processes(Fatih Demirbas 2006; Demirbas 2016). According to researchers, sulphide has the potential to impede the process of methane creation. Additionally, it has the ability to generate hydrogen sulphide gas, which possesses poisonous and corrosive properties. In order to achieve this objective, it is possible to extract and transform sulphur compounds present in the sludge into elemental sulphur. Two distinct anaerobic methods are employed for this objective. Initially, organic molecules containing sulphur undergo a conversion process to yield sulphur compounds. The sulphide compounds present in the mixture are removed by a process known as stripping, which involves the use of a stripping gas such as carbon dioxide. Subsequently, these compounds are transformed into elemental sulphur by utilising a basic adsorbent.

## 3 Fundamental Relationship in Sludge Management

To gain a comprehensive understanding of the characteristics of sludge and its generation in relation to quantity, there must be knowledge of approximately key affiliations. The previously indicated significant components delivered as:

- The correlation among magnitudes of solids and liquid present in a given system.
- The quantification of the deliberation of dry solids inside the system.
- The relationship among flow rate, concentration, and load.

The present section encompasses supplementary topics, namely the measurement and analysis of total, volatile, and fixed solids, solids capture, destruction of volatile solids as well as the assessment of sludge density.

### 3.1 TS, VS, and Fixed Solids

Sludge consists of a combination of solid particles and water. The concept of "total solids" (TS) can be further classified into two distinct categories: suspended solids (SS) and dissolved solids (DS). The majority of solid components in sludge consist of particles that are suspended. SS and DS can be further classified into two important categories of inorganic or fixed solids (FS) and organic or volatile solids (VS). The ratio between VS and TS is a dependable indicator for estimating the proportion of organic matter present in sludge solids and the extent to which it has undergone digestion. The VS/TS ratio in undigested sludges generally lies in between 0.75 and 0.80. On the other hand, it is worth noting that digested sludges exhibit a more limited spectrum, with a VS/TS ratio ranging from 0.60 to 0.65. Table 2 presents the customary ranges of volatile solids to total solids ratios (VS/TS) for sewages obtained from diverse wastewater management units. In this particular segment of the literature, the phrases DS, TS, and SS are employed alternatively for the persistence of assessing the quantity of solids present in the sludge treatment process. It is widely recognised that a considerable proportion of the overall solids present in the sludge comprises SS only (Andreoli et al. 2007).

### 3.2 Density and Specific Gravity of the Sludge

According to Crites and Tchobanoglous (2000), the specific gravity (S) of the FS elements is roughly 2.3–2.5, while for VS, the specific gravity is approximately 1.0. The value assigned to water is 1.0.

The estimation of the specific gravity of SS can be conducted according to the methods proposed by:

$$S = \frac{1}{\left(\frac{(FS/TS)}{2.5} + \frac{(VS/TS)}{1.0}\right)} \quad (1)$$

### 3.3 Destruction of Volatile Solids

The digestion process aids in the elimination of digestible organic matter out of sewage. Hence, it can be deduced that the volatile solids (VS) were either eliminated or decomposed. The quantity of FS remains consistent. The average

**Table 2** Characteristics of different sludge kinds, focusing on their density, specific gravity, VS/TS ratio, and percentage of dry solids

| Types of sludge | VS/ST ratio | % dry solids | Specific gravity of solids | Specific gravity of sludge | Density of sludge (kg/m$^3$) |
|---|---|---|---|---|---|
| Primary sludge | 0.75–0.80 | 2–6 | 1.14–1.18 | 1.003–1.01 | 1003–1010 |
| Secondary anaerobic sludge | 0.55–0.60 | 3-6 | 1.32–1.37 | 1.01–1.02 | 1010–1020 |
| Stabilisation pond sludge | 0.35–0.55 | 5–20 | 1.37–1.64 | 1.02–1.07 | 1020–1070 |
| Primary thickened sludge | 0.75–0.80 | 4–8 | 1.14–1.18 | 1.006–1.01 | 1006–1010 |
| Secondary thickened sludge | 0.75–0.80 | 2–7 | 1.14–1.18 | 1.003–1.01 | 1003–1010 |
| Thickened mixed sludge | 0.75–0.80 | 3–8 | 1.14–1.18 | 1.004–1.01 | 1004–1010 |
| Digested mixed sludge | 0.60–0.65 | 3–6 | 1.27–1.32 | 1.007–1.02 | 1007–1020 |
| Dewatered sludge | 0.60–0.65 | 20–40 | 1.27–1.32 | 1.05–1.1 | 1050–1100 |

**Table 3** Assortments of SS capture in sewage treatment

| Types of sludge | Thickening | | Digestion | | Dewatering | |
|---|---|---|---|---|---|---|
| | Procedure | Capture (%) | Procedure | Capture (%) | Procedure | Capture (%) |
| Primary sludge | Gravity | 85–92 | Secondary digester | 95 | Drying bed | 90–98 |
| | | | | | Filter press | 90–98 |
| | | | | | Centrifuge | 90–95 |
| | | | | | Belt press | 90–95 |
| Secondary sludge | Gravity | 75–85 | Secondary digester | 90–95 | Drying bed | 90–98 |
| | Flotation | 80–95 | | | Filter press | 90–98 |
| | Centrifuge | 80–95 | | | Centrifuge | 90–95 |
| | | | | | Belt press | 90–95 |
| Mixed sludge | Gravity | 80–90 | Secondary digester | 90–95 | Drying bed | 90–98 |
| | Centrifuge | 85–95 | | | Filter press | 90–98 |
| | | | | | Centrifuge | 90–95 |
| | | | | | Belt press | 90–95 |

efficiencies of VS removal during the digesting process are as follows:

$$E = 0.40 \text{ to } 0.55 \text{ (40 to 55\%)} \quad (2)$$

The calculation of the solids load (kg/d) prior to and subsequent to incorporation determined using the following equation:

$$TS_{\text{influent}} = VS_{\text{influent}} + FS_{\text{influent}} \quad (3)$$

$$TS_{\text{effluent}} = (1 - E)VS_{\text{influent}} + FS_{\text{influent}} \quad (4)$$

### 3.4 Solids Capture

In the various stages of sludge treatment, such as thickening and dewatering, it is crucial to acknowledge that complete separation of solids and liquids may not always be achieved, resulting in the retention of certain solids within the liquid phase as it progresses through subsequent stages of the sludge treatment process. A fraction of these solid substances remains present in the leftovers, drainage contaminants, and filtrates of the separating device. Due to the presence of residual solids, specifically particulate BOD, it is necessary to redirect these flows back to the initial stage of the treatment process. This allows for their integration with the plant influent and subsequent exposure to further treatment procedures (Andreoli et al. 2007).

The process of introducing solid materials into sludge is commonly referred to as solids capture, or alternatively, solids recovery. The efficiency of incorporating particles into the sludge intended for further processing stages is commonly represented as a percentage (%). Hence, the quantities of solid loads (expressed in kilogrammes of suspended solids per day) are as follows (Table 3):

$$\text{Effluent SS load in sludge} = \text{solids capture} \times \text{influent SS load} \quad (5)$$

$$\text{SS load in drained liquid} = (1 - \text{solids capture}) \times \text{influent SS load} \quad (6)$$

## 4 Quantification of Sludge Generation

### 4.1 Preliminary Sludge Generation

The amount of sludge produced during the initial treatment process, referred to as primary sludge, is dependent on the effectiveness of removing SS in the initial clarifiers. The notion of efficiency can also be construed as the capacity to capture solid substances. The average removal effectiveness of SS through capture in initial clarifiers is:

$$E = 0.60 \text{ to } 0.65 \quad (7)$$

### 4.2 Secondary Sludge Generation

The estimation of secondary sludge formation is conducted by taking into account the kinetic and stoichiometric coefficients of the specific biological wastewater management process employed. The sludge produced consists of the following fractions:

- Biological solids refer to the solid materials that are generated inside a system due to the process of organic matter clearance.
- The inert solids found in raw sewage refer to non-biodegradable solid materials that have collected within the sewage system.

The net development of suspended biological compounds is determined by the difference between total production, also known as synthesis or anabolism, and mortality, which refers to decay or catabolism.

When estimating the quantity of biological sludge that requires treatment, it is common practise to subtract a fraction from the overall amount generated. The proportion in question represents the quantity of solids that are inadvertently discharged with the final effluent. This occurs because the removal efficiency of SS in the final clarifiers are inherently less than 100%. If this adjustment is implemented in the calculation, it is important to note that the quantity of solids processed is equivalent with the difference between quantity of solids generated and the quantity of solids escape from final run-off.

## 5 Energy Positive Wastewater and Sludge Management

The global rise in population and the improvement of living ideals have caused in an increased need for resources in order to sustain a clean and safe environment. The treatment of wastewater is necessary to eliminate various forms of biological, chemical, and physical impurities prior to its release into natural water sources, including rivers, creeks, lakes, and oceans. Presently, wastewater treatment technologies, particularly those involving aerated systems, exhibit a high demand for energy. Nevertheless, wastewater is seen as a valuable energy resource. If harnessed effectively, this energy has the potential to surpass the energy demands for therapy by a factor of ten (Gude 2015). The process of wastewater treatment invariably results in the generation of substantial amounts of biomass, sometimes referred to as sludge, which serves as a significant source of energy feedstock (Rulkens 2008). A comprehensive treatment plan typically consists of initial primary processes, which involve physical and chemical methods such as the separation of inorganic solids through sedimentation, as well as adjustments to aeration and pH levels. This is then followed by secondary treatment, which encompasses both chemical and biological processes. In this stage, organic substrates are degraded by microorganisms through biochemical reactions. The treatment process then proceeds with successive steps of filtration, disinfection, and aeration. The treatment regimen in question is commonly known as secondary or biological treatment, which is widely recognised as the global standard for all wastewater treatment facilities. Although wastewater treatment is known to need a significant amount of energy, there exist various alternatives that can transform this process into one that generates energy rather than consumes it.

Wastewater typically consists of approximately 60% organic compounds on a dry basis. These organic compounds are primarily biodegradable and include nutrient compounds like $N_2$ and P. Additionally, wastewater contains both inorganic and organic pollutants like silicates, aluminates, and compounds containing calcium and magnesium along with halocarbon, hydrocarbon, phenols, pesticides, and microbiological pollutants (Rulkens 2008; Tyagi and Lo 2013). The organic component comprises carbon in the range of 50–55%, oxygen in the range of 25–30%, nitrogen in the range of 10–15%, hydrogen in the range of 6–10%, phosphorus in the range of 1–3%, and sulphur in the range of 0.5–1.5%.

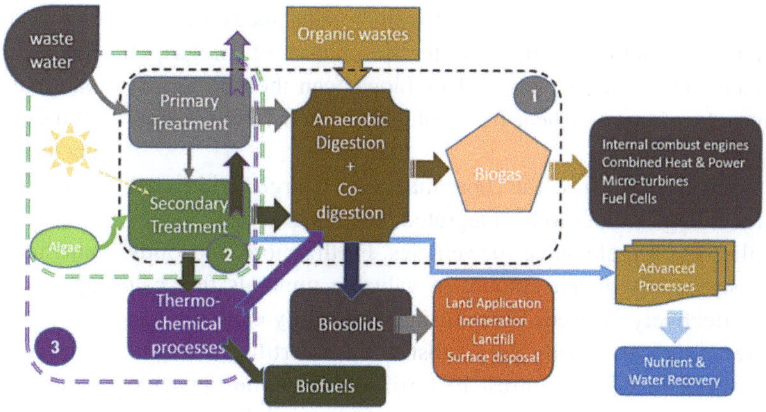


**Fig. 2** Methodologies of sludge management plants

Non-toxic organic and nutritional chemicals can be employed to cultivate the microbial biomass, facilitating the production of sludge that can subsequently undergo processing for the purpose of energy recovery. The recovery of energy from wastewater can be accomplished through three feasible designs, as illustrated in Fig. 2. One of the approaches involves the utilisation of anaerobic digestion to treat the sludge obtained from both primary and secondary treatment units to cover the energy costs related with the treatment process. Nevertheless, relying solely on this approach is sufficient to meet the energy demands of wastewater treatment, as there are other technical and systematic obstacles that exist inside these systems. The inclusion of additional organic wastes for co-digestion may be contemplated. Strategy 2 entails the utilisation of mixotrophic systems, which involve the treatment of wastewater using a combination of bacteria and algae. This approach aims to improve the efficiency of carbon utilisation, nutrient removal, and biomass generation. One potential strategy, referred to as Strategy 3, involves utilising secondary effluents obtained from wastewater treatment facilities as a means to develop algae for the purpose of biofuel production. This cultivation process would involve employing thermo-chemical techniques, either on-site or off-site.

The collective resolution of sludge management challenges in wastewater treatment facilities can be achieved through the implementation of the aforementioned three solutions. A variety of methods are currently accessible for the management of sludge, including anaerobic digestion, hydrothermal processes, pyrolysis, incineration, and gasification (Rulkens 2008; Tyagi and Lo 2013). Anaerobic digestion is a process that generates biogas to transform into other types of valuable energy, including heat, natural gas, and electricity. Anaerobic digestion (AD) is predominantly preferred over other methods due to its economic advantages and feasibility (Wiley et al. 2011). There are alternative choices that might be considered in certain circumstances where cost is not the key determining factor, but rather other concerns such as environmental and site-specific considerations hold greater significance. The process of anaerobic digestion is a biological phenomenon in which microorganisms break down organic materials in the absence of oxygen (Wiley et al. 2011). The presence of methanogens, which are microorganisms that produce methane, within anaerobic digesters facilitates the conversion of organic matter into $CO_2$, $CH_4$, and organic acids. It also exhibits reduced energy requirements and possess the ability to withstand solids characterised by elevated water content. This paper provides a detailed discussion of methods 1 and 2, focusing on the utilisation of anaerobic technologies as the primary energy recovery unit.

In order to achieve energy neutrality or positivity, wastewater treatment facilities must ascertain appropriate processing methods for energy retrieval, particularly from the sludge generated during primary and secondary treatment units, including anaerobic digestion (AD) (Gude 2015). The exclusive utilisation of anaerobic digestion for wastewater sewage does not possess the capacity to produce a positive energy impact on the environment. The present wastewater management units depend on these particular methodology which have the capability to attain energy retrieval levels of 50%. This achievement also considered to be at the upper limit and can only be accomplished by large-scale systems that have treatment capabilities beyond 10 MGD (~37,850 $m^3$/d) (Gude 2015). The inclusion of different organic wastes in the diet can also results in anaerobic digestion being energy positive. The aforementioned procedure is commonly referred to as co-digestion.

Historically, wastewater treatment facilities have been engineered to comply with regulatory requirements by eliminating various pollutants before releasing the treated water back into the ecosystem. The management of sewage sludge, an unavoidable by-product of wastewater treatment, is a multifaceted process that can contribute significantly to the overall expenses associated with wastewater treatment, ranging from 25 to 50% of the total costs (Murthy et al. 2006). Sludge possesses energy to harnessed either directly through mono or cofiring with other substances (and/or fuels), or indirectly through the formation of biogas, a sustainable form of energy. This biogas can then be turned into electricity and heat by employing cogeneration equipment.

The biogas has the potential to undergo additional processing in order to be integrated into gas grid networks or utilised as a fuel source for vehicles. Furthermore, biosolids possess a significant amount of valuable minerals that can be effectively utilised. Biosolids are frequently employed as a fertiliser, so serving as a substitute for fertilisers reliant on fossil fuels. In contrast to fertilisers, biosolids provide additional advantageous properties for soil enhancement, including carbon content, organic matter, moisture retention, and structural support to facilitate drainage. The utilisation of biosolids has the potential to significantly mitigate carbon emissions. According to a recent study conducted by ANZBP in 2012, it was found that there are offsets above 1 tonne of carbon dioxide equivalent per tonne of dry biosolids used, with the specific amount varying depending on the application (Chang et al. 2002; Pathak et al. 2009).

Nevertheless, despite the substantial expenses associated with the processing of sludge, there is a lack of focus on the influence that wastewater treatment has on the characteristics and volume of sludge, as well as the resulting biosolids. In the usual course of events, wastewater, industrial effluent, and rainwater run-off are introduced into a sewage works facility, where they undergo an initial screening process before proceeding to primary treatment. During this phase, the process of sedimentation occurs, resulting in the separation and accumulation of solids and other substances. The liquid portion obtained from this particular phase is thereafter directed to a secondary treatment stage, wherein the extraction of nutrients from the wastewater takes place. This process is conducted by a range of facultative and aerobic bacteriological processes in atmosphere oxygen, and it can be carried out in many configurations. The solid material that is discharged from the secondary treatment stage mostly consists of bacteriological matter. It is referred to by several names, such as secondary, bacteriological, waste, or surplus activated sludge (Murthy et al. 2006).

In instances where treatment levels necessitate solely the elimination of biochemical oxygen demand (BOD), it is customary for the retention period (sludge degradation time) during secondary treatment were to be roughly 6-8 days of degradation. The production yield and $O_2$ requirements resulting from a 6-day retention period, by $O_2$ utilised as an electron acceptor and for nitrogen donor ammonia have been computed. If there are further tightening of wastewater treatment standards, such as the implementation of the Urban Waste Water Treatment Directive-91/271/EEC, and nitrification becomes a necessity, there will be a significant increase in the oxygen demands. The data presented indicates $O_2$ requirement of 4.2 kg of $O_2$ per kg of ammonia oxidation, resulting in the production of 4.32 kg of $NO_3$ and the generation of 0.15 kg of sludge per kg of $NH_3$ oxidation. Due to the comparatively slower kinetics exhibited by nitrifying bacteria in comparison to heterotrophic organisms responsible for BOD removal, an increase in sludge ages becomes necessary in order to accommodate the slower growth of nitrifiers. Equation (8) was reassessed at extended sludge ages, revealing a significant escalation in ammonia release by a factor of ten after 10 days and nearly 23-fold after 20 days. Hence, it can be observed that with an increase in sludge age, there is a decrease in BOD levels and a rise in ammonia levels, both of which contribute to the promotion of nitrification kinetics.

$$\begin{aligned} & C_{10}H_{16}O_3 + 7.57\ O_2 \rightarrow 6.01\ H_2O + 5.06\ CO_2 \\ & \quad + 0.014\ HCO_3^- + 0.014\ NH_3 \\ & \quad + 0.986\ C_5H_7O_2N\ NH_3 + 1.84\ O_2 \\ & \quad + 0.09\ CO_2 + 0.023 HCO_{3^-} \rightarrow 0.953\ H_2O \\ & \quad + 0.977 NO_{3^-} + 0.023\ C_5H_7O_2N + 1.953\ H^+ \end{aligned} \tag{8}$$

Consequently, with the implementation of stricter regulations, there is a reduction in the generation of secondary sludge. If the removal of phosphorus is deemed necessary, more processing of the wastewater is then required. The process of removal is primarily dependent on the typical absorption of substances into the sludge, which is subsequently discarded, as well as the utilisation of chemicals or the promotion of improved biological absorption. The amount of phosphorous that is eliminated by typical absorption is dependent on the age of the sludge and the amount of Biochemical Oxygen Demand (BOD) that is extracted. In general, it is seen that a range of 20–30% of phosphorous can be effectively eliminated by this method. If more elimination is necessary, the utilisation of chemical precipitation may be employed. The process entails the addition of different Ca, Fe, Al salts to create insoluble phosphates, which are subsequently eliminated by the settling method.

Although the addition of precipitates does increase the volume of sludge, these precipitates are mostly inactive. Hence, in the event that chemical dosing has been implemented in a facility that already possesses sludge processing methodologies, the efficacy of the subsequent plant operations may be hindered due to the introduction of extra sludge, resulting in a reduction of its processing capacity. The introduction of chemical substances leads to a decrease in the volatile component of the sludge, resulting in a decline in the energy content of the sludge. A study conducted by (Barber 2014) examined the acceptability of digested sludges for burning. The findings revealed that the addition of ferric enhanced the ash level by 10 percentage points in comparison with unconditioned digested sludge. The results of elemental analysis indicated that the application of chemical treatment to the sludge resulted in a notable increase in the oxygen content of the volatile fraction, estimated to be around 15%. This increase is likely attributed to the precipitation of salts. Additionally, the carbon content of the sludge was seen to decrease by roughly 10% following the chemical treatment. The data analysis reveals that the decrease in calorific value, resulting from a rise in ash content, only explains half of the observed loss in energy content.

## 6 Conclusion

The implementation of more stringent wastewater standards results in an elevated ratio of secondary and chemical sludge in comparison to primary sludge. The processing of sludge is anticipated to become progressively more challenging in the coming years. The lipid and fibre content of primary sludge is higher compared to secondary sludge, whereas secondary sludge has a higher concentration of protein, nitrogen, and phosphorus. The observed disparities lead to primary sludge possessing a naturally higher energy content, a 20% increase in biogas production per kilogramme destroyed, and a lower quantity of nutrients requiring removal. On the other hand, secondary sludge exhibits reduced suitability for anaerobic digestion and dewatering compared to primary sludge. Additionally, its responsiveness diminishes as the sludge ages further. The elimination of first treatment for nutrient removal results in several adverse consequences downstream. These factors encompass an increase in the demand for aeration by more than 20%, as well as a decrease in biogas generation of around 60%. These effects arise from the utilisation of easily degradable carbon during the removal of nutrients, the loss of primary sludge, and the presence of nutrient-deprived circumstances in the digestion stage.

It is conceivable that a positive energy balance can be achieved in systems specifically designed for carbon removal or systems that utilise enhanced primary treatment. The implementation of enhanced primary treatment has been observed to result in a reduction of over 20% in downstream aeration requirements and a 25% increase in biogas production compared to standard primary treatment. The implementation of stricter wastewater rules has several negative consequences. Some of the factors that can be observed in this context are: a rise in energy consumption during the aeration process, a decrease in the production of biogas during digestion, a decline in the ability to remove water from the sludge, and the formation of cake with a greater moisture content.

## References


Abad E, Martínez K, Planas C, Palacios O, Caixach J, Rivera J (2005) Priority organic pollutant assessment of sludges for agricultural purposes. Chemosphere 61(9):1358–1369. https://doi.org/10.1016/j.chemosphere.2005.03.018

Andreoli CV, von Sperling M, Fernandes F (2007) Biological wastewater treatment series. IWA Publishing, London

Barber WPF (2014) Influence of wastewater treatment on sludge production and processing. Water Environ J 28(1):1–10. https://doi.org/10.1111/wej.12044

Barwal A, Chaudhary R (2014) To study the performance of biocarriers in moving bed biofilm reactor (MBBR) technology and kinetics of biofilm for retrofitting the existing aerobic treatment systems: a review. Rev Environ Sci Biotechnol 13(3):285–299. https://doi.org/10.1007/s11157-014-9333-7

Bharathiraja B, Yogendran D, Ranjith Kumar R, Chakravarthy M, Palani S (2014) Biofuels from sewage sludge—a review. Int J ChemTech Res 6(9):4417–4427

Chang JS, Lee KS, Lin PJ (2002) Biohydrogen production with fixed-bed bioreactors. Int J Hydrogen Energy 27(11–12):1167–1174. https://doi.org/10.1016/S0360-3199(02)00130-1

Demirbaş A (2003) Fuels recovery from municipal solid and liquid wastes (MSLW). Energy Sources 25(7):713–720. https://doi.org/10.1080/00908310390212372

Demirbas A (2016) Sulfur removal from crude oil using supercritical water. Pet Sci Technol 34(7):622–626. https://doi.org/10.1080/10916466.2016.1154871

Demirbas A, Taylan O, Kaya D (2016) Biogas production from municipal sewage sludge (MSS). Energy Sources, Part A Recover Util Environ Eff 38(20):3027–3033. https://doi.org/10.1080/15567036.2015.1124944

Demirbas A, Bamufleh HS, Edris G, Alalayah WM (2017a) Treatment of contaminated wastewater. Pet Sci Technol 35(9):883–889. https://doi.org/10.1080/10916466.2017.1290653

Demirbas A, Edris G, Alalayah WM (2017b) Sludge production from municipal wastewater treatment in sewage treatment plant. Energy Sources, Part A Recover Utili Environ Eff 39(10):999–1006. https://doi.org/10.1080/15567036.2017.1283551

Dentel SK, Dursun D (2009) Shear sensitivity of digested sludge: comparison of methods and application in conditioning and dewatering. Water Res 43(18):4617–4625. https://doi.org/10.1016/j.watres.2009.07.015

Fatih Demirbas M (2006) Technological options for producing hydrogen from renewable resources. Energy Sources, Part A Recover Util Environ Eff 28(13):1215–1223. https://doi.org/10.1080/009083190910488

Gude VG (2015) Energy and water autarky of wastewater treatment and power generation systems. Renew Sustain Energy Rev 45:52–68. https://doi.org/10.1016/j.rser.2015.01.055

Harhangi HR, Le Roy M, van Alen T, Hu B lan, Groen J, Kartal B, Tringe SG, Quan ZX, Jetten MSM, Op den Camp HJM (2012) Hydrazine synthase, a unique phylomarker with which to study the presence and biodiversity of anammox bacteria. Appl Environ Microbiol 78(3):752–758. https://doi.org/10.1128/AEM.07113-11

Kartal B, Kuenen JG, Van Loosdrecht MCM (2010) Sewage treatment with anammox. Science 328(5979):702–703. https://doi.org/10.1126/science.1185941

Karvelas M, Katsoyiannis A, Samara C (2003) Occurrence and fate of heavy metals in the wastewater treatment process. Chemosphere 53(10):1201–1210. https://doi.org/10.1016/S0045-6535(03)00591-5

Kim J, Park C, Kim TH, Lee M, Kim S, Kim SW, Lee J (2003) Effects of various pretreatments for enhanced anaerobic digestion with waste activated sludge. J Biosci Bioeng 95(3):271–275. https://doi.org/10.1263/jbb.95.271

Li C, Fang HHP (2007) Fermentative hydrogen production from wastewater and solid wastes by mixed cultures. Crit Rev Environ Sci Technol 37(1):1–39. https://doi.org/10.1080/10643380600729071

Logan BE, Regan JM (2006) Microbial fuel cells—challenges and applications. Fuel Cells Technol Fuel Process 10:1–32

Murthy S, Higgins M, Chen YC, Peot C, Toffey W (2006) High-solids centrifuge is a boon and a curse for managing anaerobically digested biosolids. Water Sci Technol 53(3):245–253. https://doi.org/10.2166/wst.2006.103

Ødegaard H, Paulsrud B, Karlsson I (2002) Wastewater sludge as a resource: sludge disposal strategies and corresponding treatment technologies aimed at sustainable handling of wastewater sludge. Water Sci Technol 46(10):295–303. https://doi.org/10.2166/wst.2002.0358

Pathak H, Jain N, Bhatia A, Mohanty S, Gupta N (2009) Global warming mitigation potential of biogas plants in India. Environ Monit Assess 157(1–4):407–418. https://doi.org/10.1007/s10661-008-0545-6

Rulkens W (2008) Sewage sludge as a biomass resource for the production of energy: overview and assessment of the various options. Energy Fuels 22(1):9–15. https://doi.org/10.1021/ef700267m

Shahbaz M, AlNouss A, Ghiat I, Mckay G, Mackey H, Elkhalifa S, Al-Ansari T (2021) A comprehensive review of biomass based thermochemical conversion technologies integrated with $CO_2$ capture and utilisation within BECCS networks. Resour Conserv Recycl 173:105734. https://doi.org/10.1016/j.resconrec.2021.105734

Szabó L, Soria A, Forsström J, Keränen JT, Hytönen E (2009) A world model of the pulp and paper industry: demand, energy consumption and emission scenarios to 2030. Environ Sci Policy 12(3):257–269. https://doi.org/10.1016/j.envsci.2009.01.011

Tyagi VK, Lo SL (2013) Sludge: a waste or renewable source for energy and resources recovery? Renew Sustain Energy Rev 25(71):708–728. https://doi.org/10.1016/j.rser.2013.05.029

Weiland P (2010) Biogas production: current state and perspectives. Appl Microbiol Biotechnol 85(4):849–860. https://doi.org/10.1007/s00253-009-2246-7

Wiley PE, Campbell JE, McKuin B (2011) Production of biodiesel and biogas from algae: a review of process train options. Water Environ Res 83(4):326–338. https://doi.org/10.2175/106143010x12780288628615

# Optimizing Sludge Dewatering Techniques

Abhishek Kumar, Hara Mohan Jena, and Krunal M. Gangawane

**Abstract**

The effective removal of moisture from sludge is a critical procedure in the treatment of wastewater and the management of the environment. Dehydrating sewage sludge involves reducing the moisture content of sediments and increasing the proportion of solid matter, hence acting as a crucial subsequent treatment stage for further processing. Concurrently, this technique facilitates a decrease in the quantity of sewage sludge. Eliminating moisture from sewage sludge relies on its composition and is accomplished by compaction, drainage, and drying techniques. Different techniques are used for efficient elimination depending on the design of the water in question. Various sludge forms possess distinct properties crucial in determining the appropriate conditioning chemicals, including cationic flocculants, ferric chloride, and lime. Additionally, these features influence the selection of dewatering equipment, such as filtration and centrifugation. The decision-making process is also influenced by the anticipated final use of the sludge, such as incineration, agricultural utilization, or other applications, in addition to the existing equipment resources. This abstract comprehensively reviews the essential factors to consider and the many ways for sludge dewatering. It emphasizes the significance of these approaches in promoting sustainable and efficient practices for managing sludge.

**Keywords**

Sludge dewatering · Sewage sludge · Dehydration · Water content · Dry matter content · Wastewater treatment

## 1 Introduction

Wastewater sludge is a residual substance produced from sewage or industrial effluents, and it comprises solid residues that are separated during the treatment process. It includes various organic and inorganic elements, microbes, and possibly hazardous compounds. Wastewater sludge is a complicated and challenging substance due to its varying composition, which is influenced by the source and treatment procedures. The importance of sludge management is in its immediate influence on environmental, public health, and economic dimensions. Insufficient managerial techniques may result in the pollution of the environment, which in turn poses hazards to both ecosystems and human well-being. Additionally, it is essential to have efficient methods for handling wastewater sludge to adhere to environmental norms and standards. Effective management, processing, and elimination of sludge are necessary to avoid the emission of hazardous compounds and guarantee the overall durability of wastewater treatment procedures (Novak 2006; To et al. 2016; Kumar and Gangawane 2022a).

Dewatering is an essential step in managing sludge, where the surplus water is extracted from the sludge, leading to a more concentrated and accessible substance. Effective dewatering is crucial for several reasons (Cao et al. 2021; Kumar and Gangawane 2022b):

A. Kumar (✉) · H. M. Jena
Department of Chemical Engineering, National Institute of Technology Rourkela, Rourkela 769008, Odisha, India
e-mail: abhisheknitrkl@outlook.com; ak170192@gmail.com

*Present Address:*
A. Kumar
Information Management & Dissemination Centre (IMDC), Council of Scientific and Industrial Research–National Metallurgical Laboratory (CSIR–NML), Jamshedpur 831007, Jharkhand, India

K. M. Gangawane
Department of Chemical Engineering, Indian Institute of Technology Jodhpur, Jodhpur 342030, Rajasthan, India

© The Author(s), under exclusive license to Springer Nature Switzerland AG 2025

A. Mandpe and M. P. Shah (eds.), *Wastewater Sludge Nexus*, Sustainable Landscape Planning and Natural Resources Management, https://doi.org/10.1007/978-3-031-45758-6_2

I. *Volume Reduction*: Dewatering dramatically decreases the amount of sludge, resulting in a more economically efficient method for transportation and subsequent processing.
II. *Transportation Costs*: Decreasing the moisture content of sludge results in a reduction in both its weight and volume, which subsequently leads to cheaper transportation expenses. It is essential to reduce the environmental impact of sludge disposal.
III. *Resource Recovery*: The process of removing water from sludge typically enables the extraction of essential resources such as organic matter, nutrients, and energy, which promotes a more sustainable approach to managing sludge.
IV. *Facilitates Further Processing*: Dewatered sludge may be more easily subjected to different processing techniques such as incineration, composting, or land application. It improves the overall effectiveness of sludge treatment and usage.

Effective dewatering methods tackle the difficulties related to controlling the moisture level of sludge, guaranteeing that it can be managed, processed, and disposed of in an ecologically sustainable and economically feasible way. Therefore, it is crucial to develop and use improved dewatering methods to optimize the overall objectives of sludge management.

## 2 Fundamentals of Sludge Dewatering

Different varieties of sludge possess unique attributes, which have a crucial impact on important choices made throughout the sludge management procedure. The choice of conditioning chemicals, such as cationic flocculants, ferric chloride, lime, and others, is closely linked to the unique characteristics of the sludge. These conditioning chemicals facilitate the treatment and preparation of sludge for further processing processes. Furthermore, selecting dewatering equipment, such as filtration or centrifuge, depends on the sludge's specific characteristics. The compatibility between the kind of sludge and the equipment chosen directly influences the efficiency and efficacy of dewatering procedures.

Furthermore, these choices are not made independently but are closely linked to the anticipated ultimate use of the sludge. The selection of conditioning chemicals and dewatering equipment is influenced by whether the sludge will be incinerated or used for agricultural spreading. It ensures that the strategy for managing the sludge aligns with its intended function (La Heij et al. 1996; Scales et al. 2004; Kumar et al. 2022; Kumar and Gangawane 2023a).

### 2.1 Sludge Formation in Water Treatment

During the water treatment, contaminants are eliminated, and the purified water is then released into the environment. Several by-products resulting from the process of pollution removal may be detected, including:

- *Subatomic particles*: These particles may either naturally settle or be generated during the physico-chemical treatment used in the water treatment process.
- *Surplus Microorganisms*: These microorganisms are produced as a separate byproduct during the treatment of dissolved organic waste.
- *Inorganic Substances that Cannot Be Broken Down by Natural Processes*: Another kind of byproduct consists of minerals impervious to biodegradation.

These residual substances are present in different amounts, floating inside the liquid. The phrase used to describe the resultant liquid, which contains suspended particles and chemicals, is sludge. Hence, the water treatment procedure guarantees the removal of contaminants and produces sludge, which requires further consideration and control in the overall treatment and disposal plan (Ahmad et al. 2016).

### 2.2 Sludge Categories

Sludge categories refer to classifications based on the origin, composition, and characteristics of sludge generated during wastewater treatment. These categories include primary sludge, biological sludge, mixed sludge, digested sludge, physicochemical sludge, and mineral sludge. Each class has unique properties influencing the selection of treatment methods, such as conditioning chemicals and dewatering equipment. Understanding these categories is essential for efficient sludge management in wastewater treatment processes (Riding 2002).

#### 2.2.1 Primary Sludge

Primary sludge is derived from the settling process and is composed of suspended particles that can be readily separated, often both big and dense. Having a relatively low volatile solids concentration, typically 55–60%, it has exceptional capability for dewatering. Pre-dewatering static thickening is sufficient. However, it is prone to rapid fermentation.

#### 2.2.2 Biological Sludge

Biological sludge is a combination of microorganisms, primarily bacteria, produced by natural wastewater treatment. Bacterial aggregates, produced by the production of exo-polymers, may be readily isolated from treated water by decantation. The notable characteristics of this substance are its high Volatile Solids content, ranging from 70 to 80%, and its moderate capacity to remove water, which sometimes necessitates dynamic thickening processes.

### 2.2.3 Mixed Sludge

Mixed sludge comprises 35 to 45% primary sludge and 55 to 65% biological sludge, resulting from combining these two types. The blending process enables more efficient dewatering by combining the characteristics of primary and biological sludges.

### 2.2.4 Digested Sludge

The process of digestion, which occurs during biological stabilization, results in the mineralization of digested sludge, leading to a reduction in the concentration of volatile solids to around 50%. It has an adequate dewatering capacity within a dry solids concentration range of 20–40 g/l. Digestion may occur at various temperatures and in the presence or absence of oxygen.

### 2.2.5 Physico-Chemical Sludge

This sludge is formed as a result of the physico-chemical treatment of wastewater. It consists of aggregates called flocs, produced using chemical substances such as coagulants and flocculants. The features of the water are directly influenced by the specific chemicals and contaminants present in it.

### 2.2.6 Mineral Sludge

Mineral sludge mainly comprises mineral particles, such as clays, generated during quarrying or mining beneficiation. It has a pronounced capacity to settle due to gravity, often reaching high concentration levels.

## 2.3 Factors in Sludge Dewatering

Crucial concerns in wastewater treatment operations revolve around the factors influencing sludge dewatering. The sludge content, quantified in grams per liter (g/l), directly impacts the assimilation and use of flocculant. A higher amount of organic matter, namely volatile solids (VS), creates difficulties in attaining efficient dewatering. Therefore, it is recommended to include a thickening process for sludge with high VS content to optimize the dewatering process. The dewatering performance is strongly affected by the colloidal character of sludge, which is governed by factors such as its origin, freshness, wastewater origin, and sludge return. Comprehending these elements is crucial for efficient sludge management and enhancing dewatering operations in wastewater treatment (Hyrycz et al. 2022).

### 2.3.1 Concentration (g/l)

The concentration of sludge, measured in g/l, plays an essential role in dewatering. Key considerations include:

- *Incorporation of Flocculant*: Higher sludge concentration makes mixing with viscous flocculant solutions challenging. Solutions include post-dilution, upstream injection, multiple injection points, or using an on-line mixer.
- *Flocculant Consumption*: If incorporated correctly, higher sludge concentration generally leads to lower flocculant consumption.

### 2.3.2 Organic Matter Content (%)

The organic matter content, comparable to volatile solids (VS), significantly impacts dewatering:

- The dewatering process becomes more challenging as the VS increases. Low levels of dehydration will be attained, accompanied by common mechanical properties and high flocculant consumption.
- It is advisable to include a thickening stage when the sludge's volume solids (VS) is high to improve the dewatering process.

### 2.3.3 Colloidal Nature of Sludge

The colloidal nature of sludge is a critical factor affecting dewatering performance, influenced by:

- *Sludge Origin*: Varies between primary, digested primary, fresh mixed, digested mixed, and biological sludge, ranging from low to high colloidal nature.
- *Freshness*: Colloidal nature increases with fermentation, particularly in septic sludge.
- *Wastewater Origin*: Sludge from dairy or brewery wastewater exhibits an increased colloidal nature.
- *Sludge Return*: Poorly controlled sludge return can elevate colloidal nature, emphasizing the need for proper management.

# 3 Mechanical Dewatering Techniques

The process of sewage sludge dewatering is efficiently carried out using mechanical techniques rather than relying on natural circumstances considered ineffective. This section explores the characteristics of several mechanical dewatering methods. These procedures are essential for achieving the sludge's optimal dry matter content, ranging from 19 to 40%.

Comprehending and enhancing these mechanical methods for removing water from sewage sludge is crucial for efficiently managing wastewater treatment systems (Mahmoud et al. 2012).

## 3.1 Belt Filter Press

Belt filter presses dehydrate primary and waste-activated sludges (WAS) from aerobic or anaerobic digestion processes.

### 3.1.1 Operation Principles

Belt filter presses are designed in two main configurations: two-belt or three-belt units. The prevailing preference is for the three-belt configuration, allowing the gravity zone to function independently from the pressing zones. Three distinct zones characterize these presses:

- *Gravity Zone*: The primary region where solid materials experience gravity dehydration.
- *Medium Pressure Zone (Wedge Zone)*: This portion enables dewatering at medium pressure after the gravity zone.
- *High-Pressure Zone (Shear Zone)*: This is the ultimate area where water is removed under high pressure.

Conditioning solids with a polymer is necessary before applying them to a belt filter press. An essential function of the polymer is to help dissolve solids, which allows for the removal of a considerable amount of water in the gravity zone. Afterward, the pressing zones are where the water is extracted. The gravity zone should eliminate about 60–70% of the water, while the wedge and shear zones should be used for the remaining water extraction. Belt filter presses are more effective in dewatering sewage sludge when set up and operated in this way.

### 3.1.2 Advantages and Limitations

There are some advantages and limitations of the belt filter press:

*Advantages*

- *Processing Special Sludge*: The belt filter press excels in processing specific sludge types such as railway, oil, and diatomite-treated wastewater sludge. It outperforms other filter presses in handling these challenging sludges, yielding favorable results.
- *Low Energy Costs*: With comparable treatment efficiency to other filter presses, the belt filter press stands out for its significantly lower water and power consumption, approximately 1/3 that of other equipment under similar conditions.
- *Space Efficiency*: The unique structure of the belt filter press allows it to occupy only half the space of other equipment while achieving equivalent treatment results. This space-efficient design is advantageous in various operational settings.
- *Stable Operation and Long Service Life*: The filter belt's high strength, short length, and particular adjustment structure contribute to its durability and stability. Using high-strength stainless steel bearings enhances corrosion resistance, extending the equipment's service life to 10 years.
- *Ease of Operation*: The belt filter press boasts a simple structure and rational design, facilitating easy operation. Even ordinary workers can operate it without the complexity of other filter presses.
- *Continuous Operation*: Unlike intermittent operation in some filter presses, the belt filter press ensures continuous 24-h operation, enhancing operational efficiency.

*Limitations*:

- *Suitability for High Sludge Moisture Content*: The belt filter press is most suitable for sludges with high moisture content, potentially limiting its efficiency for drier sludges.
- *Maintenance Challenges*: Maintenance of the belt filter press may pose challenges, potentially requiring specialized attention compared to other filter presses.

## 3.2 Centrifuge Dewatering

Centrifuges use the persistent application of centrifugal forces, often exceeding several thousand g-forces, to aid in dewatering (To et al. 2016; Mumbi et al. 2017).

### 3.2.1 Working Mechanism

The centrifuge, also known as a centrifugal decanter, operates on the principle of utilizing centrifugal force to expedite the separation of solid and liquid components. In a simplified representation, envision the centrifuge as a horizontally turning conical cylinder decanter with clarified water overflowing and dewatered sludge extracted by an Archimedean screw. Flocculated sludge is introduced into the centrifuge bowl through an injection pipe. The bowl achieves high rotation speeds, typically around 3500 rpm, causing the solid particles to be pressed against the bowl's sides in the clarification zone. Subsequently, an Archimedean screw propels the particles toward the end of the bowl's cone in the sludge spin-dry zone, as shown in Fig. 1.

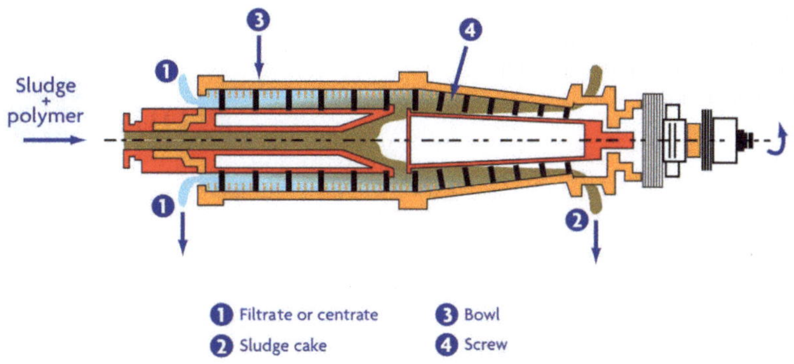


**Fig. 1** Centrifuge dewatering working mechanism (Mumbi et al. 2017)

The clarified liquid, referred to as centrate, is then evacuated through the overflow at the opposite end of the bowl. Several parameters specific to centrifuge dewatering play a crucial role in polymer selection during lab tests and industrial trials:

- *Diaphragm Diameter*: This corresponds to the diameter of the centrate exit hole. Controlled by adjustable plates, the diaphragm diameter influences the thickness of the liquid ring inside the bowl. Optimal diaphragm diameter selection balances centrate clarification and sludge dryness.
- *Relative Speed*: The relative speed denotes the difference in rotation speeds between the bowl and the central Archimedean screw. Higher relative speeds result in a faster extraction of sludge.
- *Torque*: Torque measures the pressure of the sludge on the screw, causing torsion on the screw axis. Higher sludge pressure yields higher torque. Torque and relative speed are interconnected, as an increase in relative speed reduces torque, enabling quicker sludge extraction.

It is crucial to note that torque measurement is just one example, and its relationship with relative speed impacts the efficiency of sludge extraction. High-performance centrifuges, designed with a distinct structure for the conveying screw, enhance sludge dryness by allowing for a longer residence time in the centrifuge.

### 3.2.2 Advantages and Limitations

Centrifuge dewatering has several advantages and limitations in wastewater treatment.

*Advantages*

- The centrifuge technique for sludge dewatering has many benefits, making it a favored option for private contractors. Adopting the sludge centrifuge technique has led to significant cost reductions, including around 75% in transportation expenses and an astonishing drop of 80% in disposal costs. Placing the equipment near the contractor's base of operation improves operational convenience and increases overall productivity.
- The resulting cake, a secondary outcome of this procedure, may be conveniently transported to a landfill or combined with green trash for composting. Moreover, the filtrate produced by the centrifuge process has several uses, including enhancing soil quality during droughts, recycling for use in jetting systems, or safely releasing it into a sewage system.
- This dewatering technique shows its adaptability by efficiently operating with various forms of sludge, which augments the concentration of solid particles. Significantly, it removes water from wastewater sludge without needing a filtering material, enhancing its effectiveness and suitability for various sludge management situations.

*Limitations*

Despite its advantages, the method comes with several noteworthy limitations that should be considered:

- *High Energy Consumption*: One significant drawback is the method's demand for high energy consumption in wastewater treatment processes. This aspect may contribute to increased operational costs.
- *Noise Issues*: Another notable limitation involves the generation of noise during the method's operation. This factor could pose concerns, especially when noise pollution is critical.
- *Standby Unit Requirement*: In certain instances, the method may necessitate the presence of a standby unit. This requirement adds an extra layer of complexity to the system, potentially impacting operational and capital considerations.
- *Specialized Maintenance*: While maintenance needs are generally minimal, the technical nature of the required maintenance can lead to repair delays and present operational continuity challenges.

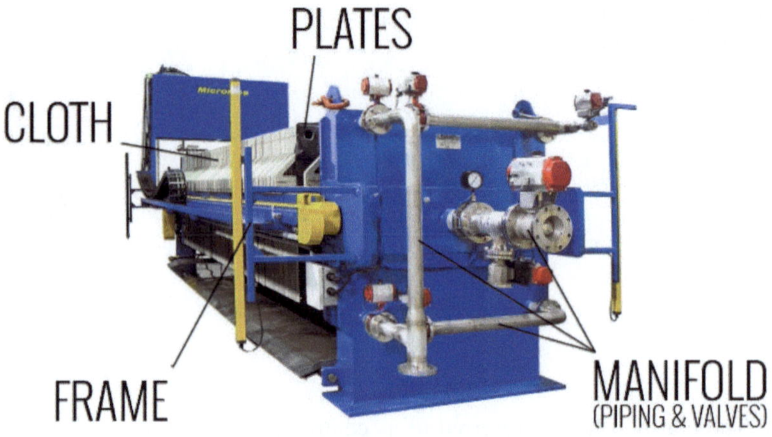


**Fig. 2** Filter press dewatering working mechanism ([CSL STYLE ERROR: reference with no printed form.])

Acknowledging these limitations is crucial for comprehensively understanding the method's feasibility and suitability for specific wastewater treatment scenarios.

## 3.3 Filter Press Dewatering

A filter press is an essential device used to separate liquids and solids. It utilizes pressure filtration to separate the two components efficiently. The operating concept entails the injection of a slurry into the filter press, which then initiates the dewatering process by applying pressure. Each filter press is custom-designed to suit the precise amount and composition of the slurry targeted for dewatering. The filter press is a flexible solution that effectively removes explicit moisture from different kinds of slurry (Shahbaz et al. 2023).

### 3.3.1 Working Mechanism

A filter press comprises four essential components, each playing a distinct role in the optimization of its operations:

1. *Frame*: The foundational structure supports and stabilizes the entire filter press system.
2. *Plates*: Plates are strategically arranged within the press to form chambers where the slurry is processed, and filter cakes are accumulated.
3. *Manifold (Piping and Valves)*: A network of pipes and valves facilitating the controlled flow of slurry and filtrate throughout the filter press.
4. *Cloth*: A critical element in filter press efficiency, the filter cloth aids in the separation process by allowing the passage of filtrate while retaining solid particles.

The operating procedure entails introducing the slurry into the chambers, where filter cakes accumulate. As the filter cakes' thickness increases, the filter's resistance increases. The filtering process stops once the ideal pressure differential is reached when the chamber becomes full. The liquid that passes through the filter cloth is collected by pipes and kept in the filter tank. The filtration process involves partitioning suspended materials inside the chambers, known as filter cakes. Subsequently, the chamber filter press is disassembled to permit the removal of the filter cakes from the plates. These cakes are detached and guided to the ultimate gathering location, as shown in Fig. 2.

Filter presses operate based on a systematic procedure that separates solids and liquids. The slurry is pushed into the machine during the filling cycle, and the solids are uniformly dispersed across the filter cloth. The particulate matter progressively builds up on the filter cloth, forming the filter cake. At the same time, the purified water, also known as filtrate, is discharged from the filter plates via corner ports and collected in the manifold. Filter presses use a pressure filtering technique, in which the feed pump incrementally increases pressure, accumulating materials inside the chambers until they are filled, generating solid filter cakes. The filter cakes are prepared for release once the partitions are filled, indicating that the filtering process is finished. The systematic methodology highlights the filter press as a dependable and efficient instrument for solid-liquid separation.

### 3.3.2 Advantages and Limitations

Filter press dewatering has several advantages and limitations in wastewater treatment.

*Advantages*

- *Efficient Solid-Liquid Separation*: Filter presses excel in achieving effective solid-liquid separation, producing drier filter cakes by removing a significant amount of water from sludge.
- *High Dewatering Efficiency*: The pressure-driven filtration mechanism of filter presses results in high dewatering efficiency, making them suitable for applications with varying sludge characteristics.

- *Versatility in Application*: Filter presses are versatile and adaptable, catering to various slurry types and diverse solid-liquid separation projects.
- *Reduced Transportation Costs*: By significantly reducing the water content in filter cakes, filter press dewatering minimizes the weight and volume of the sludge, leading to reduced transportation costs.
- *Available in Manual and Hydraulic Types*: Filter presses are available in both manual and hydraulic types, providing flexibility in operation based on specific requirements and scale.

*Limitations*:

- *Energy Consumption*: The process of filter press dewatering can demand substantial energy consumption, impacting operational costs, especially in scenarios where energy efficiency is crucial.
- *Noise Generation*: Filter presses may generate noise during operation, which can be a consideration in environments where noise pollution is a concern.
- *Standby Unit Requirement*: In certain situations, using a standby unit may be necessary, adding complexity to the system and potentially impacting operational and capital considerations.
- *Specialized Maintenance*: While maintenance needs are generally minimal, the technical nature of required maintenance can lead to repair delays and pose challenges for operational continuity.

## 4 Chemical Conditioning in Sludge Dewatering

In the preparation phase before thickening and dewatering sludge, conditioning is a common practice to improve its treatability. There are two main categories of conditioning chemicals employed for this purpose (Novak and Park 2004):

I. Mineral Chemicals:

- Examples include iron salts and lime.
- Frequently utilized in applications involving filter presses.

II. Organic Chemicals:

- Includes coagulants and flocculants.
- Cationic flocculants are particularly prevalent in these applications.

The mineral chemicals, such as iron salts and lime, find regular application, especially in scenarios involving filter presses. On the other hand, organic chemicals, notably coagulants and flocculants, are integral to the conditioning process. Cationic flocculants, in particular, are commonly encountered due to their effectiveness in enhancing the treatability of sludge during subsequent thickening and dewatering stages.

### 4.1 Mineral Chemicals

#### 4.1.1 Iron Salts

Iron salts, specifically Ferric Chloride and Iron Chloro-Sulfate, are commonly used with lime for sludge conditioning before a filter press. This combination enhances filterability by coagulating colloids, thereby reducing linked water content. Additionally, micro-flocculation of sediments, particularly hydroxides, is facilitated. Dosages for iron salts typically range between 3% and 15% of the dry content, with a growing tendency to combine them with cationic organic flocculants. This hybrid approach decreases sludge volume compared to the traditional iron salts + lime process.

#### 4.1.2 Lime

Lime is a conditioning agent that works exclusively in tandem with iron salts, specifically in filter press applications. It imparts a mineral nature to the sludge, enhancing its mechanical properties, particularly a higher specific resistance to filtration. Lime dosages typically fall between 15 and 40% of the dry content. Notably, lime can also be employed post-dewatering to stabilize the sludge. The specific resistance to filtration (r) is influenced by the size, shape, and degree of aggregation of solid particles constituting the filter press cake, and it remains independent of sludge concentration.

### 4.2 Organic Chemicals

The bulk of chemicals used in sludge dewatering are cationic flocculants.

#### 4.2.1 Flocculation in Sludge Treatment

Flocculation is an essential step in the sludge treatment process when destabilized particles come together to form larger aggregates called flocs. Flocculants, which have substantial molecular weights consisting of long chains of monomers and various ionic charges, are crucial in this process as they capture destabilized particles along their chains. Consequently, the size of particles in the water phase gradually increases throughout the flocculation process, resulting in the formation of flocs. Floc aggregation triggers water liberation, which may be effectively removed during dewatering. Employing flocculants strategically enhances the efficacy of sludge dewatering by facilitating the efficient removal of water from the sludge.

### 4.2.2 Selection of Flocculant Charge Based on Sludge Characteristics

The diversity in the source of destabilized particles is intricately connected to the characteristics of the sludge undergoing treatment. The choice of the flocculant dosage is a crucial factor influenced by the characteristics of the destabilized particles in the sludge. The choice of this option depends on the sludge's features, which may be classified into four types: biological, digested, physicochemical, or mineral.

The suggested flocculant dosage patterns, determined by the sludge type, are as follows:

- Low to medium anionic: Appropriate for mineral sludge.
- Low anionic to low cationic: Suitable for physicochemical sludge.
- Low cationic: Suitable for digested and primary sludge.
- Medium cationic: This treatment is very effective for sludge with a mixture of different components.
- Highly cationic: Suggested for the treatment of biological sludge.

This systematic approach guarantees that the flocculant dosage is tailored to the particular properties of the sludge, resulting in a highly effective and optimized flocculation procedure.

## 4.3 Organic Chemical Parameters Affecting Dewatering

Organic flocculants exhibit distinct characteristics determined by five key parameters, each of which significantly impacts the quality of flocculation and, consequently, the efficiency of dewatering processes:

### 4.3.1 Type of Charge (+ or −)

The flocculant's charge type selection is based on the nature of the particles present. Typically, the following pattern is observed:

- Anionic (−) flocculant for capturing mineral particles.
- Cationic (+) flocculant for capturing organic particles. A laboratory test is crucial for precise charge adaptation.

### 4.3.2 Charge Density (%)

Charge density, representing the quantity of + or – charge needed for optimal flocculation at the lowest dosage, is contingent on the sludge type. In municipal sludge, charge density is mainly influenced by the organic matter content (OM), often assimilated to volatile solids content (VS). Higher VS necessitates a greater cationic charge.

### 4.3.3 Molecular Weight (MW)

The choice of molecular weight, indicating the length of the polymer chain, depends on the dewatering equipment:

- For centrifuges: High to very high molecular weight is suitable due to the applied high shearing.
- For filtration: Low to medium molecular weight is preferred for adequate drainage.

### 4.3.4 Molecular Structure

Flocculant molecular structure is tailored to dewatering requirements for cationic flocculants, offering:

- *Linear systems*: Low dosage with good performance.
- *Branched structures*: Medium dosage with excellent drainage performance.
- *Cross-linked networks*: High dosage with exceptional drainage performance and shear resistance.

### 4.3.5 Type of Monomer

The choice of monomer used in flocculant synthesis also influences flocculation. Typical cationic monomers include ADAM-MeCl and APTAC, each exhibiting specific characteristics. Sodium acrylate is a frequent anionic monomer. The selection of monomers is crucial for achieving desired flocculation outcomes.

# 5 Emerging Technologies in Sludge Dewatering

Emerging technologies in sludge dewatering technologies include ultrasonic waves, magnetic fields, and microwaves. Each approach alters sludge characteristics, affecting dewatering. Each unorthodox approach is explained (Kamizela and Kowalczyk 2019):

1. *Ultrasound Field*: Ultrasonic waves through liquids cause coagulation and electrokinetic events. Sludge particle size and sound intensity affect impacts. Sludge qualities may be seen after a short sonication duration. Sludge thickening and free water removal during vacuum filtration are accelerated by ultrasonic particle packing. Sludge dewatering is greatly improved by ultrasonic exposure and polyelectrolytes before treatment.
2. *Magnetic Field*: Electric charges and magnetic poles are affected by electromagnetic fields. During raw sludge conditioning, a magnetic field reduces capillary suction time (CST) with exposure time. The magnetic field reduces treated sludge water content. Using a magnetic field and polyelectrolytes together lowers specific filtration resistance.
3. *Microwaves*: In a water-rich medium, electromagnetic microwaves behave differently. These include heating-induced non-thermal effects and thermal effects connected to increasing medium temperature. Volumetric heating in the irradiated material layer converts microwave energy into particle motion animation via chemical interactions with the electromagnetic field. Polarization of local electric charges causes energy dissipation, heat release, and diffusion-induced heating in the material layer beyond microwave reach.

These novel technologies accelerate sludge dewatering and decrease chemical reactant use.

## 6 Environmental and Economic Considerations

Examining sludge management's environmental and economic dimensions is essential for devising sustainable and economically efficient solutions. This section discusses the consequences of different sludge management strategies on the environment and financial resources (Nielsen 2015).

### 6.1 Environmental Impact of Sludge Dewatering

Sludge dewatering is an essential process in wastewater treatment that provides advantages such as decreased biosolid volume and improved water recycling. Nevertheless, the procedure has environmental implications that need careful study (Wu et al. 2020; Kumar and Gangawane 2023a, b).

- *Life Cycle Assessment (LCA)*: The life cycle environmental implications of sewage sludge treatment were evaluated, considering energy consumption, emissions, and ecotoxicity. The study uncovered possible ecological issues linked to the different phases of sludge treatment.
- *Risks of Inappropriate Treatment*: The hazards linked to inadequate treatment techniques emphasize the environmental consequences of sludge dewatering. Improper treatment of sludge may result in the release of hazardous substances such as heavy metals, persistent organic pollutants (POPs), and microorganisms, which can cause significant secondary contamination.
- *Technologies for Treatment and its Effects*: Selecting treatment technology in sewage sludge management substantially impacts environmental outcomes. Various methods, such as dewatering, stabilization, and sanitization, might exhibit diverse environmental consequences contingent upon their capacity and ultimate disposal.
- *Effects of Carbonization*: The environmental impacts of carbonization, a technique used for treating sludge, were investigated. The research highlighted the need to tackle elevated moisture levels in processed sludge to achieve optimal energy conversion.
- *Issues with Sludge Management*: The comprehensive sludge management process, including dewatering, is crucial in ensuring environmental sustainability. The challenges associated with managing sludge, from its generation to its disposal, emphasize the need for implementing safe and efficient techniques.
- *Advantages of Sludge Dewatering*: Although sludge dewatering is widely recognized as a successful technique for separating biosolids from treated wastewater, carefully considering its environmental benefits and possible disadvantages is crucial. The potential for water recycling and the reduction of biosolids favorably contribute to sustainability.

### 6.2 Economic Evaluation of Dewatering Techniques

Assessing the economic viability of dewatering methods is crucial for optimizing resource usage. Techno-economic assessments have been carried out in the context of microalgae harvesting and sludge dewatering (Spinosa et al. 2011).

- *Microalgae Harvesting*: A comprehensive quantitative techno-economic study was conducted to evaluate the efficiency and cost-effectiveness of large-scale systems for harvesting and dewatering microalgae. The research primarily examined the expenses associated with processing, the energy used, and the overall viability of various approaches.

- *Biofuel Production*: An analysis was performed to compare the technological and economic aspects of several dewatering methods used to manufacture biofuels from microalgae. The research conducted a comparative study of a clarifier-centrifuge system's efficiency and economic feasibility compared to an alternative approach.
- *Sludge Dewatering*: Techno-economic analyses have been used to assess sludge dewatering techniques, highlighting the significance of inventive and economical ways. An economic study is essential for maximizing the efficiency of sludge treatment operations.
- *Construction Dewatering*: Although not directly associated with microalgae or sludge, economic factors are vital in building dewatering techniques. Construction projects incur cost repercussions using several strategies, including healthy spots, deep wells, and sump pumps.

### 6.3 Sustainability in Sludge Management

The focus of sustainable sludge management is to use efficient and ecologically sound techniques that optimize the extraction of resources or generation of energy from sludge. This method aims to reduce adverse effects and foster sustainable ecological equilibrium. Key elements of sustainable sludge management encompass (Rulkens 2004; Spinosa et al. 2011; Sharma et al. 2021):

- *Resource Recovery*: Sustainable methods prioritize extracting valuable materials or energy from sludge, promoting a circular economy and minimizing waste.
- *Minimization of Environmental Impact*: The objective is to mitigate any detrimental impacts of sewage sludge on the environment. Constructed wetlands are widely acknowledged as a sustainable method for treating sludge, primarily because they have little maintenance and energy needs.
- *Alternative Treatment Options*: The sustainability of sludge management is improved by investigating alternate treatment alternatives, such as using sludge for fuel or feedstock production.
- *Soil Fertility and Crop Production*: It is essential to carefully consider the effects of sludge on soil fertility and crop productivity when sustainably managing sewage sludge.
- *Innovative Solutions*: Current research is investigating novel and environmentally friendly approaches to handle sewage sludge to tackle the difficulties both industrialized and developing nations face.

## 7 Future Trends and Innovations

The continuous progress in sludge dewatering technologies influences future patterns, emphasizing enhancing efficiency, sustainability, and overall efficacy (Rulkens 2004).

- *Preconditioning for Enhanced Dewaterability*: Preconditioning techniques are gaining attention to enhance the dewaterability of sludge, playing a crucial role in subsequent mechanical dewatering processes.
- *Global Market Growth*: The global market for sludge dewatering systems and equipment is experiencing substantial growth, with reports outlining current market trends and prospects, providing insights into the evolving landscape.
- *Mobile and Containerized Solutions*: Innovations in mobile or containerized solutions for sludge dewatering are becoming prominent, offering flexibility and adaptability in various environmental and operational contexts.
- *Integration of Ultrasonic Treatment*: Ultrasonic treatment is a promising technique for optimizing sludge dewatering efficiency. Research explores the impact of ultrasonic treatment on the dewatering process.
- *Focus on Minimizing Volume and Enhancing Transportability*: Recent advancements in sludge dewatering and drying technologies aim to minimize sludge volume, improve transportability, and enhance the calorific value of the processed sludge.

## 8 Conclusion

In conclusion, sustainable economic growth, public health, and environmental preservation depend on efficient wastewater sludge management. One crucial phase is dewatering, which has advantages, including resource recovery and volume reduction. Despite difficulties, a thorough understanding of sludge composition is essential for effective mechanical dewatering. Treatability is improved by chemical conditioning, and organic and mineral flocculants decrease the amount of sludge. New technologies have potential, such as ultrasonic waves. Environmental and economic factors emphasize the requirement for safe and effective procedures. Recovery of resources and alternate forms of treatment are the main goals of sustainable sludge management. Upcoming trends include increasing dewaterability, expanding worldwide markets, and creative approaches to sustainability and efficiency. Practices for treating wastewater will develop as long as research is conducted.

## References

Ahmad T, Ahmad K, Ahad A, Alam M (2016) Characterization of water treatment sludge and its reuse as coagulant. J Environ Manage 182:606–611. https://doi.org/10.1016/j.jenvman.2016.08.010

Cao B, Zhang T, Zhang W, Wang D (2021) Enhanced technology based for sewage sludge deep dewatering: a critical review. Water Res 189:116650. https://doi.org/10.1016/j.watres.2020.116650

Hyrycz M, Ochowiak M, Krupińska A et al (2022) A review of flocculants as an efficient method for increasing the efficiency of municipal sludge dewatering: mechanisms, performances, influencing factors and perspectives. Sci Total Environ 820:153328

Kamizela T, Kowalczyk M (2019) Sludge dewatering: processes for enhanced performance. In: Industrial and municipal sludge. Elsevier, pp 399–423

Kumar A, Gangawane KM, Ramanjaneyulu B (2022) Ferrofluids for waste-water treatment. In: International conference on chemical, bio and environmental engineering. Springer, pp 723–744

Kumar A, Gangawane KM (2022a) Effect of precipitating agents on the magnetic and structural properties of the synthesized ferrimagnetic nanoparticles by co-precipitation method. Powder Technol 401:117298. https://doi.org/10.1016/j.powtec.2022.117298

Kumar A, Gangawane KM (2022b) Synthesis and effect on the surface morphology & magnetic properties of ferrimagnetic nanoparticles by different wet chemical synthesis methods. Powder Technol 410:117867. https://doi.org/10.1016/j.powtec.2022.117867

Kumar A, Gangawane KM (2023a) Advanced computational approaches for water treatment. Advanced computational approaches for water treatment. CRC Press, New York, pp 1–12

Kumar A, Gangawane KM (2023b) Molecular dynamics for the membrane process. Advanced computational approaches for water treatment. CRC Press, New York, pp 13–38

La Heij EJ, Kerkhof PJAM, Herwijn AJM, Coumans WJ (1996) Fundamental aspects of sludge filtration and expression. Water Res 30:697–703. https://doi.org/10.1016/0043-1354(95)00205-7

Mahmoud A, Olivier J, Vaxelaire J, Hoadley AFA (2012) Advances in mechanical dewatering of wastewater sludge treatment. In: Wastewater reuse and management. Springer, pp 253–303

Mumbi A, Fengting L, Mwarania F, Uuganchimeg B (2017) an assessment of multi-plate screw press in dewatering process of sludge treatment (the best option?). Int J Adv Res 5:740–747. https://doi.org/10.21474/IJAR01/6014

Nielsen S (2015) Economic assessment of sludge handling and environmental impact of sludge treatment in a reed bed system. Water Sci Technol 71:1286–1292. https://doi.org/10.2166/wst.2015.072

Novak JT, Park C (2004) Chemical conditioning of sludge. Water Sci Technol 49:73–80. https://doi.org/10.2166/wst.2004.0612

Novak JT (2006) Dewatering of sewage sludge. Dry Technol 24:1257–1262. https://doi.org/10.1080/07373930600840419

Riding R (2002) Structure and composition of organic reefs and carbonate mud mounds: concepts and categories. Earth-Sci Rev 58:163–231. https://doi.org/10.1016/S0012-8252(01)00089-7

Rulkens WH (2004) Sustainable sludge management—what are the challenges for the future? Water Sci Technol 49:11–19. https://doi.org/10.2166/wst.2004.0597

Scales PJ, Dixon DR, Harbour PJ, Stickland AD (2004) The fundamentals of wastewater sludge characterization and filtration. Water Sci Technol 49:67–72. https://doi.org/10.2166/wst.2004.0610

Shahbaz M, Rashid N, Saleem J et al (2023) A review of waste management approaches to maximise sustainable value of waste from the oil and gas industry and potential for the State of Qatar. Fuel 332:126220. https://doi.org/10.1016/j.fuel.2022.126220

Sharma M, Yadav A, Mandal MK, et al (2021) Wastewater treatment and sludge management strategies for environmental sustainability. In: Circular economy and sustainability, volume 2: environmental engineering. Elsevier, pp 97–112

Spinosa L, Ayol A, Baudez JC et al (2011) Sustainable and innovative solutions for sewage sludge management. Water (Switzerland) 3:702–717. https://doi.org/10.3390/w3020702

To VHP, Nguyen TV, Vigneswaran S, Ngo HH (2016) A review on sludge dewatering indices. Water Sci Technol 74:1–16. https://doi.org/10.2166/wst.2016.102

What is a Filter Press and How Does it Work? Micronics, Inc. https://www.micronicsinc.com/filtration-news/filter-press-information/

Wu B, Dai X, Chai X (2020) Critical review on dewatering of sewage sludge: influential mechanism, conditioning technologies and implications to sludge re-utilizations. Water Res 180:115912. https://doi.org/10.1016/j.watres.2020.115912

# Innovative Approaches to Sludge Treatment and Dewatering

Mansi Jyani, Abhimanyu Bishnoi, and Urmila Gupta Phutela

**Abstract**

The proper management and treatment of sludge generated from industrial and municipal processes are crucial for environmental sustainability and resource recovery. This chapter delves into the complexities of sludge treatment and dewatering techniques, with a focus on methods to reduce volume, improve solid–liquid separation and enable responsible disposal or beneficial reuse of sludge. Beginning with an overview of sludge characteristics, the chapter discusses its composition and origin. It highlights the need for effective sludge management due to potential environmental and health risks. The chapter then explores primary treatment methods used to prepare sludge for dewatering. These methods include thickening and conditioning, which concentrate sludge solids and optimize subsequent dewatering processes. The core of the chapter examines different sludge dewatering techniques. Mechanical methods like centrifugation, filter pressing and belt filtration are explained, detailing their principles, operational considerations and efficiency. Thermal techniques like drying and incineration are explored as options to reduce sludge volume and eliminate pathogens. These approaches are evaluated based on energy requirements, environmental impacts and the potential for resource recovery from incinerated sludge. Amid growing sustainability concerns, the chapter emphasizes the importance of exploring eco-friendly sludge management options. It discusses approaches like anaerobic digestion and composting, which not only aid dewatering but also yield biogas and nutrient-rich soil amendments. In conclusion, the chapter provides a comprehensive overview of sludge treatment and dewatering techniques, spanning traditional mechanical methods to state-of-the-art innovations. It underscores the need for efficient, cost-effective and sustainable practices to address sludge management challenges. By equipping readers with a nuanced understanding of these techniques, the chapter aims to guide researchers, practitioners and policymakers toward informed decisions for responsible sludge handling and treatment in line with evolving environmental priorities.

**Keywords**

Sludge management · Environmental sustainability · Resource recovery · Dewatering techniques · Responsible disposal · Solid–Liquid separation

## 1 Introduction

In the modern age of scientific and technological advancement, the rapid growth of industry and progress has resulted in a significant rise in the generation of sewage sludge, turning its treatment into a pressing global concern. This surge in sewage discharge can be attributed to the forces of urbanization and industrialization, ultimately leading to the inevitable production of sludge at wastewater treatment plants (WWPT). While conventional mechanical dewatering methods can reduce the water content (WC) of sludge to approximately 80%, they often suffer from suboptimal dewatering performance.

M. Jyani (✉)
Department of Microbiology, College of Basic Sciences and Humanities, Punjab Agricultural University, Ludhiana, Punjab, India
e-mail: mansijyani5@gmail.com

A. Bishnoi
Department of Biochemistry, College of Basic Sciences and Humanities, Punjab Agricultural University, Ludhiana, Punjab, India
e-mail: abhimanyubishnoi2929@gmail.com

U. G. Phutela
Department of Renewable Energy Engineering, College of Agricultural Engineering and Technology, Punjab Agricultural University, Ludhiana, Punjab, India
e-mail: urmilphutela@pau.edu

© The Author(s), under exclusive license to Springer Nature Switzerland AG 2025

A. Mandpe and M. P. Shah (eds.), *Wastewater Sludge Nexus*, Sustainable Landscape Planning and Natural Resources Management, https://doi.org/10.1007/978-3-031-45758-6_3

Recognizing the critical importance of enhancing sludge dewaterability for subsequent mechanical dewatering, this chapter seeks to address a notable gap in the current discourse. Specifically, it aims to provide a comprehensive comparison and discussion regarding the effectiveness of sludge preconditioning and advanced sludge dewatering techniques (Mahmoud et al. 2012; Zhang et al. 2017). Moreover, the chapter will endeavor to shed light on the often-overlooked aspect of cost comparisons between various conditioning and dewatering technologies.

Within the pages of this chapter, we will embark on a detailed exploration of commonly employed sludge conditioning methods, subjecting them to critical scrutiny. These methods will be rigorously examined, encompassing their reaction mechanisms, application conditions, as well as their advantages and disadvantages (Rumky et al. 2023b). In parallel, we will evaluate the factors that influence widely used mechanical dewatering techniques. While traditional methods like filter pressing and centrifugal dewatering will be dissected, the spotlight will also be directed toward emerging techniques such as electroosmotic dewatering and vacuum preloading, which have recently garnered significant attention.

In general, the prevailing approach at WWTPs revolves around the pre-treatment of sludge to bolster dewatering performance, followed by the application of various dewatering technologies (Mahmoud et al. 2012; Rumky et al. 2023a). These mechanical dewatering methods encompass pressure filtration, centrifugation, and others, with innovative approaches like vacuum dewatering also gaining prominence.

With the growing need for novel strategies to enhance dewaterability in sludge management, ongoing developments in conditioning, dewatering, and drying technologies for diverse sludge types are at the forefront of scientific inquiry. Addressing the challenge posed by the expanding volume of sewage sludge, it is paramount that we critically assess new concepts and contribute to the rapid technological progress in this field. Furthermore, this chapter will delve into recent developments in dewatering and drying technologies, offering a comprehensive examination of their advantages, disadvantages, and energy requirements.

Sewage sludge, an inescapable byproduct of biological wastewater treatment, contains various hazardous substances, including pathogens, heavy metals, and organic contaminants, which pose significant pollution risks in the absence of proper disposal. Dewatering and drying technologies play pivotal roles in reducing sludge volume, facilitating transportation, increasing calorific value, and mitigating leachate production in landfill sites. The exponential growth in sewage sludge over the past decade has spurred the emergence of new technologies and the evolution of existing ones.

This chapter, therefore, seeks to provide a comprehensive review of recent advancements in the realm of sewage sludge dewatering and drying. The dewatering technology will be categorized based on moisture content, distinguishing between common, deep, and high-dry dewatering methods, while the drying methods will be divided into convective drying, conduction drying, radiation drying, and other innovative approaches. A thorough review and comparison of present dewatering and drying technologies will be conducted, with the aim of offering valuable insights for effective sludge disposal. Additionally, the chapter will illuminate the future demands and opportunities within the domain of sludge dewatering and drying. Predominant sludge disposal methods, including sanitary landfilling, land application, and incineration, will also be discussed in the context of the evolving landscape of sewage sludge management.

## 1.1 The Diverse Composition of Sludge and Resource Recovery

The composition of sludge is a complex interplay of factors influenced by its source and the treatment processes it has undergone. Sludge, typically in a semi-solid or solid form, is a byproduct of wastewater treatment, industrial procedures, and various other applications. Its constituents vary widely due to the nature of the input materials (Lambert et al. 2022), the treatment techniques employed, and the specific contaminants involved. Within the heterogeneous mix of sludge, one finds organic contaminants, viruses, inorganic minerals, microplastics, and heavy metals (Zhang et al. 2021b). Over time, these organic molecules establish a stable floc structure through long-term biochemical reactions.

Notably, the presence of organic substances, referred to as extracellular polymeric substances (EPS), in sludge can hinder the efficient dewatering of sludge, which is a pivotal aspect of sludge management (Huang et al. 2022). Depending on how these elements are managed, they can either disrupt or contribute to the well-being of our environment and public health. The inadequate disposal of sludge can lead to adverse consequences.

Some common components encountered in sludge include:

1. Water: Sludge typically contains a significant water content, often comprising a substantial portion of its total weight. To comprehensively address sludge water management, this chapter will explore four distinct categories based on the binding energies between water and solid-sludge particles, namely free water, interstitial water, surface water, and bound water, detailing how each type can be harnessed or separated through appropriate methods.
2. Solids: The solid fraction of sludge encompasses both organic and inorganic materials, ranging from organic matter like food scraps, plant debris, and microbial

biomass to inorganic materials such as sand, grit, and various chemicals.

3. Organic Matter: Organic components in sludge consist of substances like proteins, carbohydrates, lipids, and cellulose, often biodegradable and conducive to biological treatment. However, the degradation of organic matter can lead to the emission of odors and greenhouse gases, emphasizing the importance of proper management.
4. Nutrients: Sludge may contain essential nutrients like nitrogen and phosphorus, which can lead to nutrient overload in soil and water systems if not managed correctly. This can result in issues such as algal blooms, highlighting the environmental implications while also presenting opportunities for use as fertilizers when properly treated.
5. Heavy Metals: Sludge from industrial and municipal wastewater treatment can contain heavy metals such as lead, cadmium, and mercury, which require careful handling to prevent environmental contamination. The presence of these toxic heavy metals in sludge poses significant threats to human, animal, and plant life, necessitating vigilant monitoring and management.
6. Pathogens: Sludge can harbor pathogens, including bacteria, viruses, and parasites, making proper treatment and disinfection crucial for safe disposal or reuse. This section underscores the importance of pathogen management for public health.
7. Chemicals: Sludge generated from industrial processes may contain various chemicals and contaminants based on the industry, including organic chemicals, solvents, and other substances.
8. Inert Materials: Inert materials like sand, silt, and clay may be present in sludge, especially in municipal wastewater treatment.
9. Fibers: In cases where sludge originates from paper or pulp mills, it may contain fibers from the pulp.
10. Volatile Solids: These are organic solids that can be broken down by microbial activity during biological treatment.
11. Microplastics: Even minuscule microplastic particles can be found in sludge, which can have detrimental effects on aquatic and terrestrial ecosystems, a topic explored in this chapter.

The specific composition of sludge can vary widely, contingent on its source. After its generation, sludge typically undergoes various treatment processes, such as thickening, dewatering, digestion, and, in some cases, incineration or land application, to manage and stabilize its composition (Yu et al. 2010). Proper treatment and disposal are essential to ensure that sludge does not pose environmental or public health risks. This chapter emphasizes the paramount importance of efficient sludge treatment systems in mitigating the risks associated with sewage sludge disposal. Proper treatment is not only environmentally responsible but also imperative for safeguarding public health.

In this regard, effective management of large quantities of sludge involves a three-step process: thickening, dewatering, and drying. Sludge thickening reduces its water content, typically containing 96–100% water, to around 95%. However, this state is still unsuitable for transportation, necessitating a second step, dewatering, to reduce the water content to 70–80% (Tripathy and Rees 2013). Drying, often through thermal processes, is the final step to meet the required criteria for sludge disposal. Numerous factors influence dewatering rates and extents, including extracellular polymer substances, grain size distribution, cation concentration, and bacterial activity. Emerging technologies are also explored, contributing to advancements in deep sludge dehydration.

The escalating efforts in municipal and industrial wastewater purification over recent decades have led to a significant upsurge in sewage sludge production. Regardless of the type of sludge, they share high-water content, colloidal nature, and compressibility (Huang et al. 2022). Properly managing sludge is economically and technologically feasible, reducing water content to enable potential uses, such as fuel or organic matter for composting, while also cutting transport and disposal costs.

Sludge treatment and dewatering techniques are important processes in the field of wastewater and sewage management (Mahmoud et al. 2016). They involve removing water from the sludge generated during wastewater treatment, reducing its volume and making it more manageable for disposal or further processing. Here are some common sludge treatment and dewatering techniques:

1. **Gravity Thickening**: In this method, sludge is allowed to settle under gravity in a large tank, and the solid particles gradually accumulate at the bottom. This thickened sludge can then be removed from the top of the tank.
2. **Centrifuge Dewatering**: Centrifuges use centrifugal force to separate the water from the sludge. The sludge is fed into a rotating drum, and the centrifugal force pushes the water out through the drum's pores, leaving behind drier sludge.
3. **Belt Filter Press**: This technique involves passing the sludge between two belts that squeeze out the water. As the sludge travels along the belts, the pressure applied causes the water to be separated, leaving behind dewatered sludge cakes.
4. **Plate and Frame Filter Press**: Sludge is pumped into a series of chambers with filter plates and frames. Pressure is applied, forcing the water to pass through the filter cloth while the sludge forms a cake between the plates.
5. **Vacuum Filtration**: In this method, a vacuum is applied to a porous medium, such as a filter cloth, to draw water

through the medium, leaving the sludge behind as a dewatered cake.

6. **Drying Beds**: Sludge is spread out on open beds to allow natural evaporation and drainage. The sun and wind aid in reducing the moisture content of the sludge over time.
7. **Thermal Drying**: Sludge can be dried using heat in specialized dryers. This method reduces the moisture content significantly, making the sludge easier to handle and transport.
8. **Sludge Lagoons**: Sludge is stored in large lagoons or ponds, where natural processes like evaporation and microbial activity reduce the water content over time. This technique requires significant space and time.
9. **Microbial Digestion**: This biological process breaks down organic matter in sludge by bacteria in the absence of oxygen. It produces biogas (methane) and a more stabilized sludge, reducing its volume and improving its handling properties.
10. **Composting**: This method involves mixing sludge with organic materials like wood chips or straw to encourage microbial decomposition. The process generates heat, which further aids in dewatering and stabilizing the sludge.
11. **Chemical Conditioning**: Chemicals called flocculants are added to the sludge to create larger particles that settle more readily. This can improve the efficiency of other dewatering methods.
12. **Electrochemical processes**: The energy consumption of electrochemical sludge treatment processes is contingent upon several variables. These variables include the choice of electrode materials, system configuration, applied voltage, electrolyte composition, sludge properties, treatment duration, and overall system efficiency (Chen et al. 2019). This passage highlights the importance of fine-tuning these variables, not only to minimize energy usage but also to promote environmentally sustainable and eco-conscious approaches in the management of sludge.

The choice of sludge treatment and dewatering technique depends on factors such as the characteristics of the sludge (Feijoo et al. 2023), the required level of dewatering (Dey et al. 2022), available resources, and environmental regulations. Often, a combination of methods may be used to achieve the desired results effectively.

## 1.2 Gravitational Sludge Thickening

Gravity thickening is a common sludge treatment technique used to reduce the water content and volume of sludge generated during wastewater treatment processes. This process involves allowing sludge to settle under the influence of gravity in order to separate the solid particles from the water (Zhu et al. 2020). Gravity thickening is often one of the preliminary steps in the sludge management process before further dewatering or disposal. Here's how the gravity thickening process works:

1. **Sludge Collection**: After primary and secondary treatment processes, wastewater treatment plants collect the settled sludge, which is a mixture of solids and water.
2. **Thickening Tank**: The collected sludge is then transferred to a thickening tank, which is a large, shallow basin or tank designed to facilitate settling. Inside the tank, the sludge particles are allowed to settle to the bottom due to their greater density.
3. **Gravity Settling**: As sludge particles settle, they form a layer of sludge solids at the bottom of the tank, while the relatively cleaner water rises to the top.
4. **Sludge Removal**: The thickened sludge, also known as "underflow," is periodically removed from the bottom of the thickening tank. The frequency of removal depends on the rate of sludge generation and the desired degree of thickening.
5. **Water Removal**: The water separated from the sludge, known as "overflow," is typically redirected for further treatment processes or returned to the wastewater treatment process.

Gravity thickening has several advantages:

- **Simplicity**: Gravity thickening is a straightforward and easy-to-operate process that requires minimal equipment and maintenance.
- **Low Energy Consumption**: The process relies on gravity and does not require significant energy input, making it energy-efficient.
- **Cost-Effective**: Since the process is simple and requires minimal equipment, it is generally cost-effective to implement.

However, there are also some limitations and considerations:

- **Space Requirements**: Gravity thickening requires a significant amount of space, as the thickening tanks need to be relatively large.
- **Time-Consuming**: The thickening process can be slow, especially for sludges with lower settling rates.
- **Limited Thickening**: Gravity thickening might not achieve very high levels of solids concentration. Further dewatering processes might be required for more significant volume reduction.

- **Dependence on Sludge Characteristics**: The effectiveness of gravity thickening depends on the characteristics of the sludge, including its particle size, settling rate, and solids content.

Gravity thickening is often used as a preliminary step to reduce the volume of sludge before employing other dewatering techniques such as centrifugation, filtration, or mechanical dewatering methods (Wu et al. 2016). The choice of dewatering technique depends on factors like the desired final solids concentration, available resources, and the specific requirements of the treatment facility.

## 1.3 Centrifuge Dewatering

Centrifuge dewatering is a highly effective mechanical method used to further reduce the moisture content of sludge after gravity thickening or other preliminary treatment processes (Zhu et al. 2020). Centrifuges use centrifugal force to separate the water from the sludge solids, resulting in a drier and more concentrated sludge cake. This technique is widely used in wastewater treatment plants and industrial applications for sludge management. Here's how the centrifuge dewatering process works:

1. **Feed and Conditioning**: The sludge to be dewatered is fed into the centrifuge, often after undergoing some form of conditioning. Chemical additives, known as flocculants, might be added to enhance the aggregation of sludge particles, improving the separation efficiency.
2. **Acceleration and Separation**: The centrifuge rotates at high speeds, typically thousands of revolutions per minute (RPM). This rapid rotation generates centrifugal force, which pushes the sludge solids outward toward the walls of the centrifuge drum.
3. **Formation of Cake and Liquid Separation**: As the sludge solids are forced against the inner wall of the drum, water is forced out through the pores of the drum's lining, leaving behind a layer of dewatered sludge known as the "cake."
4. **Cake Discharge**: Once the dewatering process is complete, the rotating drum is slowed down, and the dewatered cake is scraped off the drum's surface using a scraper or a mechanical device. The cake is then collected for further processing or disposal.
5. **Liquid Discharge**: The separated liquid, known as the "centrate," is discharged from the centrifuge through a separate outlet. This liquid might be recycled back into the wastewater treatment process or subjected to additional treatment before disposal.

Centrifuge dewatering offers several advantages:

- **High Dewatering Efficiency**: Centrifuges can achieve significant moisture reduction, resulting in a relatively dry sludge cake with a high solids concentration.
- **Quick Processing**: Centrifuge dewatering is a relatively fast process, allowing for the continuous treatment of sludge.
- **Reduced Volume**: The dewatered sludge cake has a reduced volume, which leads to decreased disposal and transportation costs.
- **Automation**: Modern centrifuges often come equipped with automation and control systems, making them easy to operate and monitor.

However, there are also some considerations and limitations:

- **Energy Consumption**: Centrifuges require substantial energy input due to the high rotational speeds. This can lead to higher operating costs compared to some other dewatering methods.
- **Maintenance**: Maintaining and servicing centrifuges can be complex and might require specialized knowledge and resources.
- **Initial Investment**: The purchase and installation of centrifuges can be expensive.
- **Variability**: The effectiveness of centrifuge dewatering can vary depending on the characteristics of the sludge, including its particle size, flocculation behavior, and solids content.

Centrifuge dewatering is often employed when a high level of moisture reduction is required or when the sludge's characteristics make it suitable for this method (Zhang et al. 2022). It can be used as a standalone dewatering process or in combination with other techniques to achieve the desired final sludge characteristics.

## 1.4 A Belt Filter Press (BFP)

A Belt Filter Process (BFP) is a mechanical dewatering device commonly used to further reduce the moisture content of sludge that has undergone preliminary treatment processes such as gravity thickening. The BFP operates by passing sludge between two tensioned belts, which squeeze out the water, resulting in a dewatered sludge cake (Saveyn et al. 2010). This technique is popular in wastewater treatment plants and various industrial applications due to its efficiency and ability to handle a wide range of sludge types (Zhang et al. 2022). Here's how the Belt Filter Press treatment process works:

1. **Sludge Preparation**: The sludge to be dewatered is typically conditioned with polymer flocculants or other chemicals to aid in particle aggregation and improve dewatering efficiency.
2. **Belt Loading**: The conditioned sludge is introduced onto the lower belt of the BFP. As the belts move forward, the sludge is evenly spread across the width of the belts.
3. **Gravity Drainage**: The sludge on the lower belt goes through a gravity drainage phase as it moves along the initial section of the press. In this phase, some free water drains off due to gravity.
4. **Compression Zone**: As the sludge progresses through the press, it enters the compression zone where the upper and lower belts come together. The belts are then compressed against each other, and the sludge is further squeezed between them.
5. **Dewatering**: The compression of the belts creates pressure, causing water to be forced out of the sludge through the belt pores. The released water, known as the filtrate, is collected and discharged.
6. **Dewatered Cake Formation**: As water is squeezed out, the sludge forms a dewatered cake between the belts. The cake gradually builds up as the sludge moves along the press.
7. **Cake Discharge**: Once the dewatering process is complete, the dewatered cake is scraped off the belts using a scraper or mechanical device. The cake is then collected for further treatment or disposal.
8. **Belt Cleaning**: After the cake is removed, the belts are cleaned to prepare them for the next cycle. This can involve spraying water or cleaning solutions on the belts to remove residual sludge particles.

Belt Filter Press treatment offers several advantages:

- **High Dewatering Efficiency**: BFPs can achieve a significant reduction in moisture content, producing a drier sludge cake with a higher solid concentration.
- **Continuous Operation**: Belt Filter Presses are designed for continuous operation, making them suitable for handling large volumes of sludge.
- **Automated Control**: Modern BFPs often have automated control systems that help optimize the dewatering process and monitor performance.
- **Compact Design**: BFPs are relatively compact compared to some other dewatering methods, making them suitable for facilities with limited space.

However, there are also considerations and limitations:

- **Energy Consumption**: BFPs require energy to operate, particularly due to the compression and belt movement mechanisms.
- **Maintenance**: Regular maintenance and cleaning of the belts are required to ensure optimal performance.
- **Initial Investment**: The cost of purchasing and installing a Belt Filter Press can be relatively high.
- **Belt Wear**: Over time, the belts can experience wear and require replacement, which adds to operating costs.

Belt Filter Press treatment is commonly chosen when a relatively high degree of dewatering is required and when the sludge characteristics are compatible with the process. It can be used as a standalone dewatering method or combined with other techniques for further moisture reduction.

## 1.5 A Plate and Frame Filter Press

A plate and frame filter press is a mechanical dewatering device commonly used in various industries, including wastewater treatment, to further reduce the moisture content of sludge and separate solids from liquids (Cao et al. 2021). This type of filter press consists of a series of vertical plates and frames, which are arranged alternately and held together under pressure. Here's how the Plate and Frame Filter Press treatment process works:

1. **Sludge Preparation**: The sludge to be dewatered is often conditioned with chemicals to enhance particle aggregation and improve dewatering efficiency.
2. **Assembly**: The Plate and Frame Filter Press consists of a series of plates and frames. Each frame has a filter cloth or filter media on both sides. The plates and frames are aligned alternately, creating a series of chambers between them.
3. **Filling**: The conditioned sludge is pumped into the chambers created by the plates and frames. As the sludge fills the chambers, the filter media on the plates retains the solid particles while allowing the liquid to pass through.
4. **Filtration**: Once the chambers are filled with sludge, the pressure is applied. The pressure forces the liquid, or filtrate, to pass through the filter media and flow out of the press.
5. **Cake Formation**: As the liquid is forced out, the solid particles in the sludge build up on the filter media, forming a dewatered cake.

6. **Cake Removal**: After the dewatering process is complete, the pressure is released, and the plates and frames are separated. The dewatered cake can then be removed from the filter press, often with the help of scrapers or mechanical devices.
7. **Plate Cleaning**: The filter cloths or media on the plates and frames may need to be cleaned or replaced after each cycle to ensure optimal performance.

Plate and Frame Filter Press treatment offers several advantages:

- **High Dewatering Efficiency**: This method can achieve significant moisture reduction, resulting in a drier sludge cake with a high solids concentration.
- **Flexibility**: Plate and Frame Filter Presses can handle a wide range of sludge types and sizes.
- **Effective Solid–Liquid Separation**: The filter media effectively separate solids from liquids, producing a clear filtrate and a dewatered cake.
- **Reliability**: Once the filter press is set up and operational, it can run reliably and continuously.

However, there are also considerations and limitations:

- **Labor-Intensive**: Operating Plate and Frame Filter Presses can require more manual intervention compared to some other dewatering methods.
- **Cycle Time**: The dewatering process can take longer due to the filling, filtration, and cake removal steps.
- **Maintenance**: Regular maintenance, including cleaning the plates and frames, is necessary to ensure consistent performance.
- **Initial Investment**: The cost of purchasing and setting up a Plate and Frame Filter Press can be relatively high.

Plate and Frame Filter Press treatment is often chosen when a high degree of dewatering is required, and when the sludge characteristics are compatible with this method. It can be used as a standalone dewatering process or combined with other techniques for further moisture reduction.

## 1.6 Vacuum Filtration

Vaccum filteration is a dewatering technique used to separate solids from liquids by applying a vacuum to draw the liquid through a porous filter medium. This process is commonly employed in various industries, including wastewater treatment, to further reduce the moisture content of sludge and produce a drier solid cake. Here's how the vacuum filtration treatment process works:

1. **Sludge Preparation**: Similar to other dewatering methods, the sludge is often conditioned with chemicals or polymers to improve particle aggregation and enhance filtration efficiency.
2. **Filter Medium Setup**: The sludge is placed onto a filter medium, which can be a filter cloth, filter paper, or other porous material. The filter medium is supported by a rigid structure or a porous plate.
3. **Application of Vacuum**: A vacuum source is applied to the underside of the filter medium. The vacuum creates a pressure differential between the sludge side and the underside of the filter medium. This pressure difference draws the liquid (filtrate) through the filter medium, leaving behind a dewatered cake.
4. **Cake Formation**: As the filtrate is drawn through the filter medium, the solid particles in the sludge build up on the surface of the medium, forming a dewatered cake.
5. **Cake Removal**: Once the dewatering process is complete, the vacuum is released, and the dewatered cake can be removed from the filter medium.
6. **Cleaning and Maintenance**: The filter medium may need to be cleaned or replaced after each cycle to maintain optimal filtration performance.

Vacuum filtration offers several advantages:

- **Effective Solid–Liquid Separation**: Vacuum filtration provides effective separation of solids from liquids, producing a clear filtrate and a dewatered cake.
- **High Dewatering Efficiency**: This method can achieve significant moisture reduction, resulting in a drier cake with a high solids concentration.
- **Flexibility**: Vacuum filtration can be used with a variety of filter media and sludge types.
- **Automation**: Modern vacuum filtration systems often come with automated controls, making them easier to operate and monitor.

However, there are also considerations and limitations:

- **Vacuum Equipment**: Setting up and maintaining the vacuum source and associated equipment can be complex and may require skilled operators.
- **Cycle Time**: Vacuum filtration cycles can take time, as the process relies on the gradual passage of liquid through the filter medium.

- **Maintenance**: Regular maintenance of the equipment, including cleaning the filter medium and maintaining the vacuum system, is necessary for consistent performance.
- **Initial Investment**: The cost of purchasing and setting up vacuum filtration equipment can be relatively high.

Vacuum filtration treatment is often chosen when a high degree of dewatering is required, and when the sludge characteristics are compatible with this method (Huang et al. 2022). It can be used as a standalone dewatering process or combined with other techniques for further moisture reduction.

## 1.7 Drying Beds

Drying beds are a natural method of sludge dewatering that involve spreading sludge on open, shallow beds to allow moisture to evaporate and drain away. This process is widely used for dewatering sewage sludge and industrial wastewater sludge (Huang et al. 2022; Zhang et al. 2021a). Drying beds are particularly useful in areas with sufficient space and where climate conditions are favorable for evaporation. Here's how the drying beds treatment process works:

1. **Bed Preparation**: Drying beds consist of a prepared bed surface, often lined with a layer of fine gravel or sand to support the sludge and aid in drainage. The beds may be divided into smaller compartments to facilitate the drying process.
2. **Sludge Application**: The dewatered sludge is evenly spread on the surface of the drying bed using mechanical equipment or by hand.
3. **Drying and Drainage**: Over time, as the sludge sits exposed to the air, sunlight, and wind, moisture starts to evaporate from the sludge and drain through the gravel or sand layer beneath. This results in the gradual reduction of the sludge's moisture content.
4. **Crust Formation**: As the surface of the sludge dries, a crust or skin can form on top. This crust helps prevent the loss of finer sludge particles due to wind and also reduces odor emissions.
5. **Periodic Turnover**: Depending on the specific conditions and characteristics of the sludge, the sludge may need to be periodically turned or mixed to expose moist portions to the air, facilitating more even drying.
6. **Final Dewatered Cake**: After the sludge has undergone sufficient drying and dewatering, the result is a dewatered cake with a reduced moisture content and increased solids concentration.

Drying beds offer several advantages:

- **Low Energy Consumption**: Drying beds rely on natural processes such as evaporation and drainage, so they have minimal energy requirements.
- **Cost-Effectiveness**: Drying beds are relatively inexpensive to construct and maintain, making them suitable for locations with limited resources.
- **Environmental Compatibility**: The process is environmentally friendly, as it relies on natural processes and doesn't involve complex machinery or chemicals.
- **Minimal Operation and Maintenance**: Drying beds require less ongoing maintenance compared to mechanical dewatering methods.

However, there are also considerations and limitations:

- **Space Requirements**: Drying beds require a significant amount of space, so they may not be suitable for urban areas with limited land availability.
- **Weather Dependence**: The effectiveness of drying beds is highly dependent on weather conditions, including sunlight, temperature, wind, and humidity. Rain or adverse weather conditions can slow down the drying process.
- **Sludge Characteristics**: Drying beds are best suited for sludges with higher solids content and better dewatering properties.
- **Longer Dewatering Time**: Drying beds generally require longer dewatering times compared to some mechanical methods.

Drying beds treatment is often chosen in situations where space is available, and the climate is conducive to the natural evaporation of moisture from the sludge. It can be used as a standalone dewatering method or in combination with other techniques for further moisture reduction.

**General Sludge Drying Techniques**

Thermal drying of dewatered sludge becomes necessary under specific circumstances. This includes situations where the sludge must attain a particular heat value for efficient and stable combustion, or when the final water content (WC) needs to be reduced to a level suitable for fertilizer production and landfilling (Qasem et al. 2021). The ideal WC for dried sludge depends on the requirements of subsequent disposal methods, often guided by national standards and technical regulations governing sludge management. The choice of sludge drying methods should align with the specific requirements, considering factors such as WC, particle size, and pH, which are associated with the subsequent disposal pathways.

**Fig. 1** Indirect drying by vertical multi-tray dryer. Reprinted from Nayak et al. (2018) and is an open-access article (copyright©2023 by authors) distributed under the terms and conditions of the Creative Commons Attribution (CC BY) license

Concerning heat and mass transfer, sludge drying methods are commonly categorized into three types: direct drying, indirect drying, and combined drying (Wei et al. 2018). Multiple sludge dryers are available to increase the solid content of dewatered sludge, achieving levels of up to 55% or even 95%.

**Direct drying**, as the name implies, involves drying the sludge through direct thermal contact with the heating medium, like hot air, flue gas, or vapor. While direct drying methods are relatively straightforward in design, managing the evaporated water (Sha et al. 2021), particularly when recycling the heat source for energy conservation, can pose a challenge.

On the other hand, **indirect drying** offers the advantage of generating minimal exhaust gas and facilitating the reutilization of the heating medium. However, indirect drying often exhibits lower heat transfer efficiency and drying rates compared to direct drying (Solé-Bundó et al. 2019). The heating process involves the sludge being heated from the top and descending through smaller heating plates, as illustrated in Fig. 1.

**Combined drying** incorporates both convection and conduction, resulting in enhanced drying efficiency. However, this improved efficiency is accompanied by increased facility complexity and maintenance costs.

In addition to these conventional sludge drying methods, emerging techniques exhibit promise for industrial utilization. For example, sludge solar drying is evaluated in a pilot-scale greenhouse facility covering an area of 66 $m^2$. The duration required to achieve a dry sludge product with a solid content of up to 95% varies between 8 and 31 days, depending on weather conditions. Despite its economic and energy efficiency, **solar drying** faces limitations due to its significant space requirements and susceptibility to weather-related constraints.

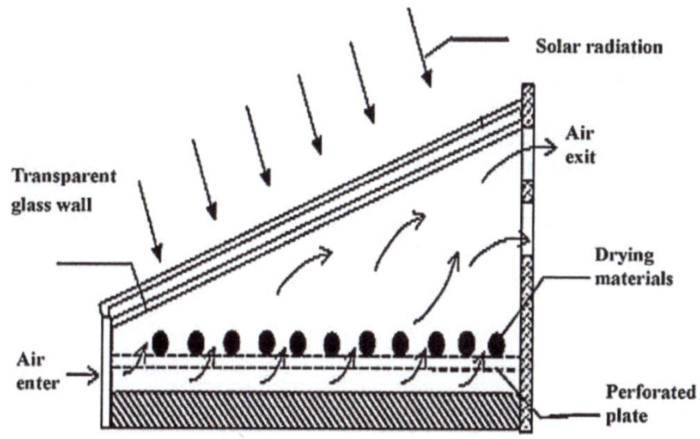


**Fig. 2** Solar drying. Reprinted from Prakash and Kumar (2013) and is an open-access article (copyright©2023 by authors) distributed under the terms and conditions of the Creative Commons Attribution (CC BY) license

The greenhouse features a triangular heap of sludge, with a key component being an automated turnover system to refresh the heating surface, as depicted in Fig. 2.

**Microwave drying**, known for its potential and efficiency, offers significantly shorter drying times. Reports indicate that microwave drying can reduce drying durations to at least one-tenth of those required by air convective drying (Chang et al. 2023). Nonetheless, there is room for improvement in enhancing the disposal capacity of microwave drying for practical industrial use.

**Fry drying**, another innovative method, can rapidly reduce the WC to below 10% in just 15 min. After fry drying, the sludge' heat value exceeds 5300 kcal/kg, making it suitable for use as solid fuel. Additionally, **electromechanical dewatering** has been applied to sludge dewatering, resulting in a WC reduction to 60%.

Among these drying technologies, **indirect drying** offers several notable advantages, including minimal exhaust gas

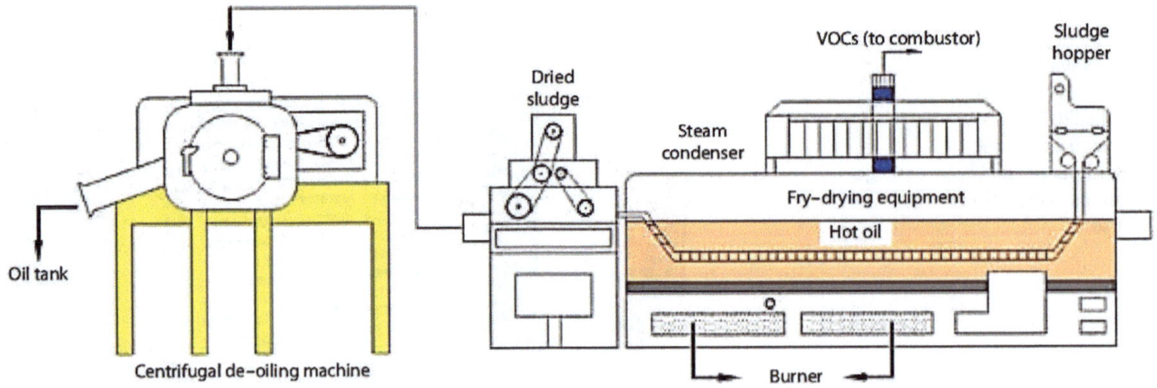


**Fig. 3** Fry drying. Reprinted from Ohm et al. (2009) and is an open-access article (copyright© 2023 by authors) distributed under the terms and conditions of the Creative Commons Attribution (CC BY) license

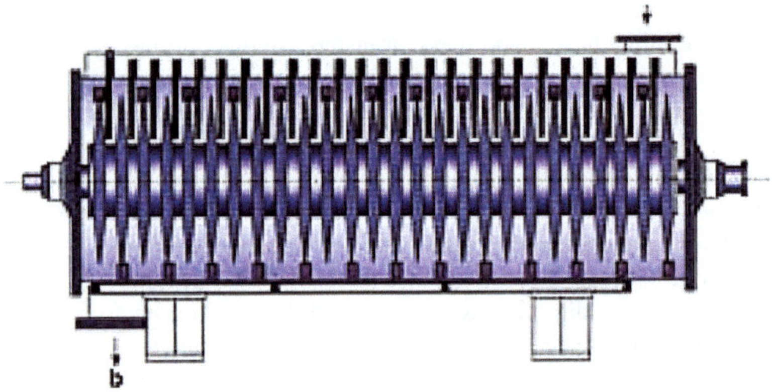

**Fig. 4** Rotary disc dryer. Reprinted from Park et al. (2022) and is an open-access article (copyright© 2023 by authors) distributed under the terms and conditions of the Creative Commons Attribution (CC BY) license

production, high heat exchange efficiency, and stable operational conditions. In contrast, **direct drying techniques** that rely on flue gas as the heat medium can encounter various intricate challenges and risks.

In the process depicted in Fig. 3, sludge enters waste oil at a high temperature, where the water in the sludge evaporates rapidly to complete the drying process (Chang et al. 2023).

### Rotary Disc Drying Techniques for Sludge

The rotary disc dryer consists of a shell, rotor, and motor. Sludge is slowly conveyed through the entire dryer and heated by a rotating disc at temperatures ranging from 200 to 300 °C in thermal oil or steam. It is widely used, either as a standalone unit or as the second stage following a thin film dryer. Figure 4 illustrates the key components and operational principles of rotary disc dryers. A typical rotary disc dryer includes a stator equipped with scrapers and a rotor housing a drive motor and agitators. The interaction between the rotor and scrapers promotes thorough mixing of the sludge. Saturated steam condenses into water after undergoing complete heat exchange with the sludge through the rotating disc. Notable suppliers of disc dryers include US Filter (USA), Altas-Stord (Denmark), Mitsubishi (Japan), and TDG-MT (China). The water content (WC) of the sludge after disc drying is reported to range between 15 and 60% under various operating conditions (Wang et al. 2018), as depicted in Fig. 4.

### Belt Drying Systems for Sludge

Belt dryers are available in various configurations, including open, semi-open, and closed-loop systems, depending on the desired operational setup (Cao et al. 2021). These dryers employ various energy sources like hot air, steam, and hot oil to provide the necessary heating fluid for the heat exchanger in the drying air system.

In the operation of a belt dryer, hot air is directed from the supply plenum downward onto the sludge within the belt passageway. The dewatered sludge is then transported along a slow-moving, horizontal stainless-steel belt, passing through an enclosed chamber where hot air circulates (Wu et al. 2016). Heat exchange between the hot air and the sludge occurs through a combination of convection and conduction.

One notable example is the BioCon™ belt dryer, which adopts a dual-belt, low-temperature drying approach. This system employs depositors to lay down thin strings of dewatered sludge on the first drying belt. The dryer operates within a temperature range, starting at 180 °C on the first belt

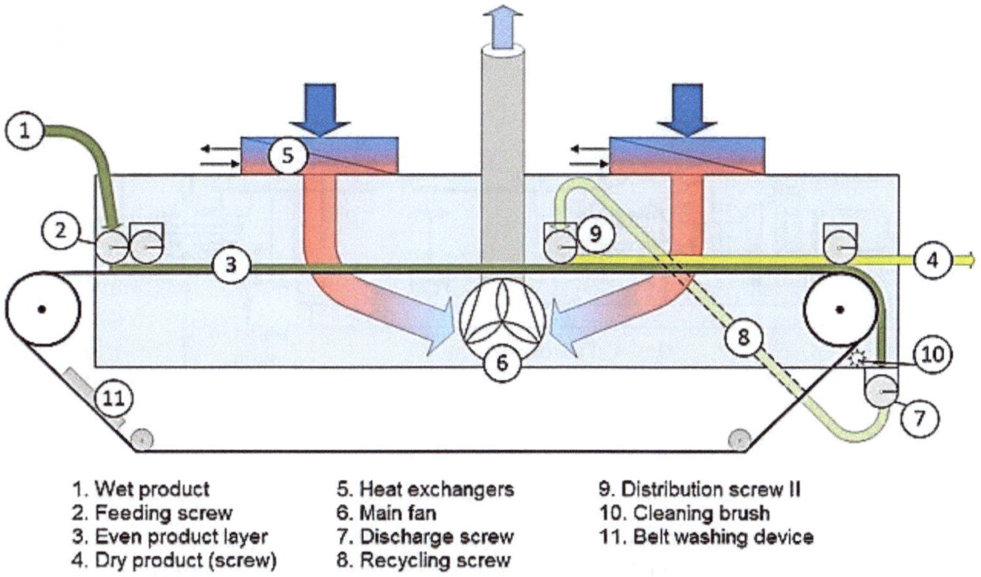


**Fig. 5** Belt dryers. Reprinted from Turunen (2017) and is an open-access article (copyright© 2023 by authors) distributed under the terms and conditions of the Creative Commons Attribution (CC BY) license

and gradually reducing to 80–100 °C on the final belt. The sludge remains within the dryer for over 60 min, ensuring the complete elimination of harmful pathogens.

Due to its user-friendly operation and energy efficiency, the belt dryer is widely utilized in sludge drying applications, particularly in processes involving sludge co-processing, such as combined drying and incineration (Usack et al. 2019). This integrated approach has already been successfully implemented in numerous large-scale applications.

In the drying process depicted in Fig. 5, sludge on the belt is heated by hot air from bottom to top.

### Paddle Drying for Sludge Treatment

In the realm of indirect drying techniques, recuperative methods are frequently utilized for sludge dewatering, with paddle drying standing out as a notable example among them. In comparison with other indirect drying approaches, the paddle dryer, which employs saturated steam or hot oil as the heat medium, offers several exceptional advantages (Rao et al. 2021). The use of low-pressure steam, typically obtained from coal-fired power plants and ranging from 0.4 to 0.6 MPa, significantly reduces energy consumption. The unique design of wedge-shaped blades lends itself to self-cleaning, thereby enhancing heat exchange efficiency. As the hollow blades stir the sludge and transform it into pellets, it becomes more suitable for combustion in fluidized bed furnaces. Furthermore, the paddle dryer, as an indirect drying method, brings the benefits of compactness and reduced exhaust volumes.

However, from a process perspective, sewage sludge poses a challenge due to its transition from an initial pasty state to a sticky one during drying (Pilli et al. 2011). Paddle drying, being one of the most complex unit operations, involves concurrent heat and mass transfer (Lagnika et al. 2018). Existing research has primarily focused on aspects such as drying kinetics, heat transfer efficiency, and the development of drying models for paddle dryers. Nevertheless, providing a rheological description of sewage sludge remains a scientifically significant endeavor. The ability to simulate the drying process in a paddle dryer is crucial for the industrial application of continuous paddle dryers.

Currently, primary contact drying models related to heat and mass transfer include the penetration model, pore network model, population balance model, and discrete element method. Concerning flow modeling (Pilli et al. 2011), the Markov chain model has been found suitable for single-shaft paddle dryers. In this system, different states within the dryer represent positions where a unit of sludge might be located, with each corresponding to the vicinity of a paddle. The matrix of transition probabilities in this system reflects the sludge's ability to move between various cells (Li et al. 2019). The model's reliability can be further confirmed by comparing test results with the simulated results of the sludge tracer's residence time distribution within the dryer.

Each of these models has its own set of advantages and disadvantages. Nevertheless, the contact drying process remains incompletely understood due to the coexistence of several physical phenomena (Li et al. 2017). Consequently, some combined methods, such as the Markov chain model and penetration model, have been proposed to simulate the penetration and drying process of sludge in a single-axis continuous paddle dryer (Chowdhury et al. 2022). Despite the simplifications introduced, experimental and simulated drying kinetics consistently demonstrate a high degree of agreement under various conditions.

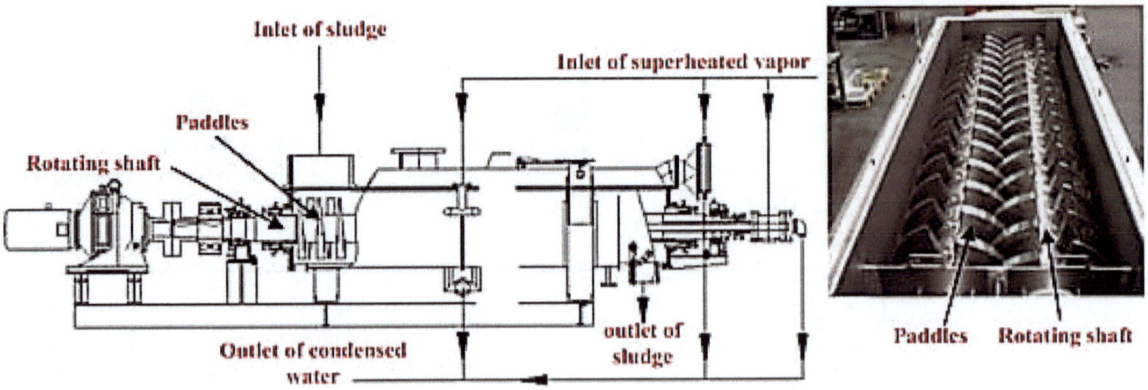


**Fig. 6** Paddle dryers. Reprinted from Liu et al. (2023) and is an open-access article (copyright© 2023 by authors) distributed under the terms and conditions of the Creative Commons Attribution (CC BY) license

As seen in Fig. 6, the process involves stirring and heating the sludge with a paddle and jacket, resulting in a constantly changing drying surface to enhance heating efficiency.

**Thermal Drying Treatment of Sludge**

Thermal drying is a sludge dewatering technique that involves using heat to evaporate moisture from sludge, resulting in a significant reduction in volume and weight. This method is particularly effective for achieving a high level of moisture reduction, making the sludge more manageable for disposal or reuse (Cao et al. 2021). Thermal drying can be accomplished through various technologies, each with its own advantages and considerations. Here are some common methods of thermal drying for sludge treatment:

1. **Direct Dryers**: Direct dryers, such as rotary dryers and drum dryers, expose sludge to direct contact with hot air or gases. The sludge is agitated and moved through the drying chamber as it comes into contact with the hot air. Evaporation occurs as moisture is vaporized from the sludge.
2. **Indirect Dryers**: Indirect dryers, like paddle dryers and belt dryers, use heated surfaces or heated gases to indirectly transfer heat to the sludge. The sludge doesn't come into direct contact with the heating element, reducing the risk of overheating or burning the sludge.
3. **Fluidized Bed Dryers**: In fluidized bed dryers, sludge is suspended on a bed of hot particles, usually sand or inert material. As hot air or gases flow through the bed, the sludge particles become fluidized and heat is transferred, causing moisture to evaporate.
4. **Belt Dryers**: Belt dryers consist of a conveyor belt that transports sludge through a drying chamber. The sludge is spread in a thin layer on the belt, and hot air is passed over and through the sludge, facilitating evaporation.
5. **Solar Drying**: Solar drying uses sunlight to heat the sludge and facilitate evaporation. Sludge is spread out on drying beds, similar to the natural drying bed method, but solar collectors or reflective surfaces enhance the drying process.

Benefits of thermal drying include:

- **High Dewatering Efficiency**: Thermal drying can achieve very low moisture levels, resulting in a significantly reduced volume of dewatered sludge.
- **Pathogen Reduction**: The high temperatures involved in thermal drying can contribute to pathogen reduction in the sludge.
- **Stabilization**: The heat applied during thermal drying can lead to the reduction of volatile organic compounds (VOCs) and other potentially harmful substances.
- **Energy Recovery**: Some thermal drying methods allow for the recovery of energy from the process, such as the use of waste heat from other industrial processes.

Considerations and limitations include:

- **Energy Consumption**: Thermal drying requires substantial energy input, which can result in high operating costs.
- **Odor Control**: Thermal drying can produce odors if not properly controlled, especially for sludge with high organic content.
- **Equipment Maintenance**: The drying equipment requires regular maintenance and cleaning to ensure optimal performance.
- **Capital Investment**: The initial investment for thermal drying equipment can be relatively high.
- **Environmental Impact**: The energy consumption and potential emissions associated with thermal drying need to be considered from an environmental standpoint.

Thermal drying is often chosen when a high level of moisture reduction is required and when the sludge characteristics make it suitable for this method. The choice of thermal drying technology depends on factors like available resources, the desired final solids concentration, and environmental regulations.

## 1.8 Sludge Lagoons Treatment of Sludge

Sludge lagoons, also known as sludge drying beds or sludge drying lagoons, are natural evaporation and drying systems used to reduce the moisture content of sludge through exposure to air, sunlight, and wind (Corrêa et al. 2011). Sludge lagoons are large, open-air basins, or ponds designed to facilitate the natural drying and dewatering of sludge (Dong et al. 2013). This method is often used as an economical and environmentally friendly way to manage sludge generated from wastewater treatment processes.

Here's how the sludge lagoons treatment process generally works:

1. **Sludge Collection**: The sludge, often previously treated through primary and secondary wastewater treatment processes, is collected and transported to the sludge lagoons.
2. **Lagoon Design**: The sludge lagoons are designed to provide a shallow, flat surface area to spread the sludge evenly. The bottom of the lagoon may be lined to prevent seepage into the ground.
3. **Sludge Spreading**: The sludge is spread evenly across the surface of the lagoon. This allows for maximum exposure to air, sunlight, and wind, which are the key factors driving the drying process.
4. **Drying and Evaporation**: Over time, as the sludge is exposed to the elements, moisture within the sludge begins to evaporate. This process reduces the moisture content of the sludge and gradually dries it out.
5. **Crust Formation**: As the surface layer of sludge dries, a crust or skin may form on top. This helps to prevent wind erosion and retain moisture deeper within the sludge layer.
6. **Drying Period**: The duration of the drying period can vary depending on factors such as weather conditions, sludge composition, and lagoon design. It can take weeks to several months to achieve the desired level of dewatering.
7. **Completion and Removal**: Once the sludge has reached the desired moisture content, it is scraped or removed from the lagoon's surface, leaving behind a dewatered sludge cake.

Benefits of sludge lagoons treatment include:

- **Low Energy Consumption**: Sludge lagoons rely on natural processes of evaporation and drying, requiring minimal energy input.
- **Cost-Effective**: Constructing and maintaining sludge lagoons is generally less expensive than some other dewatering methods.
- **Environmental Compatibility**: The process is environmentally friendly as it utilizes natural elements, avoiding the need for complex machinery or chemicals.
- **Minimal Operation and Maintenance**: Sludge lagoons require less ongoing maintenance compared to mechanical dewatering methods.

Considerations and limitations include:

- **Space Requirements**: Sludge lagoons require a significant amount of space, making them suitable for areas with available land.
- **Weather Dependence**: The effectiveness of sludge lagoons is highly dependent on weather conditions, including sunlight, temperature, wind, and humidity.
- **Drying Time**: The drying process can take longer compared to some mechanical dewatering methods.
- **Odor Management**: Depending on the sludge composition and local conditions, odor management might be necessary to prevent nuisance odor.

Sludge lagoons treatment is commonly chosen in locations with ample space and where climate conditions are conducive to natural evaporation and drying. It can be used as a standalone dewatering method or in combination with other techniques for further moisture reduction. Proper design and management are essential to ensure effective drying and minimize potential environmental impacts.

## 1.9 Biological Digestion Treatment of Sludge

Anaerobic digestion is a biological treatment process involving the decomposition of organic matter in sludge by microorganisms in an oxygen-deprived environment. This method not only reduces sludge volume but also yields valuable biogas, primarily in the form of methane, as a byproduct. Anaerobic digestion is a widely employed sludge treatment technique in wastewater treatment plants and similar facilities, offering advantages that extend beyond mere dewatering.

The diversity of microorganisms plays a crucial role in enhancing the efficiency of sewage sludge treatment, which, in turn, benefits soil quality, aquatic ecosystems, and the environment. The application of microbes in sewage sludge treatment significantly contributes to promoting water self-purification (Chen et al. 2019). Both aerobic and anaerobic microorganisms are instrumental in eliminating organic and inorganic pollutants from sewage sludge, and their strategic utilization can lead to cost-effective, efficient, and sustainable solutions for sludge treatment.

Microbes act as primary agents and universal catalysts for ecological transformations. Leveraging microbes provides a valuable approach to recovering essential nutrients, metals, and organic components from wastewater and sewage sludge while minimizing costs and energy consumption (Chen et al. 2019). Furthermore, there is considerable potential to enhance microbial structures, functionality, and metabolic pathways through the application of molecular and biotechnological methods.

Here's how the microbial digestion treatment process works:

1. **Sludge Collection**: The sludge generated from wastewater treatment processes, often after primary and secondary treatment, is collected and fed into the anaerobic digester.
2. **Digestion Tanks**: Microbial digesters are large tanks or vessels where the sludge is processed. These tanks are designed to maintain an oxygen-free environment to facilitate the growth of anaerobic microorganisms.
3. **Microbial Breakdown**: Anaerobic microorganisms, primarily bacteria, break down the organic matter in the sludge through a series of biochemical reactions. This digestion process generates biogas, which consists of methane, carbon dioxide, and trace amounts of other gases.
4. **Biogas Collection**: The biogas produced during anaerobic digestion is collected from the digester and can be used for various purposes, such as generating energy, heating, or as a fuel source.
5. **Sludge Stabilization**: Microbial digestion not only reduces the volume of sludge but also stabilizes it by reducing the organic content and pathogen levels. This leads to a more manageable and less odorous sludge product.
6. **Dewatering**: While microbial digestion doesn't primarily focus on dewatering, it does reduce the water content of the sludge due to the microbial breakdown of organic matter. Some additional dewatering might be needed depending on the desired solids concentration.

Benefits of microbial digestion treatment include:

- **Biogas Generation**: The production of biogas offers an additional energy source that can be used for electricity generation or heat production.
- **Stabilization**: The digestion process reduces the organic content and pathogens in the sludge, making it less odorous and easier to handle.
- **Volume Reduction**: The digestion process leads to a reduction in the volume of sludge, minimizing disposal and transportation costs.
- **Reduced Environmental Impact**: Microbial digestion contributes to reducing the environmental impact of sludge disposal by stabilizing it and utilizing biogas.

Considerations and limitations include:

- **Digestion Time**: The microbial digestion process can take days to weeks, depending on the specific system and conditions.
- **Temperature Sensitivity**: The efficiency of anaerobic digestion is influenced by temperature. Optimal digestion occurs within a certain temperature range.
- **Equipment and Maintenance**: Microbial digesters require proper design, maintenance, and monitoring to ensure optimal performance and prevent issues.
- **Initial Investment**: The setup and construction of microbial digestion facilities can involve significant capital investment.

The utilization of microbial digestion treatment is frequently preferred due to its capacity for energy generation and the advantages it offers in terms of sludge stabilization. It can be employed as a stand-alone treatment method or in conjunction with other approaches to create a more comprehensive sludge management strategy. The decision to apply microbial digestion hinges on factors such as available resources, facility objectives, and compliance with regulatory standards.

This approach effectively reduces the impacts of harmful substances on human health, aquatic ecosystems, and the environment. Moreover, the text also explores current trends and future prospects regarding the adoption of enhanced microbial techniques for sewage sludge treatment. This discussion aims to offer a structured and comprehensive insight into sustainable sewage sludge management through the use of microbial methods.

## 1.10 Composting Treatment of Sludge

Composting is a natural biological process that transforms organic materials, including sludge, into a stable, nutrient-rich material called compost. Composting involves the controlled decomposition of organic matter by microorganisms under aerobic conditions. Composting sludge can be an effective way to further stabilize the material, reduce its volume, and create a valuable end product that can be used as a soil amendment or fertilizer. Here's how the composting treatment process for sludge works:

1. **Sludge Preparation**: The sludge to be composted is often mixed with bulking agents such as wood chips, straw, sawdust, or other carbon-rich materials. These bulking agents help create aeration, improve the structure of the compost pile, and balance the carbon-to-nitrogen ratio.
2. **Composting Piles or Bins**: The mixture of sludge and bulking agents is arranged in composting piles or bins. These piles or bins are designed to allow air to circulate within the material, promoting aerobic microbial activity.
3. **Aeration and Turnover**: During the composting process, the piles or bins are periodically turned or aerated to ensure even decomposition and to maintain aerobic conditions. Turning helps to provide oxygen to the microorganisms and prevent anaerobic conditions.
4. **Microbial Decomposition**: Under aerobic conditions, microorganisms like bacteria, fungi, and other decomposers break down the organic matter in the sludge and bulking agents. This process generates heat as a byproduct, contributing to pathogen reduction and further decomposition.
5. **Temperature Monitoring**: The composting process generates heat, which can increase the temperature of the compost pile. Monitoring the temperature is important, as certain temperature ranges help in killing pathogens and weed seeds present in the sludge.
6. **Maturation Phase**: After the active composting phase, the compost is allowed to mature or cure. This stage involves less active microbial activity and allows for the breakdown of remaining organic matter, as well as the development of stable humus-like compounds.
7. **Final Product**: Once the composting process is complete, the end product is a stabilized, nutrient-rich material that resembles dark, crumbly soil. This compost can be used as a soil conditioner, fertilizer, or amendment to improve soil structure and fertility.

Benefits of composting treatment include:

- **Pathogen and Weed Seed Reduction**: Composting can significantly reduce the levels of pathogens and weed seeds present in the sludge, making the end product safer for use.
- **Stabilization**: Composting stabilizes the organic matter, resulting in a more predictable and manageable end product.
- **Nutrient Recycling**: The resulting compost is a valuable source of nutrients that can be used to enrich soils and improve plant growth.
- **Waste Reduction**: Composting reduces the volume of sludge, minimizing disposal costs and environmental impact.

Considerations and limitations include:

- **Proper Management**: Composting requires proper monitoring and management to ensure optimal conditions for decomposition and pathogen reduction.
- **Odor Management**: Proper aeration and turning of the compost pile help minimize odors.
- **Time**: Composting is a relatively slow process and can take several weeks to months to produce mature compost.
- **Volume Reduction**: While composting reduces the volume of the material, it might not achieve the same level of volume reduction as some other dewatering methods.

Composting treatment can be a sustainable and environmentally friendly option for managing sludge, especially in situations where the end product can be utilized for soil improvement (Wang et al. 2018). It is important to follow guidelines and regulations for compost quality and application to ensure its safe and effective use.

## 1.11 Chemical Conditioning Treatment of Sludge

Chemical conditioning is a sludge treatment method that involves the addition of chemicals to sludge in order to modify its physical and chemical properties. This treatment is primarily used to improve the dewatering characteristics of sludge, making it easier to remove water and reduce the overall moisture content (Rumky et al. 2020). Chemical conditioning can be applied to various types of sludge generated from wastewater treatment processes to enhance the efficiency of subsequent dewatering methods. Here's how the chemical conditioning treatment process works:

1. **Selection of Chemicals**: Different types of chemicals are used for conditioning sludge, including flocculants, coagulants, polymers, and lime. The choice of chemicals depends on the characteristics of the sludge and the desired treatment outcomes.

2. **Addition of Chemicals**: The selected chemicals are mixed with the sludge, either before or during the dewatering process. The chemicals interact with the sludge particles to improve their aggregation and settleability.
3. **Flocculation**: Flocculants and polymers encourage the aggregation of fine sludge particles into larger flocs. These flocs are easier to settle or dewater due to their increased size and density.
4. **Coagulation**: Coagulants destabilize the repulsive forces between sludge particles, causing them to come together and form larger aggregates.
5. **Neutralization**: Lime is often used to raise the pH of acidic sludge, promoting flocculation and aiding in the dewatering process.
6. **Mixing**: The mixture of sludge and chemicals is thoroughly mixed to ensure proper dispersion of the chemicals throughout the sludge.
7. **Dewatering**: The conditioned sludge is then subjected to a dewatering method such as centrifugation, filtration, or gravity thickening. The improved aggregation and settling properties facilitated by chemical conditioning result in more efficient dewatering.

Benefits of chemical conditioning treatment include:

- **Improved Dewatering**: Chemical conditioning enhances the dewatering efficiency of sludge, leading to a reduction in moisture content and increased solids concentration.
- **Faster Dewatering**: The use of chemicals can accelerate the dewatering process, reducing the time required for water removal.
- **Consistency**: Chemical conditioning can help stabilize the dewatering performance by mitigating the impact of variations in sludge characteristics.
- **Reduced Disposal Costs**: Dewatered sludge with higher solids content occupies less volume, reducing disposal and transportation costs.

Considerations and limitations include:

- **Chemical Selection**: The choice of chemicals must be based on the sludge characteristics, desired outcomes, and environmental considerations.
- **Dosage Optimization**: Finding the appropriate dosage of chemicals is important to achieve effective conditioning without excess usage.
- **Environmental Impact**: The use of chemicals should be managed to prevent adverse effects on the environment.
- **Regulatory Compliance**: Chemical usage should comply with local regulations and guidelines.

Chemical conditioning treatment can be an effective step in sludge management when dewatering efficiency needs to be improved. It can be applied as a standalone treatment or in combination with other methods to achieve the desired level of moisture reduction and solid content.

## 1.12 Electrochemical Process

The electrochemical process provides an efficient means of eliminating pollutants, including heavy metals and organic compounds, from sludge. These methods are highly selective in targeting specific contaminants for removal. Approaches like electrocoagulation and electrooxidation have demonstrated their effectiveness in treating sludge, thus improving its overall quality. Moreover, electrochemical treatment opens avenues for recovering valuable resources from sludge, especially compounds rich in phosphorus. This reduces reliance on chemical fertilizers and supports the generation of renewable energy through methods like microbial fuel cells and anaerobic digestion. Electrochemical disinfection techniques, such as electrochlorination and electrooxidation, have proven to be successful in deactivating pathogens in sludge, ensuring its safe disposal or potential reuse.

These versatile electrochemical processes can be applied to various types of sludge, including those from wastewater treatment, industrial processes, and agriculture. They can be adapted to different scales and configurations, suitable for both centralized and decentralized treatment systems. However, successful implementation of electrochemical sludge treatment requires careful consideration of critical factors. While cost-effectiveness can be improved through further research, scaling up these processes for large-scale treatment may require additional optimization. Parameters like electrode configuration, power requirements, and hydraulic design must be meticulously addressed to ensure consistent treatment and effective interaction with the electrodes. Managing operational conditions, including pH levels and current densities, becomes crucial, necessitating robust monitoring and control systems. Tailoring these processes to specific sludge types can enhance performance but may introduce added complexity. Therefore, thoughtful design, rigorous monitoring, and stringent control measures are essential for achieving stable and efficient operation. Regulatory constraints and approval procedures should be thoughtfully integrated during the implementation phase.

Addressing these challenges will contribute to the successful integration of electrochemical processes into sludge treatment. While electrochemical methods hold significant promise for sludge management, it is essential to tackle these limitations through ongoing research, development, and practical application. Continuous innovation and optimization will play a pivotal role in promoting their wider adoption and incorporation into sustainable sludge management practices.

The primary drawback of this system is its substantial space requirement, which, when coupled with potential odor issues, poses challenges.

## 2 Factors Influencing Sludge Dewatering

### 2.1 Characteristics of Sludge

Sludge, an inescapable byproduct of sewage treatment, is composed of various elements, including pathogens, heavy metals, organic pollutants, and other noxious substances. Among its complex components, extracellular polymeric substances (EPS) play a significant role. EPS, being highly hydrophilic and compressible, pose a substantial challenge to sludge dewatering. Consequently, the development of efficient sludge dewatering technologies is pivotal in addressing the rapid proliferation of sludge. Currently, the primary methods for achieving deep sludge dewatering involve chemical preconditioning, high-pressure filtration, and electrical mechanical dewatering. The choice of chemical preconditioning method directly impacts the ultimate efficiency of the sludge dewatering process.

### 2.2 Significance of Dewatering in Sludge Treatment

Dewatering plays a vital role in the handling of sludge. It minimizes sludge volume, eases transportation, and can even reduce leachate output from sludge landfills. However, sludge, characterized as a heterogeneous colloidal system, presents a formidable challenge. The small sludge particles form a stable suspension in water and prove exceedingly resistant to separation through mechanical pressure alone. To enhance the dewaterability of sludge and achieve deep dewatering, diverse enhanced technologies have been extensively researched and developed. These technologies encompass chemical conditioning, thermal, magnetic, and electrical methods. Nevertheless, a comprehensive review addressing all facets of prior sludge dewatering technology remains notably absent, and the discussion of the ramifications of different dewatering technologies on subsequent sludge utilization is somewhat limited.

### 2.3 Water Distribution in Wastewater Sludge

A conceptual representation illustrates the water distribution in wastewater sludge, offering various interpretations of water types within fine-particle suspensions or flocculated systems. Wastewater sludge classification typically identifies four water categories: (a) free water, unattached to particles; (b) interstitial water, trapped within flocs; (c) surface (vicinal) water, held on solid particles' surfaces through adsorption and adhesion; and (d) intercellularly and chemically bound water. This classification, while simplifying the complex behavior of wastewater sludge, generally suffices for dewatering considerations. Mechanical dewatering primarily targets the removal of free and interstitial water, and the spatial distribution of water inherently influences the energy needed for its removal.

Sludge can be categorized into three distinct types of water content:

- Free Water: This constitutes 65–85% of the total water content in sludge, representing moisture freely existing between sludge particles. It is relatively easy to separate, making it a primary target for sludge thickening.
- Interstitial Water: Making up about 10–25% of the total water content, this water type is primarily found in the interstitial spaces between sludge flocs. It doesn't flow freely on the surface of solid-sludge particles due to surface tension, requiring significant mechanical forces for separation.
- Bound Water (BW): Accounting for only 10% of the total water content, BW is water held within sludge EPS and intracellular water. It is particularly challenging to remove mechanically. The elimination of BW and intracellular water typically requires structural modification of EPS or cell membrane disintegration to convert them into free water. Common techniques include coagulation, biological decomposition, high-temperature heating, freezing, and oxidation.

## 2.4 Influence of Physicochemical Properties on Dewatering Performance

### 2.4.1 Influence of Floc Characteristics on Sludge Dewatering Performance

- **Sludge Solid Concentration**: The concentration of sludge solids significantly varies during the dewatering process. An increase in total suspended solids (TSS) content was observed to influence sludge dewaterability.
- **Sludge Floc Size Distribution**: The size distribution of sludge flocs plays a crucial role in determining specific surface area, porosity, and filtration resistance of sludge cake. It significantly impacts dewatering performance.
- **Sludge Floc Strength**: Low floc strength results in poor dewaterability, as fragile flocs easily break into smaller particles that obstruct channels in the sludge cake and filter cloths.
- **Surface Charge**: The zeta potential value affects the aggregation and sedimentation of sludge particles. A stronger negative charge increases the stability of flocs, making dewatering more challenging. This can be mitigated through processes like electric neutralization and double electric layer compression.

### 2.4.2 Solution Chemistry Property

The dewatering efficiency of sludge is markedly affected by solution chemistry properties such as pH and ionic strength. Sludge dewaterability tends to improve in acidic conditions, while alkaline conditions can worsen it by causing significant EPS dissolution and filter membrane blockage. Proper management of these properties is crucial to enhance dewatering processes.

In addition, there has been limited research focused on the integration of various disciplines during the sludge dewatering process. Table 1 in the accompanying materials offers a summary of the advantages and disadvantages of each dewatering technique.

## 3 Conclusion

In conclusion, this chapter provides a foundational understanding of the multifaceted field of sludge management and treatment, acknowledging its crucial role in environmental sustainability and resource reclamation. It commences by delving into the composition, origins, and potential risks associated with sludge, emphasizing the necessity for effective management practices. The chapter then comprehensively explores primary treatment methods, highlighting the importance of techniques such as thickening and conditioning in preparing sludge for subsequent dewatering. Mechanical methods like centrifugation, filter pressing, and belt filtration are examined, along with technological advancements such as advanced polymers and additives, all aimed at optimizing dewatering efficiency. The chapter also underscores the importance of selecting the right combination of force fields to optimize the dewatering process, combining multiple force fields holds promise for enhancing the dewatering capabilities of conventional processes. By employing heat, shear, acoustic, magnetic, microwave, and electrical fields in conjunction with mechanical force, it is possible to intensify the dewatering process, leading to higher levels of water removal and greater final solids content. This intensified dewatering approach can be effective for a wide range of challenging slurries that are difficult to dewater using conventional methods or would require extreme conditions with standard dewatering devices.

Furthermore, the chapter investigates thermal techniques like drying and incineration, offering insights into their potential for volume reduction and pathogen elimination while considering their energy requirements and environmental impacts. Emerging techniques like electro-dewatering and

**Table 1** Recapitulation of the advantages and disadvantages of each dewatering technique (Wang et al. 2018)

| Dewatering methods | Advantages | Disadvantages |
|---|---|---|
| Natural dewatering | Easy manipulation, low operation cost | Odor, large occupied area needed, unstable performance because of climate, long drying period |
| Centrifugal dewatering | Compact structure, small occupation space, odor confinement | Damageable equipment, high operation cost |
| Belt dewatering | Continuous feeding, low operation cost | Sensitive to sludge properties, addition of polymer |
| Plate-frame dewatering | Easy manipulation stable operation, high dewatering efficiency | Non-continuous feeding, heavy consumption of press cloth |
| Electrical dewatering | High dewatering efficiency, removal of charged pollutants | High energy consumption, corrosion of anode material |

freeze–thaw dewatering are also presented, shedding light on their experimental and practical applications. In response to mounting sustainability concerns, the chapter underscores the significance of eco-friendly sludge management alternatives. Anaerobic digestion and composting are introduced as effective dewatering methods and sources of biogas and nutrient-rich soil amendments. It reinforces the urgency for efficient, cost-effective, and sustainable practices to address the challenges of sludge management. By equipping readers with a detailed understanding of these methods, the chapter aims to guide researchers, practitioners, and policymakers in making informed decisions that align with evolving environmental priorities, ultimately promoting responsible sludge handling and treatment.

## References


Cao B, Zhang T, Zhang W, Wang D (2021) Enhanced technology based for sewage sludge deep dewatering: a critical review. Water Res 189:116650

Chang H, Zhao Y, Bisinella V, Damgaard A, Christensen TH (2023) Climate change impacts of conventional sewage sludge treatment and disposal. Water Res 240:120109

Chen Y, Chen H, Li J, Xiao L (2019) Rapid and efficient activated sludge treatment by electro-Fenton oxidation. Water Res 152:181–190

Chowdhury SD, Bandyopadhyay R, Bhunia P (2022) Reutilization of sludge as fertilizer. In: Clean energy and resource recovery. Elsevier, pp 423–434

Corrêa JLG, Dev SRS, Gariepy Y, Raghavan GSV (2011) Drying of pineapple by microwave-vacuum with osmotic pretreatment. Dry Technol 29:1556–1561

Dey AR, Nekrasevits A, Semken RS, Mikkola A, Sihvonen T (2022) Applying ultrasound plus electrokinetics to enhance sludge dewatering. Dry Technol 40:2990–3002

Dong B, Liu X, Dai L, Dai X (2013) Changes of heavy metal speciation during high-solid anaerobic digestion of sewage sludge. Bioresour Technol 131:152–158

Feijoo S, Estévez S, Kamali M, Dewil R, Moreira MT (2023) Scale-up modelling and life cycle assessment of electrochemical oxidation in wastewater treatment. Chem Eng J 455:140627

Huang S, Zhou X, Zhou L, Huang Z, Shen J (2022) Constant-current electro-dewatering of sewage sludge: effect of anthracite modification on dehydration performance and economic benefit. J Env Chem Eng 10:107087

Lagnika C, Huang J, Jiang N, Li D, Liu C, Song J, Wei Q, Zhang M (2018) Ultrasound-assisted osmotic process on quality of microwave vacuum drying sweet potato. Dry Technol 1–13

Lambert N, Van Aken P, Smets I, Appels L, Dewil R (2022) Performance assessment of ultrasonic sludge disintegration in activated sludge wastewater treatment plants under nutrient-deficient conditions. Chem Eng J 431:133979

Li ZS, Fleureau JM, Tang LS (2017) Aspects of compaction and drying-wetting curves of a subgrade clayey soil. Geotech 67:1–1126

Li ZS, Tang LS, Zhang LJ, Luo ZG (2019) Dewatering sludge by osmotic technique—a comparative experimental study. Dry Technol 37(6):680–690

Liu W, Gui M, Zha Y, Li Z (2023) Numerical investigation of the effect of surface wettability and rotation on condensation heat transfer in a sludge dryer vertical paddle. Energies 16(2):901

Mahmoud A, Hoadley AF, Conrardy JB, Olivier J, Vaxelaire J (2016) Influence of process operating parameters on dryness level and energy saving during wastewater sludge electro-dewatering. Water Res 103:109–123

Mahmoud A, Olivier J, Vaxelaire J, Hoadley AF (2012) Advances in mechanical dewatering of wastewater sludge treatment. Wastewater Reuse Manage 253–303

Nayak PK, Mohan CC, Radhakrishnan K (2018) Effect of microwave pretreatment on the color degradation kinetics in mustard greens (*Brassica juncea*). Chem Eng Commun 205(9):1261–1273

Ohm TI, Chae JS, Kim JE, Kim HK, Moon SH (2009) A study on the dewatering of industrial waste sludge by fry-drying technology. J Hazard Mater 168(1):445–450

Park G, Eom T, Kwak H, Kim C (2022) Optimization of screw mixer to improve drying performance of livestock manure dryer using CFD analysis. Appl Sci 12(6):2872

Pilli S, Bhunia P, Yan S, LeBlanc RJ, Tyagi RD, Surampalli RY (2011) Ultrasonic pretreatment of sludge: a review. Ultrason Sonochem 18:1–18

Prakash O, Kumar A (2013) Historical review and recent trends in solar drying systems. Int J Green Energ 10(7):690–738

Qasem NAA, Mohammed RH, Lawal DU (2021) Removal of heavy metal ions from wastewater: a comprehensive and critical review. NPJ Clean Water 4:36

Rao B, Pang H, Fan F, Zhang J, Xu P, Qiu S, Su J (2021) Pore-scale model and dewatering performance of municipal sludge by ultrahigh pressurized electro-dewatering with constant voltage gradient. Water Res 189:116611

Rumky J, Bandina E, Repo E (2023a) Behavior of sludge dewaterability and nutrient contents after treatment with cellulose-based flocculants with combined PTS and catalytic behavior of sludge towards tetracycline degradation. Resources 12:17

Rumky J, Visigalli S, Turolla A, Gelmi E, Necibi C, Gronchi P, Sillanpää M, Canziani R (2020) Electro-dewatering treatment of sludge: assessment of the influence on relevant indicators for disposal in agriculture. J Environ Manage 268

Rumky J, Deb A, Shim MJ, Laakso E, Repo E (2023) A review on the recent advances in electrochemical treatment technologies for sludge dewatering and alternative uses. J Hazard Mater Adv 100341

Saveyn H, Curvers D, Jacobsen R, Meeren PV (2010) Improved dewatering by freeze–thawing of predewatered sludge cakes. Asia Pac J Chem Eng 5:798–803

Sha L, Wu Z, Ling Z, Liu X, Yu X, Zhang S (2021) Dewaterability and energy consumption of electro-dewatered sludge near the anode and the cathode during electro-dewatering process. J Environ Chem Eng 9:105729

Solé-Bundó M, Garfí M, Matamoros V, Ferrer I (2019) Co-digestion of microalgae and primary sludge: effect on biogas production and microcontaminants removal. Sci Total Environ 660:974–981

Tripathy S, Rees S (2013) Suction of some polyethylene glycols commonly used for unsaturated soil testing. Geotech Test J 36:20120041–20120780

Turunen K (2017) Energy efficiency improvement potential in a paper mill by means of heat load balancing

Usack JG, Van Doren LG, Posmanik R, Tester JW, Angenent LT (2019) Harnessing anaerobic digestion for combined cooling, heat, and power on dairy farms: an environmental life cycle and techno-economic assessment of added cooling pathways. J Dairy Sci 102:3630–3645

Wang K, Wu Y, Wang Z, Wang W, Ren N (2018) Insight into effects of electro-dewatering pretreatment on nitrous oxide emission involved in related functional genes in sewage sludge composting. Bioresour Technol 265:25–32

Wei H, Gao B, Ren J, Li A, Yang H (2018) Coagulation/flocculation in dewatering of sludge: a review. Water Res 143:608–631

Wu Y, Zhang P, Zhang H, Zeng G, Liu J, Ye J, Fang W, Gou X (2016) Possibility of sludge conditioning and dewatering with rice husk biochar modified by ferric chloride. Bioresour Technol 205:258–263
Yu Q, Lei H, Li Z, Li H, Chen K, Zhang X, Liang R (2010) Physical and chemical properties of waste-activated sludge after microwave treatment. Water Res 44:2841–2849
Zhang Q, Hu J, Lee DJ, Chang Y, Lee YJ (2017) Sludge treatment: current research trends. Bioresour Technol 243:1159–1172
Zhang W, Tang M, Li D, Yang P, Xu S, Wang D (2021a) Effects of alkalinity on interaction between EPS and hydroxy-aluminum with different speciation in wastewater sludge conditioning with aluminum-based inorganic polymer flocculant. J Environ Sci 100:257–268
Zhang S, Wang F, Mei Z, Lv L, Chi Y (2021b) Status and development of sludge incineration in China. Waste Biomass Valori 12:3541–3574
Zhang D, Hou R, Wang W, Zhao H (2022) Recovery and reuse of floc sludge for high-performance capacitors. FESE 16:78
Zhu Y, Xiao K, Zhou Y, Yu W, Tao S, Lu C, Lu D, Yu Z, Liang S, Hu J, Hou H, Liu B, Yang J (2020) Profiling of amino acids and their interactions with proteinaceous compounds for sewage sludge dewatering by Fenton oxidation treatment. Water Res 175:115645

# Managing Sludge from Textile and Leather Industries

Ankita Sharma, Ankita Kumari, and Sudesh Kumar

**Abstract**

Textile industry is booming all over the globe, with India ranking second and the United States coming in third in terms of textile manufacturing after China. The expansion of textile industry has repercussions on the global economy. The textile industry has become one of the foundations of the economy. Among world's textile producers, China and India holds top positions in the list. India is the world's second-largest supplier of textiles and apparel, accounting for 5% of global trade (Chand et al. in J Mater Cycles Waste Manag 25(6), 3132–3143, 2023). Global textile production of fiber has replicated from 58 million tons in 2000 to 109 million tons in 2020, and by 2023, the production is expected to reach 145 million tons. As the technologies that could support clothes to be recycled into virgin fibers are now emerging, only 1% of used clothes are recycled into fresh clothes (E. Union in The impact of textile production and waste on the environment (infographics), 2020). From 1960 to 2018, the total amount of textiles generated (in tons), recycled, transformed into compost, burned in energy recovery, and landfilled was recorded and represented in (United Nations Environmental Protection Agency in Textiles: material-specific data, 2024).

**Keywords**

Textile · Leather · Sludge · Management · Sustainable

A. Sharma (✉) · A. Kumari
Banasthali Vidyapith, Banasthali, India
e-mail: ankitadinesh117@gmail.com

S. Kumar
National Council of Educational Research and Training (NCERT), New Delhi, India

## 1 Introduction

From Fig. 1, it can be elucidated that the generation has increased drastically since 1990s. The recycling and combustion into energy has increased since 2000s. It is obvious to note that with increase in generation of textile products, the landfill waste has also increased parallelly (United Nations Environmental Protection Agency 2024).

The leather industry deals with the production of leather through a process known as tanning, in which animal skins are treated with chemicals such as sodium sulfide and chrome salts. Because of the significant generation of chemical and organic pollutants, primarily in effluent discharge, this process, particularly the conventional method, poses a significant environmental threat (Patel et al. 2021). Thousands of tanneries can be found throughout India and the world. It was evaluated that over 2161 leather industries annually handle about 314 tons of skins and 500,000 tons of hides. The leather manufacturing process generates 30 L of wastewater for every kilogram of animal skin, resulting in Indian industries emitting 50,000 $m^3$ of effluent daily. Tanning permanently changes the protein structure of the rawhide/skin. Toxic elements such as chromium, arsenic, heavy metals, and organic chemicals like dyes are produced in high concentrations in leather industry effluents. In the chrome tanning and dyeing processes, chromium salts and organic dyes are discharged into the environment. It is critical to remove and recover these hazardous chemicals from effluent as soon as possible for reuse in industrial applications (Kumar et al. 2023).

Circular economy is defined as a method for tackling the resources the concept of circularity of resources, effectiveness, and optimization promotes the utilization of waste as a resource to produce value products. The world's present economic system relies on a linear economic model. This is based on use, make, and dispose model. Alternatively,

© The Author(s), under exclusive license to Springer Nature Switzerland AG 2025

A. Mandpe and M. P. Shah (eds.), *Wastewater Sludge Nexus*, Sustainable Landscape Planning and Natural Resources Management, https://doi.org/10.1007/978-3-031-45758-6_4

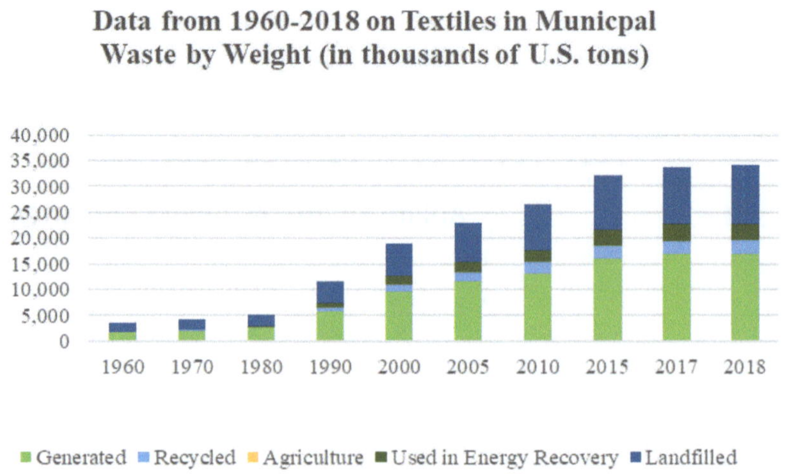


**Fig. 1** Textiles sludge in municipal waste by weight (in thousands of U.S. tons) between 1960 and 2018 (United Nations Environmental Protection Agency 2024)

manufacturing, consumption, and waste disposal results in pollution by generation of huge amount of waste and inefficient resource use. While in circular economy model the production, consumption, and waste disposal to recycle it for future production is involved, and it is primarily governed by the Recycling, Reduce, and Reuse (3Rs) concept. Thus, we can say that circular economy model is one of the most environmentally friendly and long run development strategies (Ahmed et al. 2022).

## 2 Management of Textile Sludge

The manufacturing of textile fiber generates sludge and scrapes in every stage. The fiber recycling technologies are classified as primary, secondary, tertiary, and quaternary. The primary recycling methods concentrate on scraps discharged from industrial, whereas the post-consumer products are mechanically processed in secondary recycling. Tertiary recycling processes plastic waste using pyrolysis and hydrolysis Fig. 2 to create monomers, chemicals, or fuels. Quaternary recycling entails burning of fibrous waste and reusing the generated heat (Juanga-Labayen et al. 2022).

### 2.1 Vermicomposting

The vermicomposting is a sustainable way to combat the textile and leather sludge. In vermicomposting, different species of earthworm are used. There are different types of earthworms as mentioned in the Table 1 (Kale and Karmegam 2010).

The inclusion of organic matter in textile and leather sludge renders vermicomposting an excellent and eco-friendly technique. Earthworms release organic acids like humic acid, fulvic acid, etc. into the soil during vermistabilization. These organic acids have presence –OH and –COOH groups the environment binds with heavy metal ions and reduces their bioavailability. This a natural defense mechanism of earthworm which prevents bioaccumulation (Maity et al. 2009). However, with increase in production and industrialization as a result of global population explosion in this century, proper amendment needs to be done. A potential candidate is the cow dung which not only contains a number of microbes but also a number of –OH and –COOH groups enhancing metal-exchange process. The addition of cow dung in 1:1 ratio enhances its ability in a beneficial manner. Another interesting possibility is the poultry waste, which is rich in nitrogen and phosphorous can also be used with cow dung and earthworm litter in 1:1:2 ratio, respectively (Yuvaraj et al. 2019a).

Earthworms use a sophisticated defense mechanism in environments with heavy metal contamination to deal with metal ions that are resistant to leaching by organic acids. Metallothionein (MT), a heavy metal-binding protein, is a key player in this detoxification process. Earthworms' chloragenous cells contain the genes responsible for MT production. Metallothionein is distinguished by the presence of 12 cysteine residues that act as active metal binding sites. These cysteine residues contain specific thiol groups, which are sulfur-containing functional groups that are highly reactive. When thiol groups come into contact with metal ions, they form thiolate bonds with both essential and non-essential metals. This binding process is critical for sequestering these metals and reducing their toxicity. The formation of thiolate bonds with metal ions is a sophisticated molecular strategy used by earthworms to effectively immobilize and detoxify heavy metals. Earthworms can thrive in environments with high metal concentrations by doing so, demonstrating their remarkable adaptation to and mitigation of metal-induced stress in their ecosystems. This complex biological process not only protects earthworms but also contributes to the

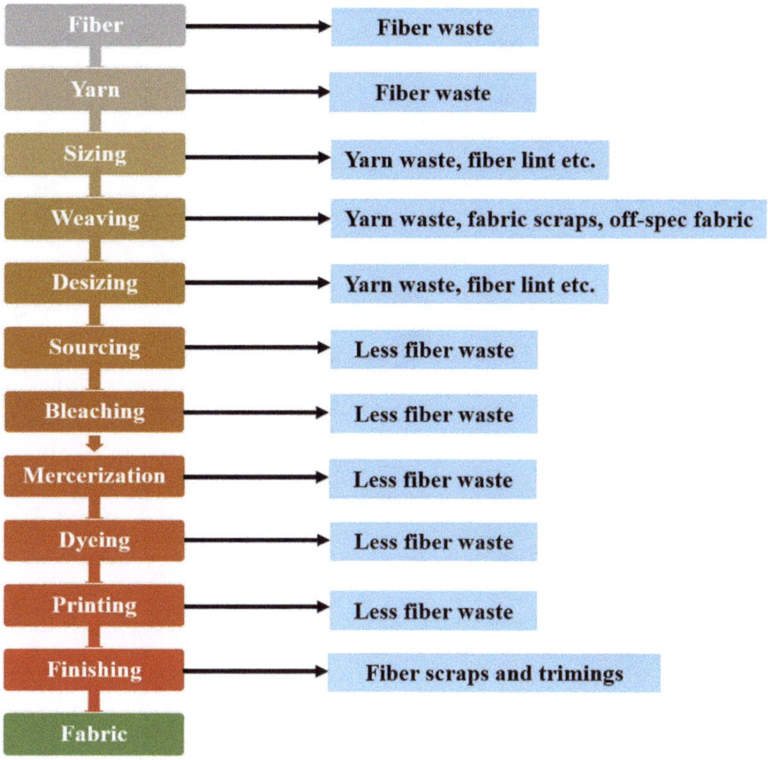


**Fig. 2** Cloth production process and type of fabric waste generated at each step (Chand et al. 2023)

**Table 1** Different categories of earthworms used for vermicomposting

| Type of earthworm | Epigenic | Endogenic | Anecic | Coprophogic | Hydrophilic |
|---|---|---|---|---|---|
| Habitat | On the surface of soil | Top soil (10–15 cm) | In soil upto 3m deep<br>Vertical burrowers | Upper layer of moist soil | Damp soil |
| Feeding habit | Plant waste | Organic matter<br>Horizontal burrowers | Plant litter | Organic waste material | Plant litter, animal dung |
| Example | *Eudrilus eugeniae, Lumbricus rubellus, Perionyx excavatus* | *Octochaetona thurstoni, Proctodrilus oculata* | *Lumbricus terrestris, Aporrectodea longa* | *Eisenia fetida* | *Eiseniella tetraedra* |

overall health of the ecosystem by reducing the potential negative effects of heavy metal pollution (Gómez-Brandón and Domínguez 2014; Lores et al. 2006). In the context of the metallothionein (MT) degradation process in earthworms, microbial activity plays an important role in enhancing overall degradation via a process known as mineralization. Mineralization is a biological process that converts organic matter, in this case metal ions and metallothionein, into inorganic forms, usually minerals. The process of mechanism of MT protein production and action is represented in Fig. 3.

Mineralization occurs in two stages, each involving complex microbial processes:

*Microbial Colonization*: Microorganisms colonize the organic material, which in this case includes metallothionein and associated metal ions, in the first step. Through enzymatic activities, these microorganisms, which are typically bacteria or fungi, break down the complex organic compounds in simple substances. The organic material is used as nutrient source by the microbial community, resulting in metabolic byproducts.

*Biomineralization*: It involves converting the byproducts and remaining organic material into stable inorganic compounds. Microbes are essential in facilitating this conversion process. They have the potential to cause precipitation reactions, converting soluble organic compounds into insoluble mineral

**Fig. 3** Mechanism of MT protein production and its action (Gómez-Brandón and Domínguez 2014; Lores et al. 2006)

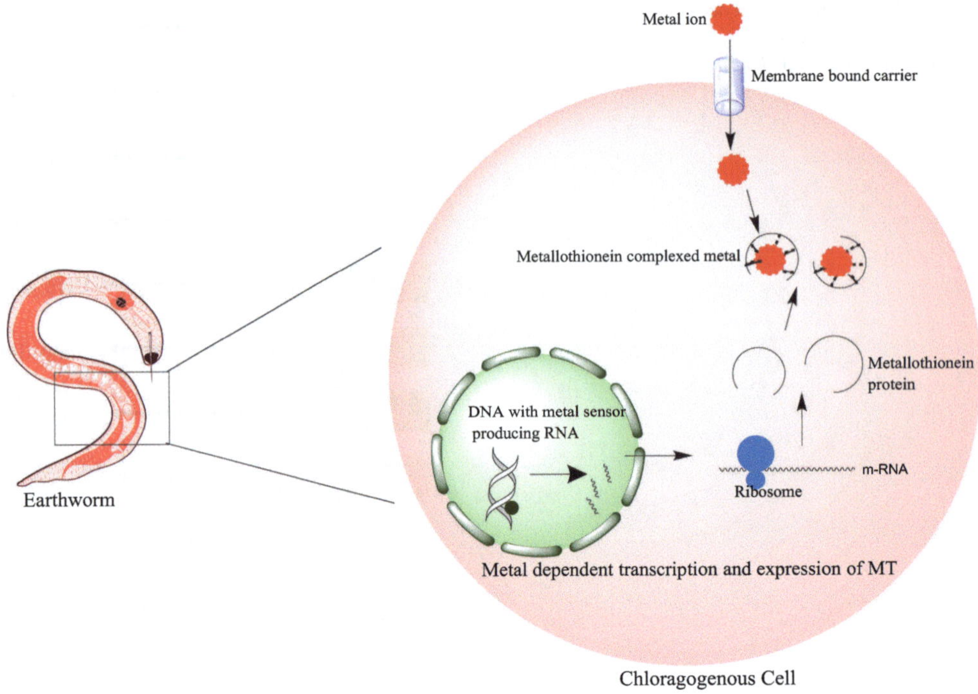


forms. This step in the biomineralization process helps to immobilize metal ions and other elements, making them less mobile and potentially less harmful to the environment.

In conclusion, microbial involvement in the degradation process extends beyond the breakdown of complex organic compounds to the transformation of these compounds into more stable, less environmentally impactful mineral forms via a two-step mechanism: microbial colonization and biomineralization.

Vermicomposting is recognized as an outstanding ecological and sustainable approach. Besides, earthworms are well-suited to thrive under metal stress conditions, making them highly advantageous for vermicomposting because of several additional benefits, such as (Yuvaraj et al. 2019b, 2020)

(i) *Degradation of Pesticides*:
The earthworms like *Eisenia fetida*, *Lumbricus terrestris*, etc. have detoxification mechanisms involving a number of enzymes, including cytochrome P450, carboxylesterase, and phosphotriesterase. These enzymes collaborate to metabolize and break down toxic substances in the environment, lowering their overall toxicity. This ability benefits not only the earthworms themselves, but also helps to improve soil quality and ecosystem health.

(ii) *Prevention from Cellular Damage*:
The antioxidants, like lipid peroxides, catalase, glutathione-S-transferase, and superoxide dismutase, help earthworms *to* combat oxidative stress, neutralize harmful reactive oxygen species, and prevent cellular damage. These defense mechanisms are critical for the overall health and survival of organisms in environments containing oxidative stressors.

(iii) *Presence of Phytohormones in Their Castings*:
Earthworm castings are a powerful natural fertilizer for plants because they are high in phytohormones such as auxins, gibberellic acids, and cytokinins. Auxins aid in seed germination and root development, whereas gibberellic acids aid in stem elongation and flowering. Cytokinins in the castings promote cell division, which influences chlorophyll synthesis and results in higher chlorophyll content. This nutrient-dense mixture promotes vigorous plant growth, increasing overall vitality. Earthworms are important in the breakdown of organic matter, and their castings act as a bioavailable fertilizer, promoting healthier and more vigorous plant growth. Earthworm castings provide an ideal environment for seedlings, positively impacting key aspects of plant growth ranging from germination to increased leaf surface area and chlorophyll production.

(iv) *Castings Make a Fresh Layer of Substrate*:
Earthworm castings help the formation of a new layer of substrate. When earthworms digest organic matter, the resulting castings are deposited on the soil surface as a nutrient-rich material. This vermicast layer serves as a natural and beneficial top dressing. It improves soil structure, water retention, and plant growth by providing essential nutrients. The fresh substrate layer created by earthworm castings promotes a healthy microbial environment, encouraging beneficial soil

organisms to thrive. This process not only adds nutrients to the soil but also improves its overall fertility. The continuous deposition of earthworm castings creates a dynamic and self-sustaining cycle that improves soil quality and stimulate the growth of a variety of plants in the ecosystem.

(v) *Increase Micro and Macronutrients in Soils*
Earthworm castings are essential for increasing soil micronutrients and macronutrients. Earthworms' digestive processes enrich the castings with essential nutrients as they consume organic matter. Microorganisms in the earthworm's digestive system break down organic materials even further, releasing nutrients in a plant-available form. Nitrogen, phosphorus, potassium, calcium, magnesium, and trace minerals are concentrated in these castings. Earthworm castings improve nutrient levels in the soil by promoting balanced and sustainable soil fertility. This nutrient enrichment promotes plant growth, increases crop yields, and benefits the ecosystem's overall health. Furthermore, earthworm castings improve soil cation exchange capacity, allowing for better retention of nutrients and availability for plant uptake.

## 2.2 Biofuel Production

The biofuel is produced by enzymatic hydrolysis and anaerobic digestion. The production of biofuel from the untreated textile waste is slow process due to high crystallinity of cellulose and presence of synthetic fibers. To make fiber more penetrable to enzymes, pretreatment of fibers is done in order to increase the surface area, porosity and change in its crystallinity.

**Pretreatment**

The cotton cloths are obtained from *Gossypium* plant and are generally cellulose-based cloth. They are crystalline due to the presence of intra and interchain hydrogen bond in cellulose structure. Mercerization is a process of treatment of cotton cloth with alkali for intra-crystalline swelling and modification of crystallographic cell from cellulose I to cellulose II. This would increase the hydrophilicity and enhance hydrolysis of the cloth making it more susceptible to digestion. The hydrolysis of cellulose chain breaks the hydrogen bond and washing of fiber rearranges and reform the hydrogen bonds but decreases the number of hydrogen bonds. The polyester cloth, due to its high crystallinity, is difficult for biodegradation. A comparative study brought by the pretreatment with 99% Alkali, concentrated phosphoric acid and non-treated textile sludge. It was elucidated that the yield of bioethanol was maximum in NaOH i.e., 99 and 66% in phosphoric acid. The best suited condition 0 °C and 12% NaOH while for phosphoric acid it was 85% phosphoric acid which recovers 79.2% sugar (Jeihanipour and Taherzadeh 2009). Hasanzadeh et al. (2017) have used $Na_2CO_3$ for hot alkaline pretreatment of jeans textile waste (which is made up of polyester and cellulose). This enhanced (i) increased water swelling capacity of cloth, (ii) hydrolysis of non-cellulosic part, (iii) partially conversion of cellulose I to cellulose II, and (iv) decreased crystallinity. The results showed increase in ethanol yield by 49% and biogas was 329 mL/gm VS. In some researches, the use of certain urea containing compounds like urea, thiourea and their mixtures with NaOH prevents the reformation of inter and intra-cellulose bonds. The dissolution of cellulose into an aqueous environment facilitated by formation of an inclusion complex (IC) involving NaOH hydrate-urea hydrate-free water-cellulose which promotes effective dispersion. The primary role of NaOH hydrate lies in disrupting both intermolecular and intramolecular hydrogen bonds. This occurs as hydrogen bonds shift between cellulose and NaOH hydrate, especially at lower temperatures. The NaOH attacks the amorphous portion of cellulose and destroys adjacent crystalline portion. Concurrently, thiourea hydrates or urea hydrates acts to deter cellulose from self-association by functioning as H-donor. In combination, these components within the IC synergistically encourage cellulose dispersion in water, ensuring a more favorable and stable aqueous system for various applications. The pretreatment of fibers with enzymes followed by alkali treatment increases efficiency. The enzymes create a smoothened fiber surface by removing fibrillars and decrease length of the fiber (Wang et al. 2008). A combination of alkali, urea, and Polyethylene Glycol (PEG) could be used for pretreatment where the PEG prevents reassociation of hydroxyl groups of cellulose by acting as a H-Bond acceptor (Mohsenzadeh et al. 2012).

A new variety of pretreatment reagents have emerged namely, Ionic Liquids (ILs) which are a combination of cations and anions having a wide range of solvation properties for cellulose, wood and also for extraction of nanocellulose. They have unique solubility and H-bonding ability between hydroxyl group of cellulose and highly electronegative IL anion. Generally, the chloride-based ILs are preferred due to small size and hydrophilic behavior of chloride ions acting as H-bond acceptor Among them, the imidazolium-based room temperature IL (RTIL) are the most suitable ILs due to low viscosity, non-volatile, and easy to recover and reuse behavior. The four major types of cations of RTIL that dissolve cellulose are EMIM (1-Ethyl-3-methylimidazolium), BMIM (1-Butyl-3-methylimidazolium), and AMIM (1-Allyl-3-methylimidazolium). Out of all, [AMIM]Cl is a better candidate for microcrystalline cellulosic structure, low crystalline substances, etc. The structure of AMIM is electrostatically stable thus, due to its chemical stability it can be reused. It also has other important properties like less corrosive, non-flammable, less viscous nature (Abushammala and Mao 2020).

Wool is also a major component of textile sludge. The degradation of wool with liquid $N_2$ enhances particle size reduction and increase solubility and makes it bioavailable for the microbial degradation. It further increases the methane production to 100%, 78%, 77% for Bluefaced Leicester, Scotch mule and Texel wool fibers respectively (Kuzmanova et al. 2018).

**Bioethanol**

The textile sludge contains a variety of cloth like wool, cotton and synthetic fiber. The synthetic fibers, due to their high crystallinity, are difficult to digest. Thus, wool and cotton can be separated from them at different levels. The wool is generally made up of proteins thus, can be hydrolyzed by proteases and cotton by cellulases. The wool and cotton enzymatic hydrolysis yields upto 95% and 85%, respectively, followed by digestion by *Saccharomyces cerevisiae* (Quartinello et al. 2018). Gholamzad et al. (2014), used Simultaneous Saccharification and Fermentation (SSF) method to analyzed a mixture of textile cloth i.e., polyester and cotton textile, and found that 12% NaOH at −20 °C not only increased enzymatic hydrolysis by 30% but also improved polyester recovery to 98% without any change in its properties. For enzymatic hydrolysis, there are some limiting factors like the presence of inhibitors like sugars, lignin, pseudolignin, water constraint. The proper working of enzymes requires mild conditions while the mineral acids provide harsh and undesirable conditions by breaking hemicellulose forming inhibitors like other sugars. Some common enzymes are listed in the Table 2 with their inhibitors (Duque et al. 2021).

The yield of bioethanol production from textile sludge can be controlled by the presence of more reducible sugars in the solution and cellulose structure prone to enzyme attack. The effect of pretreatment methods like Ultrasound-Assisted Alkali Pretreatment (UAAP), Dilute $H_2SO_4$ pretreatment (DSAP), and High Pressure-Assisted Alkali Pretreatment (HPAP) was performed by Wang et al. (2016). The fiber treated with HPAP were most suitable for bioethanol production because they had highest reducing sugar (271.70 mg/gm dry biomass), strong ability to remove lignin, more degradation of cellulose. In DSAP, the samples had more appropriate dissolution of acid-soluble lignin, hemicellulose and other compounds.

**Biogas Production**

The pretreatment has proved to improve the biogas production because it increases the availability of surface area for anaerobic digestion. A comparative study was conducted between single and two-step batch digestion for cotton/polyester and viscose/polyester. For pretreated jeans fabric the $CH_4$:$CO_2$ ratio increased without any accumulation of Volatile Fatty Acids (VFA). It was observed that for two stage batch process, the biogas production was increased, lag phase was shorter in comparison to single phase, can handle more Organic Loading Rate (OLR) with less Hydraulic Retention Time (HRT) (Jeihanipour et al. 2013). Addition of seeding material like cow dung, pig dung, and goat dung can enhance the production of methane rich biogas. The percentage of methane reaches around 78%, 68%, and 78% when pig dung, goat dung and cow dung respectively were used as seeding materials for laboratory scale biogas production. The biogas showed good results when tested in engine as a dual fuel with diesel (Saravanan et al. 2009). Food waste is another easily available seeding material. But, due to the presence of oil and spices, the proliferation of methanogens is inhibited, thus they could be used with other seeding materials like cow dung to amplify its performance. The on-site digestion of food waste with textile sludge could produce around 0.288L biogas $kg^{-1}$VS/TS added (Kumar et al. 2020).

The maintenance of temperature is one of the key factors for biogas production. The biogas is operated at mesophilic temperature range (25–40 °C) or thermophilic (40–60°C) where the activity of methanogens is maximum. The temperature of the digestor is generally kept constant during the production. With increase in the internal depth of the digestor, the temperature also increases. In order to increase production of biogas, the heat loss should be prevented to maintain constant temperature. It could be done by using water bath, saw dust etc. The results show that yield of

**Table 2** Enzymes used for bioethanol production (Duque et al. 2021)

| Properties | Celliobiohydrolase | Endo-1,4-β-D-glucanase | 1,4-β-D-glucosidase |
|---|---|---|---|
| Specifications | Exo-1,4-β-D-glucanase<br>EC 3.2.1.91 | Endo-1,4-β-D-glucanase<br>EC 3.2.1.4 | 1,4-β-D-glucosidases<br>EC 3.2.1.74 |
| Working | Separate cellobiose units from ends<br>Works linearly | Increase number of chains ends via hydrolysis of internal bonds for cellobiohydrolase<br>Works randomly | Produce monomers from disaccharide |
| Inhibitory factor | Cellobiose | Cellobiose | Glucose |

biogas was increased with high methane content of 63 and 53% in water bath and saw dust based temperature maintained digester respectively (Twizerimana et al. 2021). Besides this, the C/N ratio is also major factor for biogas production. It should be kept between 20 and 30 to ensure proper nitrogen supply to the cells for their proliferation and degradation of produced $CO_2$. Low C/N ratio could result in production of toxic $NH_4^+$ ions (Fricke et al. 2007).

**Syngas Production**

Syngas is a mixture of CO and $H_2$ gases which have less greenhouse emissions. The most common method of syngas production is gasification where carbonaceous material is burnt (T > 700°C) to form syngas. Bagheri et al. (2020) developed a model for compared the High Heat Values (HHV) of 252 solid wastes. The carbon and nitrogen are the major factors which influence the heat recovery which are second highest for textile industries. The synthetic petroleum-based fibers i.e., polyester, acrylic, and nylon have high carbon content and higher HHV values. This study explored the co-pyrolysis/gasification of cotton stalk and municipal sewage sludge using textile dyeing sludge (TDS) pyrolysis residues as catalysts for syngas production. The research employed TGA and a fixed-bed reactor, revealing that adding TDS residue from pyrolysis at 600 °C (TDS-600) effectively reduced the activation energy of the reaction, enhancing the co-pyrolysis process. When 10% TDS-600 was introduced, a maximum syngas yield of 0.808 m3/kg was achieved at 850 °C and 20% steam volume fraction, with a 32% increase by raising TDS-600 to 20%. The study also evaluated heavy metal contamination in the ash, suggesting potential agricultural use due to acceptable contamination levels (Yan et al. 2022). Nguyen and Chiang (2021) investigated the co-gasification of textile sludge (TS) and sewage sludge (SS) using a laboratory-scale gasifier at 850 °C. The addition of TS increases the CO yield and improves the heating value of the gas and eliminated tar during gasification. An integrated hot-gas cleaning system effectively reduced tar (upto 30%) and ammonia (upto 35%), highlighting the feasibility of co-gasifying these waste materials to improve the gas quality.

## 2.3 Building Materials

Infrastructure is a key factor to compare the developing stage of a country. The building materials are generally comprising of structural and non-structural. Textile waste ash is a potential candidate for its reinforcement in building material. Raw textile sludge contains high amount of water already; thus, it is more prone for odor and tedious to handle. When 10% textile sludge ash is incorporated with 40% fly ash and 30% quarry dust in brick. The brick shows 8.5% higher strength. On increasing the textile sludge, the mechanical strength decease, microstructure evolution increase, and metal leachability decrease (Ashok Kumar et al. 2019). The presence of organic or hydrophobic groups in building material like cement gets absorbed on the surface of the grains and forms a temporary protective layer resulting in delay in hydration. A positive point to note that, the textile sludge contains organic compounds like dyes. The presence of hydrophobic groups like in azodyes the $N = N$ follows the same trend (Zhan et al. 2019). Chemical resistance using 5% $H_2SO_4$ was done and found that with increase in textile sludge percentage in replacement, the weight loss increases. The textile sludge increases porosity when it is heated during the process due to escape of gases like $CO_2$. The acid sweeps into it and deteriorates it (Singh et al. 2019). The main issue with handling the textile sludge is the $NH_3$ odor. Thus, lime is used as a source of alkali to produce $OH^-$ ions that convert the $NH_4^+$ in the sludge to $NH_3$ gas. It also increases the strength of bricks when kept 10% with aggregate to amount ratio of 10 in the bricks. Also, the $NH_3$ have the tendency to inhibit cement hydration (Zhan and Poon 2015). The studies have shown that the textile sludge incorporation with higher % in the building materials increases the porosity, setting time and decreases the flexural and compressive strength. A study has shown that the textile sludge at 25% replacement of sand exhibited maximum compressive strength of 20 MPa. On replacement of Portland cement and addition of 5% textile sludge shows similar properties with mortar made up of Portland cement alone. Besides, mortar, they have conducted experiments on concrete with 30 wt% of substitution (Rahman et al. 2017).

# 3 Management of Leather Sludge

Leather sludge is produced at various stages of leather production and is a byproduct that must be managed carefully. Several steps in the tanning and leather manufacturing processes are involved in the generation of leather sludge (Kandasamy et al. 2020).

**Pretanning**

*Soaking*: The process of removing salt from raw hides/skins and soaking them in water to restore moisture and remove impurities. Remove blood, dirt, manure, preservatives, and soluble proteins Fig. 4.

Method: Multiple water soaks with bactericide addition are performed in pits or paddles. It takes 3 h to overnight.

*Green Fleshing, Dehairing, and Liming*: Fleshing to remove excess fat, dehairing with alkali solutions (lime, sodium sulfide), and liming to open pores. High COD in effluent due to the production of organic and inorganic salts, keratin, and fleshing waste.

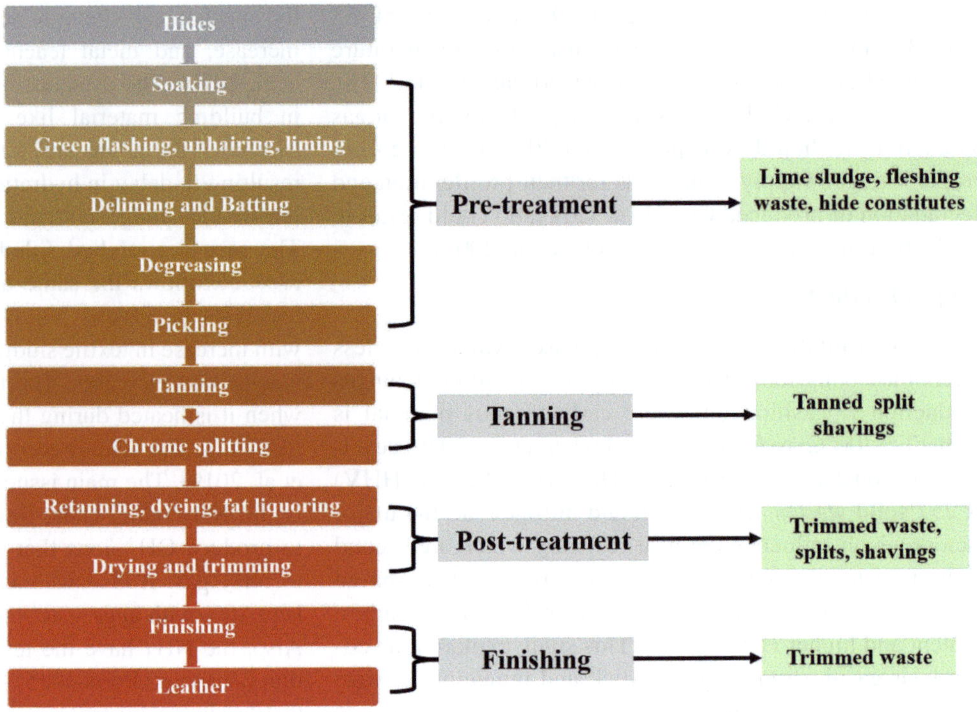


**Fig. 4** Process of leather production (Kandasamy et al. 2020)

*Lime Fleshing and Splitting*: Operations include second fleshing for cleaning and shaping, as well as splitting hides into two or three layers. As a result, organic materials containing alkaline fat are disposed of as solid waste.

*Bating and deliming*: Deliming is using ammonium salts to reduce pH after liming contributes to effluent hardness. Bating is done to exposed enzymes for texture, softness, and flexibility, proteinaceous solid waste is produced.

*Degreasing*: Process of removing excess natural fats for uniform distribution. High COD in effluent as a result of surfactants or organic solvents.

*Pickling*: Adjust the pH to 2.5–3.0 with acids, sodium chloride, and bactericides. As a result, the effluent contains a high concentration of dissolved solids, COD, and BOD.

#### Tanning

Tanning is the process of converting raw untreated hides into leather. Chrome tanning, which uses chromium salts, produces stable and flexible leather but poses disposal issues. Natural agents are used in vegetable tanning to produce less stable, water-resistant leather. Mineral-free leather with increased stability is produced using alternative methods. While important, operations such as samming and sorting contribute to increased waste and effluent hardness. Overall, the tanning process shapes the properties of leather, with distinct advantages and waste management considerations.

#### Post-tanning

*Retanning, dyeing, and fat liquefaction*: Retanning is used to achieve uniformity, which is then dyed and fat liquored for softness. Effluent is impacted by coating chemicals used in finishing.

*Drying and trimming*: The process of stretching, compressing, drying, and trimming leather to achieve an even surface.

*Finishing Methods*: With wax pigments and resins, the goal is to impart esthetic beauty and increase product value. Coating chemicals contribute to effluent by producing solid and liquid waste.

Solid residues separate and form sludge as wastewater is treated or settles. This sludge is made up of tanning agents, organic material, and other substances used in the various stages of leather processing. Sludge is typically dried to reduce its moisture content before being collected as a semi-solid waste material.

#### Biofuel Production

As represented in Fig. 5, different biofuels are produced by the treatment of leather waste. Anaerobic digestion of organic matter is a complex microbial process that occurs in the absence of oxygen and involves various bacteria and archaea. Hydrolysis, fermentation/acidogenesis, acetogenesis, followed by methanogenesis, are the four stages of this process. Hydrolytic bacteria exoenzymes degrade polymer into monomers during hydrolysis. Acidogenesis

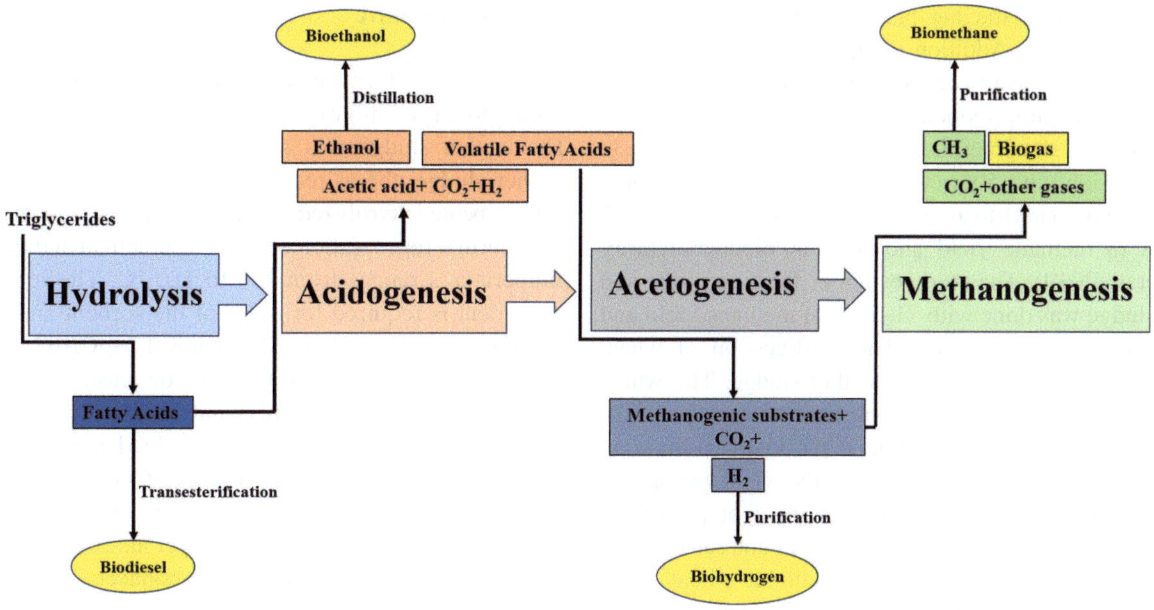


**Fig. 5** Anaerobic digestion mechanism and biofuels produced in different steps

occurs when fermentative bacteria convert the monomers produced by hydrolysis into acetic acid, carbon dioxide, hydrogen and volatile fatty acids. Acetogenesis is aided by acetogenic bacteria, which convert fermentative intermediates like volatile fatty acids into methanogenic substrates such as carbon dioxide, acetic acids, hydrogen, and monocarbon compounds. Finally, methanogenic archaea convert the acetogenesis products into biogas, which is primarily composed of methane (55–70%) and carbon dioxide (30–45%), with traces of other gases.

### Biodiesel

Biodiesel is generally produced from vegetable oil/animal fat/animal waste by transesterification. The generated product is crude which is refined and finally used as fuel. The use of animal fat like tallow, fish fat, chicken fat, etc. The fat content vary from animal to animal with 2–6% cattle, 5–30% sheep, and 2–10% goat (Aftab and Hameed 2006). The use of animal waste for biodiesel production has advantages. The biodiesel has high cetane number, non-corrosive, low Free Fatty Acid (FFA), clean and renewable source of energy Single stage transesterification was performed as the leather waste oil has less than 1% free fatty acid content (Vignesh and Barik 2019). Generally, the LS has 40% saturated fat and 57% unsaturated fat. When blend of biodiesel with diesel used as 10% (v/v) and 30% (v/v), the smoke opacity in 30% decreased due to less C/H content and less $O_2$ (Keskin et al. 2020). Calcined powder of crab shell was used as a catalyst to synthesize biodiesel from leather sludge with purity 98.6 ± 0.4 wt% with Fatty Acid Ethyl Ester (FAEE) yield above 90 wt% (Yuliana et al. 2021). Research on catalyst-free biodiesel production using single-step critical ethanol (SpCE) technique using Response Surface Methodology (RSM). Various studies like reaction time, temperature, etc. were conducted in supercritical condition. All the properties were falling under ASTM D6751 standards. With rise in temperature the H-bond between water and -OH group in ethanol weakens and increases miscibility with non-polar LS. The miscibility of LS in water and ethanol is low at room temperature. Increase in temperature, hydrolyze lipids and favors SpCE with rise in in-situ esterification. The number of $H_3O^+$/$H^+$ ions increase with rise in temperature behaving like a bifunctional catalyst resulting in high recovery of FAEE. Thus, it is a positive approach if water content is high (Yuliana et al. 2019).

The presence of high water content and less FFA content in LS oil is not favorable for acid based catalysis as water serves as a limitation for acid based transesterification (Singh et al. 2023). The sustainable conversion of LS to biodiesel with L-valine amido ethyl methyl imidazolium bromide ([L-Vaemim]Br) as a novel ionic liquid catalyst. It was found that CO emissions were less, indicating complete combustion (Ranjitha et al. 2020).

### Biogas and Biomethane

The tanning process makes the leather resistant to environmental degradation because the $Cr^{+3}$ ions form cross-linking between collagen fibers. Thus, LS should be pretreated for biogas. Hydrothermal and extrusion method is one of them which reduces the lag phase for biogas production (Gomes et al. 2020). Addition of lipase enzyme (commercial grade) i.e., Steapsin seems to increase biogas production by 15% and time reduction by 30% (Sri et al. 2011). Co-digestion of slaughterhouse sludge and leather waste sludge activation with 50% (v/v) yielded 215 mL methane/gm VS. As compared to mono-digestion, there was 156% increment in methane yield and 60% in biodegradability (Mpofu et al. 2019). The co-digestion of fleshing waste and primary sludge was done with 71–77% of methane yield and 512% volatile solid reduction. The co-digestion of wheat saw has shown good yields with leather sludge. The wheat was treated with a combination of pretreatment methods like (i) acid + thermal and (ii) alkaline + thermal (iii) thermal. The pretreatment with alkali affected the yield negatively inhibiting methanogenesis. The biogas yield of pretreated with HCl was highest with 59% higher than untreated wheat straw and almost 167% higher than without wheat straw (Simioni 2022). Sludge and shavings (hide and tanned hide) are anaerobically biodegradable and yield methane of 4.1–11.3 mL methane/g of VSS added. Experiments performed showed that low concentrations of chromium containing sludge produced about three times more biogas with 55% of $CH_4$ on average. Thus, the low concentrations of chromium is more relevant for methane and biogas production than its absence (Borges et al. 2017).

### 3.1 Building Materials

The use of 10% leather sludge in bricks for commercial purpose not only satisfies the ASTM and BDS standards but also reduce the firing energy by 15–47% for incorporation of 10–40% leather sludge (Islam et al. 2017). Leather sludge with cement increases the flexural and compressive strength, decreases water absorption and improved density due to presence of high Calcium hydrosilicate (C-S–H bond) in samples (Mortar 2019). During the liming stage of leather production, the skin is treated with $Na_2S$ and lime in order to dissolve epidermis, fat/grease, non-fibrous protein. The incorporation of 6% of liming sludge with bricks have maximum strength and more water retention capacity. The $CO_2$ escaped during heating made the bricks porous. The leather sludge, fly ash and Ground Granulated Blast Furnace Slag (GGBS), lime were used to prepare green bricks. The optimum composition selected was 20%w/w LS with binder mix of GGBS:lime:fly ash following ratio of 20:30:50. The bricks gave positive results in norms for chromium leaching in non-hazardous wastes. The elastic modulus of carbonated was high due to high porosity which helped in capturing more amount of metals as a result preventing their leaching in the environment which could be beneficial for replacement of clay (Jothilingam et al. 2022).

### 3.2 Adsorbant

A two-step pyrolyzed leather sludge waste is used as a granulated sorbent of n-butyl acetate from leather air emissions. The leather sludge was derived from a pyrolysis plant of leather sludge, the as received (PTS) treated at 700 °C and after repeated pyrolyzed (PPTS) i.e. pyrolyzed at 850 °C. The thermal regeneration of PPTS is practical and desorption temperatures ranges from 150 to 180 °C. The regeneration of sorbent is required for reuse of the sorbent and recovery of VOCs (Rossi et al. 2023). A novel MgO/BC composite was synthesized from non-tanned hide waste. The multilayered structure of collagen fibers and the strong permeability of $Mg^{+2}$ ions in collagen fibers. Besides, leather waste contained a lot of functional groups such as –COOH, $-NH_3$, –OH, etc. which made attachment of $Mg^{+2}$ ions easy and MgO/BC rich in functional groups (Tang et al. 2021). The leather sludge after pyrolysis, is rich in carbonaceous matter so can be used as a replacement of activated carbon. The ash has high thermal resistance in comparison to commercially available activated carbon, porosity with mesopores and micropores. Also, it has the greater number of surface functional groups (Palani et al. 2017). A novel approach for converting leather waste into cost-effective material for acoustics application. The protein hydrolysate was prepared from the leather waste via acid hydrolysis and mixed with polyvinyl (PVA) to produce nanofibers through electrospinning process. The sound absorption studies revealed that the bilayer material made up of coir disc and fleshing hydrolysate-PVA nanofibers is superior in sound absorbing capacity at lower frequencies in comparison with pure coir and PVA nanofibers with coir material (Selvaraj et al. 2019).

## 4 Future Prospects

The setting of regulatory restrictions on the number of chemical reagents (like chromium) used by tanneries and how much of it is included in the finished leather goods The European Union's member states are primarily involved in the aforementioned plan. To supply them with shared infrastructure (such as wastewater treatment plants), legislative authorities in any event encourage and support the central organization of leather industries in certain clusters of organized industrial regions. Many European countries follow good housekeeping practices at industry level like (Kokkinos and Zouboulis 2020):

(i) input–output analysis (via supplies) and monitoring of used chemicals, including toxicological properties and quantities.
(ii) reducing chemical use to the least possible level according to final product specifications

(iii) careful handling and safeguarding of raw materials as well as final goods to prevent leaks, accidents, and the generation of excessive wastewater
(iv) separation of waste waterways, should possible, to facilitate proper recycling
(v) careful tracking of the essential operational parameters to improve the manufacturing process
(vi) routine maintenance of sewage treatment systems
(vii) consideration of options for specific water reuse
(viii) scrutiny of the respective major waste disposal options.

Sustainability is the most important requirement for the textile industry's long-term viability. A paradigm shift toward environmentally friendly production is required for long-term success. Embracing the Circular Economy model, which focuses on waste reduction and the use of renewable resources, is critical throughout the product life cycle. Water conservation, reducing carbon footprints, supporting local artisans, ensuring cruelty-free practices, and advocating for fair trade, just wages, and safe working conditions are all initiatives. This discussion delves into novel approaches to sourcing raw materials and upcycling waste, providing valuable insights to fill research gaps. Addressing city-level textile issues, investigating national and international research, and fostering industry-institute collaborations all promise tangible solutions to pervasive pollution problems. Table 3 Sustainable-fashion promoting brands and enlists some sustainable fashion promoting industries and their respective waste raw materials and products (Aishwariya 2020).

The fast fashion's ascent has been instrumental in driving up consumption, in part because of social media and the industry's ability to spread fashion trends to a wider audience more quickly than in the past. The frequent availability of new trends at extremely low prices is known as fast fashion. The new approaches to address this issue include creating new clothing rental business models, designing products to facilitate recycling and reusing them (circular fashion), persuading customers to purchase higher-quality, longer-lasting clothing (slow fashion), and generally influencing consumer behavior toward more environmentally friendly choices (European Union 2020). The table suggests the steps to be taken by individual sectors. Table 4 enlists the major steps to be taken by individual sectors in a society for a sustainable and responsible usage of resources (Niinimäki et al. 2020).

**Table 3** Sustainable-fashion promoting brands and their respective waste raw materials and products (Aishwariya 2020)

| S. no | Waste raw material | Product | Brand |
|---|---|---|---|
| 1 | • Fish skin<br>• Natural dyes<br>• Tofu WASTE YArn | Hand bags | World of Mayu |
| 2 | Milk | Fabric | Qmilk |
| 3 | Coffee grounds waste | Fabric | Signtex |
| 4 | Waste coffee grounds | Polyester fabric | Adidas |
| 5 | Corn Husk | Shoes | Veja |
| 6 | • Milk<br>• Corn<br>• Eucalyptus<br>• Banana<br>• Bamboo | Fabric | DHURI |
| 7 | Orange peel | Fabric | • H&M<br>• Ferragmo |

**Table 4** Major steps to be taken by individual sectors in a society (Niinimäki et al. 2020)

| | Policymakers | Industry | Retailers | Consumers |
|---|---|---|---|---|
| Steps to be taken | • Legislation<br>• Rules and regulations<br>• Green taxation<br>• Tools for better equilibrium and a gradual system<br>• Extended producer accountability policy | • Prevent waste by investing in pollution-control technology<br>• Avoiding surplus manufacturing<br>• Shutting the material loop<br>• Transparency in the supply chain | • Innovative business models to encourage slow consumption and circular economy<br>• A new tag system that takes into account environmental impact of product | • Increase the lifespan of your products<br>• Slow consumption<br>• Conscious consumption<br>• Slow fashion |

## 5 Conclusion

As a byproduct of their manufacturing processes, the textile and leather industries generate significant amounts of waste sludge. Researchers and industries are actively investigating sustainable management practices to address the environmental impact of this waste. One method is to repurpose textile and leather sludge for biofuel production. The organic components of the sludge can be converted into biofuels using advanced technologies, providing a renewable and environmentally friendly energy source. Another use for these waste sludges is in vermicomposting, a process in which earthworms break down and transform organic materials into nutrient-rich compost. The resulting vermicompost acts as a natural fertilizer, improving soil fertility and promoting sustainable agriculture. Furthermore, researchers are looking into ways to incorporate textiles. Researchers are also looking into ways to incorporate textile and leather sludges into building materials. This not only reduces waste but also contributes to the development of eco-friendly building materials, which aligns with the principles of a circular economy. The emphasis on sustainable waste sludge management reflects the textile and leather industries' broader commitment to reducing their environmental footprint. Researchers hope to reduce the impact on ecosystems, promote resource efficiency, and contribute to the overall sustainability of these industries by developing innovative and eco-friendly applications for waste generated.

## References


Abushammala H, Mao J (2020) A review on the partial and complete dissolution and fractionation of wood and lignocelluloses using imidazolium ionic liquids. Polymers (Basel) 12(1). https://doi.org/10.3390/polym12010194

Aftab MN, Hameed A (2006) Biodegradation of leather waste by enzymatic treatment. Chinese J Process Eng 6(3):8–11

Ahmed Z, Mahmud S, Acet DH (2022) Circular economy model for developing countries: evidence from Bangladesh. Heliyon 8(5):e09530. https://doi.org/10.1016/j.heliyon.2022.e09530

Aishwariya S (2020) Covering the entire spectrum of circular economy and conscious clothing. Asian Text J 29(2):33–36

Ashok Kumar VT, Ravichandran PT, Kannan Rajkumar PR (2019) Use of textile effluent treatment plant sludge as sustainable material in brick manufacturing, vol 846. Springer Singapore. https://doi.org/10.1007/978-981-13-2182-5_27

Bagheri M, Esfilar R, Sina Golchi M, Kennedy CA (2020) Towards a circular economy: a comprehensive study of higher heat values and emission potential of various municipal solid wastes. Waste Manag 101:210–221. https://doi.org/10.1016/j.wasman.2019.09.042

Borges C, Spier F, Gutterres M (2018) Resources, conservation & recycling biogas production for anaerobic co-digestion of tannery solid wastes under presence and absence of the tanning agent. Resour Conserv Recycl 130:51–59. https://doi.org/10.1016/j.resconrec.2017.11.018

Chand S, Chand S, Raula B (2023) Textile and apparel industries waste and its sustainable management approaches. J Mater Cycles Waste Manag 25(6):3132–3143. https://doi.org/10.1007/s10163-023-01761-1

Duque A, Álvarez C, Doménech P, Manzanares P, Moreno AD (2021) Advanced bioethanol production: from novel raw materials to integrated biorefineries. Processes 9(2):1–30. https://doi.org/10.3390/pr9020206

E. Union, "European Union," (2020) The impact of textile production and waste on the environment (infographics). [Online]. Available: https://www.europarl.europa.eu/news/en/headlines/society/20201208STO93327/the-impact-of-textile-production-and-waste-on-the-environment-infographics

Fricke K, Santen H, Wallmann R, Hüttner A, Dichtl N (2007) Operating problems in anaerobic digestion plants resulting from nitrogen in MSW. Waste Manag 27(1):30–43. https://doi.org/10.1016/j.wasman.2006.03.003

Gholamzad E, Karimi K, Masoomi M (2014) Effective conversion of waste polyester-cotton textile to ethanol and recovery of polyester by alkaline pretreatment. Chem Eng J 253:40–45. https://doi.org/10.1016/j.cej.2014.04.109

Gomes CS, Repke J, Meyer M (2020) Investigation of different pretreatments of chromium leather shavings to improve biogas production

Gómez-Brandón M, Domínguez J (2014) Recycling of solid organic wastes through vermicomposting: microbial community changes throughout the process and use of vermicompost as a soil amendment. Crit Rev Environ Sci Technol 44(12):1289–1312. https://doi.org/10.1080/10643389.2013.763588

Hasanzadeh E, Mirmohamadsadeghi S, Karimi K (2018) Enhancing energy production from waste textile by hydrolysis of synthetic parts. Fuel 218:41–48. https://doi.org/10.1016/j.fuel.2018.01.035

Islam A, Mizan A, Ahmed T (2017) Sustainable use of tannery sludge in brick manufacturing in Bangladesh. Waste Manag 60:259–269. https://doi.org/10.1016/j.wasman.2016.12.041

Jeihanipour A, Taherzadeh MJ (2009) Ethanol production from cotton-based waste textiles. Bioresour Technol 100(2):1007–1010. https://doi.org/10.1016/j.biortech.2008.07.020

Jeihanipour A, Aslanzadeh S, Rajendran K, Balasubramanian G, Taherzadeh MJ (2013) High-rate biogas production from waste textiles using a two-stage process. Renew Energy 52:128–135. https://doi.org/10.1016/j.renene.2012.10.042

Jothilingam M, Preethi V, Chandana PS, Janardhanan G (2023) Fabrication of sustainable green bricks by the effective utilization of tannery sludge as main additive. Structures 48:182–194. https://doi.org/10.1016/j.istruc.2022.12.057

Juanga-Labayen JP, Labayen IV, Yuan Q (2022) A review on textile recycling practices and challenges. Textiles 2(1):174–188. https://doi.org/10.3390/textiles2010010

Kale RD, Karmegam N (2010) The role of earthworms in tropics with emphasis on indian ecosystems. Appl Environ Soil Sci 2010:1–16. https://doi.org/10.1155/2010/414356

Kandasamy R, Venkatesan SK, Uddin MI, Ganesan S (2020) Solid waste and production of high value-added. Elsevier Inc. https://doi.org/10.1016/B978-0-12-817951-2.00001-8

Keskin A, Mehmet Ş, Osman A (2020) Experimental studies on biodiesel production from leather industry waste fat and its effect on diesel engine characteristics. Fuel 276. https://doi.org/10.1016/j.fuel.2020.118000

Kokkinos E, Zouboulis AI (2020) The chromium recovery and reuse from tanneries : a case study according to the principles of circular economy. Springer Singapore. https://doi.org/10.1007/978-981-15-6296-9

Kumar P, Samuchiwal S, Malik A (2020) Anaerobic digestion of textile industries wastes for biogas production. Biomass Convers Biorefinery 10(3):715–724. https://doi.org/10.1007/s13399-020-00601-8

Kumar R, Basu A, Bishayee B, Chatterjee RP, Behera M, Ang WL, Pal P, Shah M, Tripathy SK, Ambika S, Janani VA, Chakrabortty S, Nayak J (2023) Management of tannery waste effluents towards the reclamation of clean water using an integrated membrane system: a state-of-the-art review. Environ Res 229:115881. https://doi.org/10.1016/j.envres.2023.115881

Kuzmanova E, Zhelev N, Akunna JC (2018) Effect of liquid nitrogen pretreatment on various types of wool waste fibres for biogas production. Heliyon 4(5):e00619. https://doi.org/10.1016/j.heliyon.2018.e00619

Lores M, Gómez-Brandón M, Pérez-Díaz D, Domínguez J (2006) Using FAME profiles for the characterization of animal wastes and vermicomposts. Soil Biol Biochem 38(9):2993–2996. https://doi.org/10.1016/j.soilbio.2006.05.001

Maity S, Bhattacharya S, Chaudhury S (2009) Metallothionein response in earthworms *Lampito mauritii* (Kinberg) exposed to fly ash. Chemosphere 77(3):319–324. https://doi.org/10.1016/j.chemosphere.2009.07.011

Mohsenzadeh A, Jeihanipour A, Karimi K, Taherzadeh MJ (2012) Alkali pretreatment of softwood spruce and hardwood birch by NaOH/thiourea, NaOH/urea, NaOH/urea/thiourea, and NaOH/PEG to improve ethanol and biogas production. J Chem Technol Biotechnol 87(8):1209–1214. https://doi.org/10.1002/jctb.3695

Mortar C (2019) A study on tannery sludge as a raw material for cement mortar

Mpofu AB, Oyekola OO, Welz PJ (2019) Co-digestion of tannery waste activated sludge with slaughterhouse sludge to improve organic biodegradability and biomethane generation. Process Saf Environ Prot 131:235–245. https://doi.org/10.1016/j.psep.2019.09.018

Nguyen VT, Chiang KY (2021) Sewage and textile sludge thermal degradation kinetic study using multistep approach. Thermochim Acta 698:178871. https://doi.org/10.1016/j.tca.2021.178871

Niinimäki K, Peters G, Dahlbo H, Perry P, Rissanen T, Gwilt A (2020) The environmental price of fast fashion. Nat Rev Earth Environ 1(4):189–200. https://doi.org/10.1038/s43017-020-0039-9

Palani Y, Jonnalagadda RR, Nishter NF (2017) Adsorption on activated carbon derived from tannery fleshing waste : adsorption isotherms. Thermodyn Kinetics 36(6). https://doi.org/10.1002/ep.12637

Patel N, Shahane S, Chauhan D, Rai D, Khan MZA (2021) Environmental impact and treatment of tannery waste. In: Water pollution and remediation: organic pollutants. environmental chemistry for a sustainable world. https://doi.org/10.1007/978-3-030-52395-4_16

Quartinello F et al (2018) Highly selective enzymatic recovery of building blocks fromwool-cotton-polyester textile waste blends. Polymers (Basel) 10(10). https://doi.org/10.3390/polym10101107

Rahman M, Rahman M, Tamez K, Islam A (2017) Textile effluent treatment plant sludge: characterization and utilization in building materials. Arab J Sci Eng 1435–1442. https://doi.org/10.1007/s13369-016-2298-9

Ranjitha J, Raghavendra SG, Vijayalakshmi S, Deepanraj B (2020) Production, optimisation and engine characteristics of beef tallow biodiesel rendered from leather fleshing and slaughterhouse wastes, pp 675–688

Rossi D, Cappello M, Antognoli M, Brunazzi E, Seggiani M (2023) Pyrolyzed tannery sludge as adsorbent of volatile organic compounds from tannery air emissions. Chem Eng J 454(P3):140320. https://doi.org/10.1016/j.cej.2022.140320

Saravanan CG, Sendilvelan S, Arul S, Raj CS (2009) Bio gas from textile cotton waste—an alternate fuel for diesel engines. Open Waste Manag. J. 2(1):1–5. https://doi.org/10.2174/1876400200902010001

Selvaraj S, Jeevan V, Jonnalagadda RR, Fathima NN (2019) Conversion of tannery solid waste to sound absorbing nano fibrous materials: a road to sustainability. J Clean Prod 213:375–383. https://doi.org/10.1016/j.jclepro.2018.12.144

Simioni T, Borges C, Dettmer A, Gutterres M (2022) Enhancement of biogas production by anaerobic co-digestion of leather waste with raw and pretreated wheat straw. Energy 253:124051. https://doi.org/10.1016/j.energy.2022.124051

Singh J, Kaur H, Khattra SK (2019) Use of textile mill sludge in concrete for waste characterization and its management. Springer International Publishing. https://doi.org/10.1007/978-3-030-02707-0

Singh D, Sharma D, Kumar P, Amit S, Dinesh J, Sharma K (2023) Characterization of homogenous acid catalyzed biodiesel production from palm oil: experimental investigation and numerical simulation. Environ Sci Pollut Res 34481–34502. https://doi.org/10.1007/s11356-022-24515-2

Sri K, Kameswari B, Kalyanaraman C, Porselvam S, Thanasekaran K, Division ET (2011) Enhancement of biogas generation by addition of lipase in the co-digestion of tannery solid wastes. Clean Soil Air Water 39(8):781–786. https://doi.org/10.1002/clen.201000408

Tang Y, Zhao J, Zhang Y, Zhou J, Shi B (2021) Chemosphere Conversion of tannery solid waste to an adsorbent for high-efficiency dye removal from tannery wastewater: a road to circular utilization. Chemosphere 263:127987. https://doi.org/10.1016/j.chemosphere.2020.127987

Twizerimana M, M'Arimi MM, Nganyi EO, Omara T, Olomo E, Kawelamzenje NA (2021) Anaerobic digestion of cotton yarn wastes for biogas production: feasibility of using sawdust to control digester temperature at room temperature. Oalib 08(07):1–15. https://doi.org/10.4236/oalib.1107654

United Nations Environmental Protection Agency (2024) Textiles: material-specific data. [Online]. Available: https://www.epa.gov/facts-and-figures-about-materials-waste-and-recycling/textiles-material-specific-data#:~:text=EPA measures the generation%2C recycling%2C composting%2C combustion with, 5.8 percent of total MSW generation that year

Vignesh G, Barik D (2019) Toxic waste from biodiesel production industries and its utilization. Elsevier Ltd. https://doi.org/10.1016/B978-0-08-102528-4.00006-7

Wang Y, Zhao Y, Deng Y (2008) Effect of enzymatic treatment on cotton fiber dissolution in NaOH/urea solution at cold temperature. Carbohydr Polym 72(1):178–184. https://doi.org/10.1016/j.carbpol.2007.08.003

Wang M et al (2016) Bioethanol production from cotton stalk: a comparative study of various pretreatments. Fuel 184:527–532. https://doi.org/10.1016/j.fuel.2016.07.061

Yan J et al (2023) Co-gasification of municipal sewage sludge and cotton stalk enhanced by metal-enriched texture dyeing sludge additives for syngas production. Fuel 341:127669. https://doi.org/10.1016/j.fuel.2023.127669

Yuliana M et al (2020) Biomass and bioenergy a one-pot synthesis of biodiesel from leather tanning waste using supercritical ethanol: process optimization. Biomass Bioenergy 142:105761. https://doi.org/10.1016/j.biombioe.2020.105761

Yuliana M et al (2021) Biomass and bioenergy efficient conversion of leather tanning waste to biodiesel using crab shell-based catalyst: waste-to-energy approach. Biomass Bioenergy 151:106155. https://doi.org/10.1016/j.biombioe.2021.106155

Yuvaraj A, Thangaraj R, Maheswaran R (2019a) Decomposition of poultry litter through vermicomposting using earthworm *Drawida sulcata* and its effect on plant growth. Int J Environ Sci Technol 16(11):7241–7254. https://doi.org/10.1007/s13762-018-2083-2

Yuvaraj A, Karmegam N, Tripathi S, Kannan S, Thangaraj R (2019b) Environment-friendly management of textile mill wastewater sludge using epigeic earthworms: bioaccumulation of heavy metals and metallothionein production. J Environ Manage 254:109813. https://doi.org/10.1016/j.jenvman.2019.109813

Yuvaraj A et al (2020) Recycling of leather industrial sludge through vermitechnology for a cleaner environment—a review. Ind Crops Prod 155:112791. https://doi.org/10.1016/j.indcrop.2020.112791

Zhan BJ, Poon CS (2015) Study on feasibility of reutilizing textile effluent sludge for producing concrete blocks. J Clean Prod 101:174–179. https://doi.org/10.1016/j.jclepro.2015.03.083

Zhan BJ, Li JS, Xuan DX, Poon CS (2019) Recycling hazardous textile effluent sludge in cement-based construction materials: physicochemical interactions between sludge and cement. J Hazard Mater 381:2020. https://doi.org/10.1016/j.jhazmat.2019.121034

# Textile and Leather Industry Residues: Managing Complex Sludge Challenges

Shelley Oberoi

**Abstract**

There has been a noticeable increase in the global population in recent decades, along with significant improvements in overall living standards. These two advancements have significantly contributed to the escalating demand for textiles, consequently leading to a surge in textile manufacturing. The annual global production of apparel and textile fibers exceeds 110 million metric tons, resulting in a significant generation of textile waste. The textile manufacturing processes necessitate substantial quantities of water. Additionally, it is important to note that this phenomenon can potentially result in detrimental impacts on both the overall ecosystem and the well-being of human populations. Based on scientific research, it has been determined that the leather industry is responsible for the production of substantial quantities of industrial sewage. The waste generated during the different stages of leather production is derived from various technological processes. It is important to note that approximately 1 metric ton of raw hide results in a total waste output of approximately 700 kg. Furthermore, it is important to note that tannery waste is known to contain chromium compounds, which are widely used as tanning agents in the leather industry. This presents an additional peril to the ecosystem. Based on the composition and levels of chemicals present in the wastewater, if left untreated, these substances have the potential to contaminate water sources. Furthermore, it is important to note that the textile and leather industries serve as significant contributors to the generation of hazardous waste. Not all of the chemical compounds that are introduced into the flotation process ultimately find their way into the end product, but rather are discharged along with the utilized process water. Hence, environmental protection necessitates effective sludge management.

**Keywords**

Global population · Textile waste · Raw hide · Leather industry · Sewage

## 1 Introduction

The textile industry is a major contributor to environmental pollution, primarily due to its dyeing and pre-treatment processes. These processes generate liquid waste and sludge that contain reactive dyes and chemicals. The leather industry produces waste from chromium-tanned leather, which is considered to be hazardous. The exploration of advanced treatment methodologies, such as adsorption and activated carbon, is aimed at reducing these wastes. The tanning industry is responsible for the release of more than 250 chemicals into the environment, which has been linked to various health issues. The utilization of vermiculture-treated leather wastes in sustainable agriculture holds potential for addressing pollution concerns. Tannery waste has the potential to be utilized in the synthesis of surfactants, biopolymers, and biodegradable re-tanning agents. Additional methods for addressing the issue include the reduction of decabromodiphenyl ether in textile sludge, the utilization of cement kiln dust and coal filters, and the incorporation of biochar derived from textile dyeing sludge. Various sustainable sludge management methods, including incineration of textile sludge ash, fluidized bed pyrolysis, and thermal plasma gasification, have the potential to effectively tackle heavy metal pollution and enhance sustainability.

S. Oberoi (✉)
Department of Humanities and Applied Sciences, KC College of Engineering and Management Studies and Research, Thane, India
e-mail: shelly.oberoi@kccemsr.edu.in

© The Author(s), under exclusive license to Springer Nature Switzerland AG 2025

A. Mandpe and M. P. Shah (eds.), *Wastewater Sludge Nexus*, Sustainable Landscape Planning and Natural Resources Management, https://doi.org/10.1007/978-3-031-45758-6_5

## 2 Residues from the Textile Industry

The textile industry, a notable source of environmental pollution, encompasses cotton, wool, and synthetic fibers. The cotton textile industry and its dyeing processes yield substantial quantities of aqueous effluent throughout the dyeing and finishing procedures. The watery solution often has bright colors because of the reactive dyes and chemicals that are still there. These chemicals and dyes have complex molecular structures and high chromaticity levels. The aqueous effluent also encompasses recalcitrant substances.

The organic compounds derived from the pre-treatment phase of pulp, cotton gum, cellulose, hemicellulose, and alkali are the primary contributors to pollution in this particular scenario. The pre-treatment of wastewater accounts for roughly 45% of the total wastewater produced, whereas the dyeing and printing procedures contribute between 50 and 55% to the overall wastewater volume. The sizing agents that are used in the weaving of fabric are removed by desizing and scouring, which are pre-treatment processes for cotton fabric. The textile industry is polluted because contamination occurs in cotton, cotton gum, hemicellulose, slurry, and alkali, which are used in the weaving process. During the pre-treatment process, the average chemical oxygen demand (COD) molar concentration is 3000 mg/L.

Terephthalic acid and ethylene glycol are products of the reduction process of pre-treatment of polyester fibers at 90 °C for 45 min in the presence of alkali to give a silky appearance to polyester fiber. This process is classified into continuous and batch configurations where the COD constituents make 60% or more of the waste water volume. The presence of dyes in waste water is also impacting the environment and human beings. When treating textile dyeing wastewater, the pH value (in the range of 10–11) is a crucial factor to take into account. The presence of total nitrogen and ammonia nitrogen is because of dyes in waste water, which have a concentration of 300 mg/L. Similarly, phosphorus is found in waste water because of phosphorous compounds in detergents (Ranade and Bhandari 2014).

According to Chen et al. (2021), CPs are named chlorophenols, which have a tendency to accumulate in sludge that is generated during the textile dyeing process. This study looked into the issues that chlorophenols in five textile dyeing plants (TDPs) were causing. The ecological screening level (ESL) of CPs was carried out to analyze their pollution levels, and potential risk was determined by the risk quotient (RQ) value and dioxin conversion rate. A reduced rate of CP augmentation was obtained by employing a concentration of 180 mmol/L with a stoichiometric ratio of hydrogen peroxide to ferrous ions of 1:1 while maintaining a pH range of 3–4 and allowing the reaction to proceed for a duration of 30 min (Chen et al. 2021).

### 2.1 Complex Textile Sludge Management

The underlying issue pertaining to industrial textile wastewater lies in its highly carcinogenic properties, which give rise to allergic reactions and engender the formation of toxic compounds, alongside various other environmental concerns. Henceforth, the aqueous effluent derived from the textile and dyeing sectors elicits significant apprehension in terms of textile wastewater management. Wang and Peng (2010) expounded upon the scientific utilization of naturally occurring zeolites as adsorption agents for the purification of sewage and water solutions originating from different sources. It was found from the study that zeolite absorbs ammonium ions up to 20 mg/L.When it is used in powdered form, with particle sizes ranging from 4 to 8 mm, it shows superior efficacy (Wang and Peng 2010). Derden and Huybrechts (2013) worked on an effective methodology to reduce the amount of decabromodiphenyl ether (Deca-BDE) released from the textile industry. It is observed, by using bond dissociation energy (BDE), the production of cotton and synthetic fibers for clothing and floor coverings has increased (Derden and Huybrechts 2013). Lian et al. (2009) also observed that brominated flame retardants can be used as fire-resistant fibers in fabricated textiles. Therefore, in the course of the finalization procedure, elevated quantities of Deca-BDE were being incorporated into the sludge, thereby posing detrimental effects on the environment and soil. Henceforth, another methodology was acknowledged as the level of emissions associated with the best approach (BAT-AEL), thereby confirming that the emission statistics within the determined BAT-AEL range were below 20 mg/L (Lian et al. 2009). Mahmoued (2010) has put forth an alternative approach to treating wastewater in the textile industry, specifically through the implementation of coal filter treatment. In the aforementioned protocols, the utilization of cement kiln dust (CKD) and CKD/coal filters enables the efficacious removal of all impurities from textile sludge. The various analytical studies, like X-ray diffraction (XRD), differential scanning calorimeter (DSC), and infrared (IR), were conducted to quantitate the colored remaining part. From the studies, it has been found that CKD is an efficient way to eliminate acid dyes from sludge in Egypt (Mahmoued 2010).

Saraya and Nassar (2015) examined the adsorption characteristics of Reactive Blue 7 dye, obtained from textile sludge, utilizing a cost-effective adsorbent referred to as CKD. The CKD exhibited a remarkable maximum adsorption capacity of 100 mg/G under ambient temperature conditions. In addition, they found that adding different solvents and acids to the CKD that was dyed caused the dye to be released in water and alcohols because the dye physically stuck to the surfaces. Therefore, the majority of researchers have shown a preference for absorbent materials that possess both excellent absorbency properties and economic viability

(Saraya and Nassar 2015). Instances of such substances encompass waste from farming (Mahmoud et al. 2012), which can function as an integrating adsorbent (Allen 1987); tire char (Mui et al. 2010); and waste from farming like rice bran and bone char (Ip et al. 2010). The investigators observed that the compound extracted from the textile wastewater consisted solely of cadmium, with an elimination capacity of 100% attained by means of the application of activated carbon and cement kiln dust (CKD). The experimental results indicate that CKD demonstrates promise as a cost-efficient and powerful sorbent for the removal of cadmium (Cd) and copper (Cu) from textile effluent when compared to activated carbon.

In their investigation, Kumar and Porkodi (2009) employed rubber wood sawdust as a substitute for CKD to facilitate the adsorption-mediated removal of Bismarck brown from water-based solutions. Nevertheless, caution must be exercised when this compound encounters wastewater or the environment, as its carcinogenic properties may pose a threat to human life (Kumar and Porkodi 2009). Malik (2003) did more research on the ability of activated carbon to remove acid dyes from water-based solutions. The activated carbon was derived from economically viable sources, namely mahogany sawdust and rice husk.

Ultimately, the researcher reached the conclusion that sawdust exhibited superior adsorption capacity in comparison to rice husk (Malik 2003). It was further noted by Charumathi and Das (2012) that the adsorption capacity was primarily influenced by the duration of contact, the quantity of adsorbent employed, and the pH level of the wastewater, respectively (Charumathi and Das 2012). Ferrero (2007) elucidated the utilization of hazelnut shells and sawdust as potential adsorbents for the efficacious removal of methylene blue, a basic dye, and acid blue 25 from water-based solutions. The researcher meticulously conducted a series of experiments to ascertain the adsorption capacity of these organic materials. The wastewater undergoes subsequent amalgamation with river water or soil water, thereby engendering significant societal ramifications in terms of both aqueous and chemical interactions with the environment. Therefore, in the present era, the prominent obstacles lie in the effective utilization of textile wastewater, either by subjecting it to various treatment methodologies for its reuse within the textile industry or by addressing the substantial expenses associated with its treatment. Upon the conclusion of the treatment process, it is noteworthy that the waste water undergoes a transformation, resulting in its conversion into regular water suitable for subsequent utilization within the textile industry.

Coincidentally, substantial volumes of textile sludge are concurrently generated as a secondary outcome of the textile industry. In the current study, investigator employed three different types of sewage waste, specifically desiccated sludge, sludge granules, and sludge ashes, to produce materials for construction. They additionally deduced that, at a weight percentage of 20, sewage sludge exhibits potential for incorporation into the production process of soft mud bricks.

Subsequently, he observed that the concrete's compressive strength reached a value of 23.55 N/mm$^2$, as per the standards set for M20-grade concrete (Ferrero 2007).

Chougule and Sonaje (2013) initially put forth the concept of conducting a cost–benefit analysis pertaining to the recycling of wastewater solely in the context of wet processing. Furthermore, they advocated for the adoption of water management methodologies across various industries, encompassing the utilization of alternative sources of pristine water throughout the entire production cycle, ranging from wet processing to the evaluation of finished products (Chougule and Sonaje 2013). Teal (1997) performed a comprehensive sensitivity analysis on the existing treatment processes and devised a cost-effective methodology by circumventing the need for pre-treatment processes in all instances. The primary emphasis of the researcher revolved around the diminished utilization of dye and fabric in terms of advantageous outcomes.

Additionally, a noteworthy rationale was proposed for the abatement of chemicals during the process, labor, variable overheads, and fixed overheads (Teal 1997). In another discussion, scientists explored the cost analysis of the processing of textile dyes, specifically focusing on the combined utilization of hydrogen peroxide ($H_2O_2$) and ultraviolet (UV) radiation. Additionally, scientists examined the efficacy of the biological treatment process when applied to higher concentrations of reactive textile dyes. It has been suggested that hydrogen peroxide ($H_2O_2$) exhibits negligible absorption capabilities. However, with advancements in ultraviolet (UV) technology, there is potential for enhanced efficiency. Additionally, the biological treatment stage holds promise for reducing energy expenditure and achieving a greater degree of mineralization. It has been reported that the use of ozone for treatment purposes incurs a significantly higher investment cost (9).

Singh and Srivastava (2022) observed in their investigation that the sludge originating from the textile dyeing industry underwent a successful conversion into biochar with remarkable efficacy. The utilization of this biochar was employed for the purpose of eradicating ofloxacin, which is a pharmaceutical pollutant. There was a lot of calcium (Ca) in the slurry that had not been reacting. This is because lime was used as a coagulating agent in the effluent treatment plant (ETP). The biochar was able to adsorb up to 21.6 mg/G when the conditions were perfect (starting concentration, 30 mg/L, temperature, of 25 $\pm$ 2 °C, pH value of 6 $\pm$ 0.2, dose of 5 g/L, and time of 240 min). A study on the thermal regeneration of biochar produced an initial removal efficiency of about 73% in the first cycle, which then gradually dropped to about 59% after five consecutive cycles (Singh and Srivastava 2022). The textile industry in Tiruppur, Tamil Nadu, serves as a significant contributor to aqueous contamination. The effluent treatment plants in the industry have partially alleviated this concern; however, the resultant precipitate presents storage and disposal challenges owing to its elevated concentrations of heavy metals, sulfates, and chlorides. These chemical species have the potential to undergo leaching processes, thereby infiltrating rainwater and surface water, thereby presenting a potential hazard to the integrity of the surrounding natural environment. Many investigations are going on for the packaging of textile waste into concrete, where waste plastics and oils can act as effective binders. But it was observed that DIETPS (compounds of sulfates, chlorides, and heavy metals) decreased the mechanical strength of both polymer and geopolymer-based concrete. These results showed that plastic can act as an effective binder (Raghunathan 2018).

This study outlines the synthesis of an important adsorbent known as textile sludge-based activated carbon (TSBAC), which is used to purify textile waste generated from dyeing. The sludge used to create this TSBAC has undergone chemical and heat treatment with phosphoric acid. This adsorbent has active functional groups and a large specific area, and it shows excellent affinity for synthetic dyes like reactive red 198 dye, reactive yellow 145 dye, and methylene blue separately and in synthetic textile waste water. According to the pseudo-second-order and isotherm models of Langmuir, the methylene blue and reactive red 198 can adsorb 101.8 mg/G and 76.6 mg/G, respectively. The reactive yellow 145 can adsorb 75.1 mg/G. In experiments, it was found that TSBAC was used to remove the dye in six cycles (Oke and Mohan 2022). This investigation was conducted at the Common Effluent Treatment Plant (CETP), situated in Karur, Tamil Nadu, to show the application of sludge from a textile effluent treatment plant (ETP) as an alternative to cement. In experiments, the physical characteristics of cement (consistency, initial and final setting times) were examined by adding different proportions of sludge. In the final part of the investigation, the compressive strength of mortar cubes was examined, and sludge was applied as a substitute for cement. Furthermore, the chemical properties of the treated water solution, such as its pH, electrical conductivity, hardness, and heavy metal concentration, have been checked at different times for different amounts of sludge in cement-mortar cubes. There were the following results: chemical sludge from textile plants' wastewater treatment processes could be used instead of cement. This method can replace up to 20% of the cement content in mortar while still maintaining its essential compressive strength (Jeevanandam et al. 2015).

On the same concept, a comprehensive investigation was conducted on the Ethiopian textile sector, revealing a noteworthy surge in its expansion. However, it was observed that the wastewater treatment facilities associated with this industry produce a substantial quantity of sludge, which frequently exhibits contamination by heavy metals. The conventional methods of disposal, such as landfilling and incineration, are deemed unsuitable due to the production of leachate and residues resulting from the incineration process. This has resulted in an increasing need for sustainable sludge management. The objective of this investigation is to ascertain an environmentally friendly resolution for managing textile sludge via incineration and its utilization as a partial alternative to cement in the production of concrete. By subjecting the substance to incineration at a temperature of 625 °C for a period of four hours, the results were obtained as follows: a moisture content of 78.1%, a volatile organic matter content of 61.2%, and an inorganic ash content of 59.6%. Silicon dioxide ($SiO_2$), iron III oxide ($Fe_2O_3$), calcium oxide (CaO), and aluminum oxide ($Al_2O_3$) were also found. By substituting the cement with incinerated textile sludge ash in different proportions, the concrete block was synthesized. As a result, there was a decrease in the hydration duration by a maximum of 20%, but the desired compressive strength was achieved. The rate of water absorption was also increased by textile sludge ash, and the substitution of cement with a 20% proportion resulted in heavy metal concentrations that followed the United States Environmental Protection Agency's (USEPA's) permissible limit. The findings of the experiment suggest that adding textile sludge ash to cement concrete can reduce heavy metal contamination linked to sludge disposal and reduce the adverse effects of cement manufacturing procedures on the environment (Kasaw et al. 2021).

Ran et al. (2019) found that fluidized bed pyrolysis (FBP) is an efficient way to keep away textile dyeing sludge (DS). The pyrolysis behavior of DS was analyzed under different experimental conditions. Fluorocarbons (FCs) efficiently acted as chelating agents for heavy metals, resulting in residue ratios varying from 64 to 89 wt% within the temperature range of around 450 to 850 °C. Researcher found that calcium oxide (CaO), kaolin, and calcium-bentonite could increase the amounts of lead (Pb), nickel (Ni), chromium (Cr), and zinc (Zn) in fuel cells. Calcium-bentonite exhibited the most pronounced recuperation efficiencies for nickel (Ni), zinc (Zn), and lead (Pb) potassium nickelate ($K_2Ni(CN)_4$), zinc aluminate ($ZnAl_2O_4$), zinc sulfide (ZnS), lead silicate, and lead sulfate ($PbSO_4$) were successfully discerned within the FCs, whereas zinc aluminate ($ZnAl_2O_4$) was specifically detected in the DS. The results of the extraction showed that higher temperatures and the addition of catalysts sped up the process of changing heavy metals from the oxidizable part to the residual part. This reduced the amount of pollution in the environment (Ran et al. 2019). Textile dyeing sludge is an expanding and perilous compound with a plethora of potential health hazards. The limitations of conventional treatment methods encompass the potential for contamination of water and soil, the inadequate degradation of hazardous organic compounds, and the inefficacious immobilization of toxic heavy metals. To treat textile dyeing sludge, thermal plasma gasification has been used by applying a custom-made rotating arc plasma reactor. This process has a carbon conversion efficiency of 99.9% and an energy conversion efficiency of 71.8%. The gas enriched with syngas has a reduced heating value of 8.91 $MJ/Nm^2$ (Wang et al. 2019).

In this experiment, an aerobic granular sludge (AGS) system was used for treating textile wastewater. The textile dyeing effluent samples were taken from a textile mill and subjected to a bioreactor. It was observed that chemical oxygen demand (COD), total suspended solids (TSS), and the removal of lubricant from textile wastewater were affected at different concentrations of mixed liquor suspended solids (MLSS). It was found that when the concentration of MLSS was increased, its effectiveness increased to extract organic matter from the sample water. This process was found to be effective in removing COD from 91.2 to 94.5%. The rate of organic decomposition was also increased by increasing microbial density in the aeration tank. The TSS removal percentage was also increased from 83.5 to 98%, and there was also an increase of 76.4 from 62.5% for oil and grease. According to the study, the AGS system works well at cleaning up complicated and heavily polluted denim textile wastewater. This lowers the risk of secondary pollution that is common in this industry. The efficacy of the AGS system in treating industrial textile effluent is remarkable (Malik et al. 2021).

## 3 Residues from the Leather Industry

Engaged in the synthesis of diverse commodities such as footwear, purses, clothing, and other articles, the leather industry has exhibited a prolonged existence. The procurement of raw materials by the leather industry primarily relies on the utilization of byproducts derived from the food sector, with a particular emphasis on the remnants generated by the meat processing industry. This byproduct is used to create leather goods that are both fashionable and functional. Globally, the trade of leather and leather goods constitutes a substantial proportion of overall commerce. The production is predicated on readily available and replenishable resources. With an increasing urban population and a surge in demand for leather goods, the current annual trade value of this industry has already surpassed the noteworthy sum of US$80 billion.

A significant number of individuals employ and ingest leather-based commodities on a day-to-day basis. The meat processing sector and slaughterhouses serve as the primary suppliers of raw materials for the leather processing industry. Tanneries employ the utilization of this raw material to undergo a chemical transformation, resulting in the production of leather. The tanning process serves as a pivotal component within the leather industry, facilitating the transformation of raw animal hides into refined leather products (Barik 2018).

In nations such as India, the leather industry serves as a significant economic contributor. The wastewater effluents may exceed 150 metric tons per day as a consequence of the substantial water utilization during the leather purification procedure. The aqueous solution, comprising malodorous compounds and particulate matter, exhibits a volumetric capacity ranging from 20,000 to 80,000 liters (L). This voluminous effluent is generated as a consequence of the treatment of a metric ton of unprocessed dermal substrate. Elevated chemical oxygen demand (COD), heightened biological oxygen demand (BOD), diminished chromium species, sodium chloride (NaCl), calcium (Ca), magnesium (Mg), sulfide compounds, oils, tannins, and biocides are among the deleterious chemicals detected in effluent originating from the leather industry (Ferreira et al. 2023). The mechanical operations employed in the leather industry produce a substantial amount of solid waste, encompassing chromium-tanned leather waste (CTLW), which is categorized as deleterious according to Brazilian regulations. Scientists have conducted investigations into methodologies for diminishing, categorizing, and harnessing these residual substances in alternative sectors, either as novel adsorbents or as secondary commodities possessing amalgamated worth. The adsorption method is used in this study to selectively remove chromaticity from dye-containing effluent from a pilot-scale leather processing facility. A complete factorial design was employed to investigate the effects of all variables on adsorption, and the feasibility of utilizing the process as a pre-treatment prior to implementing conventional treatment methods was verified (Piccin et al. 2016).

### 3.1 Residues from the Tanneries

An estimated $40 billion worth of leather is created every year by the world's leather industry, which spans 18 billion square feet. The majority of the world's leather comes from developing nations, where 65% of the total is used for footwear. Chemical and mechanical processes involved in tanning are tannins, acids, alkalis, solvents, auxiliaries, surfactants, metallorganic colors, tanning agents, sulfonated oils, salts, and leather-making machinery, all of which are necessary for the transformation of raw hide into finished leather. The process results in an effluent volume of 30 L for every kilogram of skin or hide that is treated. Indian tanneries release over 50,000 $m^3$ of wastewater each day, which contains significant amounts of organic contaminants (Doble and Kumar 2005). The tanning process produces a large amount of waste known as tanning sludge (TnS) when the wastewater from the leather industry is treated. Wastewater treatment technology, raw material composition, and skin type determine the amount and quality of TnS. The heavy metal ions, along with other contaminants like fixing agents and dyes, make this sludge very harmful. The presence of chromium salts in tanning agent sludge is also very concerning, which is used in 90% of world tanneries. From 6030 mg/Kg all the way up to 474,000 mg/Kg, their concentration fluctuates substantially. Tannins in TnS have the potential to reduce methane generation and volatile solid (VS) removal, two indicators of poor anaerobic digestion (AD) performance. How TnS is managed is dependent on its properties. The high level of chromium (Cr), which is significantly higher than the permissible limits in compost 100 mg/Kg, casts doubt on the potential effectiveness of biological approaches. Additionally, the quantities hinder the AD process. Using primary sludge from tannery waste water treatment plants, multiple studies have explored the possibility of anaerobic codigestion in lime fleshings. The results show that the process can be successful in removing VS and producing methane ($CH_4$), which is comparable to what has been reported for different types of organic waste. Incineration is another method for recovering energy from TnS. Several pre-treatment procedures are carried out on TnS before incineration in order to lessen its environmental impact, specifically by reducing chromium oxidation and the volatilization of heavy metals. It has been effectively utilized in vitrification to immobilize heavy metals in a silica matrix and pyrolysis to treat and decrease sludge and minimize volatilization. In contrast to vitrification byproducts, which can serve as alternatives to aggregates in building projects, pyrolysis byproducts are energy carriers that can generate heat and power (Grosser and Celary 2019).

According to Tadesse and Guya (2017), a multitude of harmful substances are released into the environment by the tannery industry. The more than 250 chemicals used in the production of leather release a complex combination of organic chlorinated phenols, toxic chromium (VI), and other harmful pollutants. These include sulfides, phenols, magnesium (Mg), sodium (Na), potassium (K), azo-dyes, cadmium (Cd), cobalt (Co), copper (Cu), antimony (Sb), barium (Ba) lead (Pb), selenium (Se), mercury (Hg), zinc (Zn), arsenic (As), PCB, nickel (Ni), formaldehyde resins, pesticide residues, mineral salts, dyes, and solvents such as grease and oils. The most common and harmful sewage contaminants to humans, other organisms, and ecosystems are chlorinated phenols and chromium (Cr, VI). Soil, water, and vegetable concentrations of chromium (Cr, VI) exceed 99.99% of the allowable limit set by the World Health Organization (WHO), the Food and Agriculture Organization (FAO), the Environmental Protection Agency (EPA), and other nations. Acute toxicity, mutagenicity, carcinogenicity, and hypertension are some of the health problems caused by these harmful chemicals in communities that do not treat their wastewater. There is more chromium than what is considered safe for human consumption in water samples taken from rivers and springs farther downstream 0.05 mg/L, according to the World Health Organization. The rise of cities and factories has made environmental pollution a pressing issue on a global scale. These pollutants are entering the environment, which includes water, soil, plants, animals, and humans. A major issue in the modern world is water contamination, and the next two most essential parts of the ecosystem are soil and air. Because soil is the last destination for many pollutants—both organic and inorganic—discharged into the environment by industries, soil contamination has emerged as a major issue in America's industrialized regions. The effluents released from the tannery depend on the tannery dimensions, the quantity of water used, and the chemicals used. When it comes to industrial waste, none is more polluting than tanning effluents, the most prevalent of which are from vegetable tanning. All the developing countries are heavily impacted by chromium (Cr) pollution caused by the tanning industry. Heavy metals like chromium and phenolic solutions are released into the environment in large quantities by nearly every leather industry (Tadesse and Guya 2017).

Chemicals such as sodium chloride, lime, sodium sulfide, protein hydrolysates, colors, surfactants, grease, solvents, and tanning agents (both synthetic and mineral) are abundant in tannery wastewaters. The use of precipitation to remove chromium salts from wastewaters produced by tanneries has been the subject of research. But since chromium salts are not completely depleted, using them after a concentration in chromium oxide ($Cr_2O_3$) was suggested. The waste waters are rendered unfit for further use in the tanning process after each usage due to the enrichment in neutral salts. Several techniques have been suggested for recovering chrome from depleted baths; one of these is to convert the precipitate, which is chromium hydroxide $Cr(OH)_3$, into a basic salt, such as sodium bichromate ($Na_2Cr_2O_7$) or chromium sulfate $Cr_2(SO_4)_3$. This method can be carried out using a variety of alkaline chemicals, such as sodium hydroxide, ammonium hydroxide, calcium hydroxide, magnesium oxide, etc. This study suggests three different ways to do things: a lime $Ca(OH)_2$ precipitation system, a polyelectrolyte flocculation system, and a chrome hydroxide filtration system with a rotating filter. On the basis of chemical and physical analyses, these results showed that the initial chromium oxide ($Cr_2O_3$) level of the wastewater samples treated with the calcium hydroxide polyelectrolyte precipitation system ranged from 10 to 25 mg/L during the hide tanning process (Zăinescu et al. 2018).

Chowdhury et al. (2015) have conducted a thorough investigation into the industrial sector of Bangladesh. The majority of industries in Bangladesh exhibit a predilection towards the utilization of chrome tanning. Following the aforementioned process, the sludge contains a substantial quantity of sodium chloride (NaCl), sodium sulfate ($Na_2SO_4$), organic compounds, metallic compounds, putrefying suspended matter, biological tanning materials, proteins, emulsified lipids, ammonium salts, sulfides, sulfuric acid, chrome, chlorides, sodium bicarbonate, sulfates, calcium salts, organic nitrogen, and ammonia. The research looked at a number of aspects of effluent quality, including ionic load, physicochemical parameters, and biological markers. Finally, these findings showed that there is no purpose in irrigating crops with the heavily polluted leather industry effluent (Chowdhury et al. 2015).

According to Fei and Liu (2016), one sector that causes a great deal of pollution is the leather industry, which is mostly located in poor nations (Malaiškienė et al. 2019). The tanning and post-tanning procedures, which use chemicals derived from chromium, are used by about 90% of the light leather sector. Improved intrinsic leather qualities are a result of the tanning and post-tanning processes, which stabilize proteins used to make leather and make it resistant to heat, deterioration, and thermo mechanical stress. Chrome tanning, the most commonly used method, utilizes a lot of basic chromium salts and produces more than 0.12 kg of pollutants per ton of treated leather (Fei and Liu 2016). A study was conducted on tannery sludge as a raw material for the cement mortar. The concrete samples that were subjected to testing contained 3–12% tannery sludge according to the weight of cement and 3–9% tannery sludge according to the weight of sand. The results of an X-ray diffraction (XRD) and scanning electron microscopy (SEM) of mortar and tannery sludge showed that ultrasonic pulse velocity (UPV) values, flexural and compressive strengths, and water absorption were all improved when 6% tannery sludge was included in place of cement in the mixture. These were findings that samples that had been treated with tannery sludge had more portlandite and less ettringite. This made the structure denser and gave it more C-S-H. The results showed that the chromium leaching values in the cement mortars did not surpass the limits specified in Directive 2003/33/EC. As a result of the many chemicals and freshwater resources used in the leather industry's production process, the industry's wastewater sludge, hides, buffing dust, and other solid waste products are released into the environment. Iron oxide (FeO), aluminum sulfate $Al_2(SO_4)_3$, magnesium oxide (MgO) and heavy metals are among the many dangerous compounds found in leather industry waste. Humidity, pH, temperature, and amendment materials are necessary to stimulate earthworm activity. Based on the findings, it was observed that vermi-treated leather wastes are a promising medium for sustainable agriculture since they promote plant growth. Additionally, the meticulously collected data showed in the study that vermitreatment could be an alternate approach to our environment with better agro ecosystems (Yuvaraj et al. 2020).

### 3.2 Complex Leather Sludge Management

The dominant economic paradigm, founded on a linear energy and material exchange model known as "take-make-waste," has led to a rapid increase in raw material utilization and waste generation. Unfortunately, the inadequate handling of these processes has adverse consequences for the ecosystem. Hence, it is crucial to formulate chemical solutions that can efficiently alleviate the deleterious environmental impacts of the industry while simultaneously satisfying the requirements of consumers (Sadhukhan et al. 2020). The circular economy (CE), colloquially referred to as the sustainable development paradigm, places significant emphasis on the processes of remanufacturing, repairing, and upgrading various components (Korhonen and Snäkin 2015). This involves the utilization of sustainable resources and the reutilization of waste materials as secondary inputs. The CE concept exhibits superior environmental compatibility and energy conservation while also demonstrating enhanced cost-effectiveness in comparison to the conventional economy model. This is achieved through the utilization of waste reuse strategies and the mitigation of pollutant emissions. The incorporation of novel technologies and the amplification of enterprises' reputations are two mechanisms through which their integration catalyzes business expansion (Chojnacka et al. 2020). The utilization of tannery waste, comprising precious compounds that exhibit renewability, has the potential to represent a noteworthy advancement towards the pursuit of sustainable development. $H_2O$ (up to 80% in materials) and protein (2.5–10.5% w/w) are the predominant components of unprocessed leather and processed waste, including fleshings, trimmings, splits, and scourings (Masilamani et al. 2016). The hydrolytic process of this proteinaceous waste can result in the production of collagen or gelatin, both of which serve as cost-effective feedstocks used in the pharmaceutical and cosmetic industries. The existing literature presents promising avenues for the valorization of leather waste in various chemical processes. The methodologies outlined entail the utilization of the protein component found in tanning residues for the synthesis of surfactants (Bautista et al. 2015); the extraction of activated carbon from bio-collagen leather waste for the purpose of refining biogas (Cabrera-Codony et al. 2021); the production of biopolymers derived from bovine hides; the development of sound-absorbing materials utilizing leather collagen hydrolysates (Selvaraj et al. 2021); and the acquisition of organic NPK fertilizers from poultry bones and leather (Majee et al. 2019). Furthermore, the synthesis of biodegradable re-tanning agents can be achieved through the utilization of byproducts derived from the leather industry (Sathish et al. 2019). Chromium (Cr), an element employed in the leather tanning process, serves as a notable origin of contamination, especially upon undergoing oxidation from $Cr^{3+}$ to $Cr^{6+}$. This particular chemical transformation gives rise to a variant that presents a potential mutagenic and carcinogenic hazard to human health. The leakage of solid waste treated with chromium and the presence of increased levels of chromium in wastewater possess the inherent capability to cause groundwater and soil contamination (Famielec 2020). Organic chelates possess the remarkable ability to effectively treat solid chrome leather waste, exhibiting an impressive efficacy rate of 95% (Malek et al. 2009). Mineral acids, namely hydrochloric acid (HCl), nitric acid ($HNO_3$), and sulfuric acid ($H_2SO_4$), can be employed for the purpose of chromium recovery.

Tasca and Puccini (2019), who also assessed the re-tanning, oiling, and dyeing processes of hides, deliberated upon the ecological concerns within the realm of the tanning sector (Tasca and Puccini 2019). Hu et al. (2011) conducted a study titled "Ecological Management of Tannery Waste in the CE Model," elucidating their findings in an article. The scientists presented a thorough delineation of tannery byproducts and their corresponding disposal techniques, encompassing approaches for chromium (Cr), solid waste, wastewater minimization, and ammonium and sulf salts (Hu et al. 2011). Pringle et al. (2016) talked about the problems that come up when circular economy (CE) ideas are used in the leather industry. They did a life cycle assessment (LCA) of leather and the waste that is made when it is made. Researchers also investigated the biological, chemical, and mechanical recycling mechanisms, along with the corresponding challenges (technological aspects, recovered material value, secondary market considerations, and economic feasibility) (Pringle et al. 2016).

Navarro et al. (2020) presented a novel scientific methodology that holds promise for augmenting and optimizing the leather industry towards the pursuit of sustainable development (Navarro et al. 2020). Moktadir et al. (2018) investigated the variables impacting the assimilation of ecologically sustainable methodologies within the leather technology domain in Bangladesh. As concerns pertaining to global waste levels and raw material consumption escalate, entrepreneurs face the challenge of harmonizing consumer demands with conscientious waste management. In order to accomplish this objective, further investigation is required to ascertain the optimal approach for enhancing the treatment of tannery waste and implementing the introduction of recycled materials into the market with maximum efficiency (Moktadir et al. 2018).

## 4 Synthetic Dyes in Leather and Textile Residues

The aforementioned sectors, namely the textile, leather, paint, cosmetic, and food industries, are the primary beneficiaries of dyes for the purpose of coloration. Synthetic dyes, synthesized towards the conclusion of the nineteenth century, have predominantly supplanted numerous natural hues derived from fauna, minerals, or flora. It is anticipated that dyes exhibit the characteristics of being non-toxic, non-carcinogenic, and non-allergenic. Nevertheless, textile dyes represent the most frequently recorded origins of inadvertent garment repercussions, and it is well established that specific dyes, like Butter Yellow, formerly employed in culinary applications, exhibit carcinogenic properties. In contrast to foods and beauty products, textiles rarely list the exact chemicals used in their coloring, which makes it very hard to know for sure what they are made of. Moreover, the extensive utilization of dyes originating from developing nations or the Far East, which possess a historical background and are economically more affordable, may compensate for the alterations in manufacturing practices observed in industrialized nations. This is particularly relevant in light of the prohibition of dyes that exhibit high allergenic, mutagenic, or carcinogenic properties in Europe and the United States (Le Coz 2005). Dyeing is carried out on leather within the confines of tannery drums, which are cylindrical reactors positioned horizontally and employed on a large-scale industrial level. Effluents containing dyes, oils, surfactants, synthetic and natural tanning agents, as well as colors, exhibit inherent complexity and pose potential hazards. Preliminary investigations indicate that upon employing dyeing techniques on leather with concentrations exceeding 0.5 gm/L, the resultant effluent yield may escalate to 7 gm/Kg of treated leather. The industrial sector commonly employs physical–chemical techniques for the treatment of these effluents, subsequently employing biological treatment via activated sludge. Nevertheless, the toxicity of the effluent resulting from the dyeing process can have a substantial influence on the effectiveness of biological treatment procedures. Toxicology assays have been incorporated into environmental regulations, as observed in Brazil, where the evaluation of toxicity is mandated for all ultimate effluents (Piccin et al. 2016).

The effluents from the leather industry, which consist of various dyestuffs, necessitate the implementation of advanced treatment methodologies, such as adsorption, to achieve efficient removal. Powdered commercial coal-derived activated carbon serves as the prevailing adsorbent for the purpose of dyestuff elimination. This investigation employed activated carbon derived from desiccated tannery sludge as a viable alternative to coal-based activated carbon. The investigation evaluated the effectiveness of tannery sludge and coal-derived activated carbon in the process of color removal. The experimental findings revealed that the tannery sludge exhibited a commendable dye removal efficacy of 80–90%, which can be favorably compared to that of coal-derived activated carbon. Nevertheless, the adsorption power of the dye onto the tannery sludge exhibited a lower equilibrium value compared to that of the commercial activated carbon. The Langmuir isotherm was determined to exhibit greater suitability for the isotherm data in comparison to the Freundlich isotherm (Pollutants from textile manufacturing processes; Sreedhar Reddy et al. 2008).

## References


Allen S (1987) Equilibrium adsorption isotherms for peat. Fuel 66:1171–1175. https://doi.org/10.1016/0016-2361(87)90053-6

Barik D (2018) Energy from toxic organic waste for heat and power generation. Woodhead Publishing

Bautista ME, Pérez L, García MT et al (2015) Valorization of tannery wastes: Lipoamino acid surfactant mixtures from the protein fraction of process wastewater. Chem Eng J 262:399–408. https://doi.org/10.1016/j.cej.2014.10.004

Cabrera-Codony A, Ruiz B, Gil RR et al (2021) From biocollagenic waste to efficient biogas purification: applying circular economy in the leather industry. Environ Technol Innov 21:101229. https://doi.org/10.1016/j.eti.2020.101229

Charumathi D, Das N (2012) Packed bed column studies for the removal of synthetic dyes from textile wastewater using immobilised dead *C. tropicalis*. Desalination 285:22–30. https://doi.org/10.1016/j.desal.2011.09.023

Chen X, Ning X, Lai X et al (2021) Chlorophenols in textile dyeing sludge: pollution characteristics and environmental risk control. J Hazard Mater 416:125721. https://doi.org/10.1016/j.jhazmat.2021.125721

Chojnacka K, Moustakas K, Witek-Krowiak A (2020) Bio-based fertilizers: a practical approach towards circular economy. Biores Technol 295:122223. https://doi.org/10.1016/j.biortech.2019.122223

Chougule MB, Sonaje NP (2013) Cost-benefit analysis of wastewater recycling plant for textile wet processing. Int J Comput Eng 3(1):27–31

Chowdhury M, Mostafa MG, Biswas TK et al (2015) Characterization of the effluents from leather processing industries. Environ Process 2:173–187. https://doi.org/10.1007/s40710-015-0065-7

Derden A, Huybrechts D (2013) Brominated flame retardants in textile wastewater: reducing Deca-BDE using best available techniques. J Clean Prod 53:167–175. https://doi.org/10.1016/j.jclepro.2013.04.013

Doble M, Kumar A (2005) Biotreatment of industrial effluents. Elsevier. https://doi.org/10.1016/B978-075067838-4/50013-0

Famielec S (2020) Chromium concentrate recovery from solid tannery waste in a thermal process. Materials 13:1533. https://doi.org/10.3390/ma13071533

Fei Y, Liu C (2016) Detoxification and resource recovery of chromium-containing wastes. Environ Mater Waste 265–284. https://doi.org/10.1016/b978-0-12-803837-6.00012-3

Ferreira LFR, Kumar A, Bilal M (2023) Recent advancements in waste water management: implications and biological solutions, vol 9, 1st edn. https://shop.elsevier.com/books/recent-advancements-in-waste-water-management-implications-and-biological-solutions/ferreira/978-0-443-19388-0

Ferrero F (2007) Dye removal by low cost adsorbents: Hazelnut shells in comparison with wood sawdust. J Hazard Mater 142:144–152. https://doi.org/10.1016/j.jhazmat.2006.07.072

Grosser A, Celary P (2019) 30—Biogas (methane production) and energy recovery from different sludges. Indus Municipal Sludge 705:740. https://doi.org/10.1016/B978-0-12-815907-1.00030-1

Hu J, Xiao Z, Zhou R et al (2011) Ecological utilization of leather tannery waste with circular economy model. J Clean Prod 19:221–228. https://doi.org/10.1016/j.jclepro.2010.09.018

Ip AWM, Barford JP, McKay G (2010) A comparative study on the kinetics and mechanisms of removal of Reactive Black 5 by adsorption onto activated carbons and bone char. Chem Eng J 157:434–442. https://doi.org/10.1016/j.cej.2009.12.003

Jeevanandam S, Ravikumar K, Das A, Goel M (2015) Comprehensive study on textile dyeing sludge as a substitute for cement in cement-mortar. Int J Technol 5:219. https://doi.org/10.5958/2231-3915.2015.00024.3

Kasaw E, Adane Z, Gebino G et al (2021) Incineration of textile sludge for partial replacement of cement in concrete production: a case of Ethiopian textile industries. Adv Mater Sci Eng 2021:1–6. https://doi.org/10.1155/2021/9984598

Korhonen J, Snäkin J-P (2015) Quantifying the relationship of resilience and eco-efficiency in complex adaptive energy systems. Ecol Econ 120:83–92. https://doi.org/10.1016/j.ecolecon.2015.09.00

Kumar KV, Porkodi K (2009) Equilibrium and thermodynamics of dye removal from aqueous solution by adsorption using rubber wood saw dust. Int J Environ Technol Manage 10:295. https://doi.org/10.1504/ijetm.2009.023736

Le Coz CJ (2005) Encyclopedia of toxicology, 2nd edn. https://doi.org/10.1016/B0-12-369400-0/00359-8

Lian L, Guo L, Wang A (2009) Use of CaCl2 modified bentonite for removal of Congo red dye from aqueous solutions. Desalination 249:797–801. https://doi.org/10.1016/j.desal.2009.02.064

Mahmoud DK, Salleh MAM, Karim WAWA et al (2012) Batch adsorption of basic dye using acid treated kenaf fibre char: equilibrium, kinetic and thermodynamic studies. Chem Eng J 181–182:449–457. https://doi.org/10.1016/j.cej.2011.11.116

Mahmoued EK (2010) Cement kiln dust and coal filters treatment of textile industrial effluents. Desalination 255:175–178. https://doi.org/10.1016/j.desal.2009.12.025

Majee S, Halder G, Mandal T (2019) Formulating nitrogen-phosphorous-potassium enriched organic manure from solid waste: a novel approach of waste valorization. Process Saf Environ Prot 132:160–168. https://doi.org/10.1016/j.psep.2019.10.013

Malaiškienė J, Kizinievič O, Kizinievič V (2019) A study on tannery sludge as a raw material for cement mortar. Materials 12:1562. https://doi.org/10.3390/ma12091562

Malek A, Hachemi M, Didier V (2009) New approach of depollution of solid chromium leather waste by the use of organic chelates. J Hazard Mater 170:156–162. https://doi.org/10.1016/j.jhazmat.2009.04.118

Malik PK (2003) Use of activated carbons prepared from sawdust and rice-husk for adsorption of acid dyes: a case study of Acid Yellow 36. Dyes Pigm 56:239–249. https://doi.org/10.1016/s0143-7208(02)00159-6

Malik A, Hussain M, Uddin F et al (2021) Investigation of textile dyeing effluent using activated sludge system to assess the removal efficiency. Water Environ Res 93:2931–2940. https://doi.org/10.1002/wer.1639

Masilamani D, Madhan B, Shanmugam G et al (2016) Extraction of collagen from raw trimming wastes of tannery: a waste to wealth approach. J Clean Prod 113:338–344. https://doi.org/10.1016/j.jclepro.2015.11.087

Moktadir MA, Rahman T, Rahman MH et al (2018) Drivers to sustainable manufacturing practices and circular economy: a perspective of leather industries in Bangladesh. J Clean Prod 174:1366–1380. https://doi.org/10.1016/j.jclepro.2017.11.063

Mui ELK, Cheung WH, McKay G (2010) Tyre char preparation from waste tyre rubber for dye removal from effluents. J Hazard Mater 175:151–158. https://doi.org/10.1016/j.jhazmat.2009.09.142

Navarro D, Wu J, Lin W et al (2020) Life cycle assessment and leather production. J Leather Sci Eng 2. https://doi.org/10.1186/s42825-020-00035-y

Oke N, Mohan S (2022) Development of nanoporous textile sludge based adsorbent for the dye removal from industrial textile effluent. J Hazard Mater 422:126864. https://doi.org/10.1016/j.jhazmat.2021.126864

Piccin JS, Gomes CS, Mella B, Gutterres M (2016) Color removal from real leather dyeing effluent using tannery waste as an adsorbent. J Environ Chem Eng 4:1061–1067. https://doi.org/10.1016/j.jece.2016.01.010

Pollutants from textile manufacturing processes https://arkvietnam.com/en/safest-and-most-effective-textile-dyeing-wastewater-treatment

Pringle T, Barwood M, Rahimifard S (2016) The challenges in achieving a circular economy within leather recycling. Procedia CIRP 48:544–549. https://doi.org/10.1016/j.procir.2016.04.112

Raghunathan T (2018) A study on geopolymer with dyeing industry effluent treatment plant sludge. Int J Trend Sci Res Develop 3:950–953. https://doi.org/10.31142/ijtsrd19165

Ran C, Liu Y, Siddiqui AR et al (2019) Pyrolysis of textile dyeing sludge in fluidized bed: analysis of products, and migration and distribution of heavy metals. J Clean Prod 241:118308. https://doi.org/10.1016/j.jclepro.2019.118308

Ranade VV, Bhandari VM (2014) Industrial wastewater treatment, recycling and reuse. Butterworth-Heinemann

Sadhukhan J, Dugmore TIJ, Matharu A et al (2020) Perspectives on "Game Changer" global challenges for sustainable 21st century: plant-based diet, unavoidable food waste biorefining, and circular economy. Sustainability 12:1976. https://doi.org/10.3390/su12051976

Saraya MEl-SI, Nassar HS(2015) Removal of Reactive Blue7 dye from aqueous solution using solid waste. Int J Res Eng Technol 3(7):19–36

Sathish M, Madhan B, Raghava Rao J (2019) Leather solid waste: an eco-benign raw material for leather chemical preparation—a circular economy example. Waste Manage 87:357–367. https://doi.org/10.1016/j.wasman.2019.02.026

Selvaraj S, Ramalingam S, Parida S et al (2021) Chromium containing leather trimmings valorization: sustainable sound absorber from collagen hydrolysate intercalated electrospun nanofibers. J Hazard Mater 405:124231. https://doi.org/10.1016/j.jhazmat.2020.124231

Singh V, Srivastava VC (2022) Transformation of textile dyeing industrial sludge into economical biochar for sorption of ofloxacin: equilibrium, kinetic, and cost analysis. Biomass Conversion Biorefinery. https://doi.org/10.1007/s13399-022-02554-6

Sreedhar Reddy S, Kotaiah B, Siva Prasad Reddy N (2008) Color pollution control in textile dyeing industry effluents using tannery sludge derived activated carbon. Bull Chem Soc Ethiopia 22. https://doi.org/10.4314/bcse.v22i3.61211

Tadesse GL, Guya TK (2017) Impacts of tannery effluent on environments and human health: a review article. https://core.ac.uk/download/pdf/234687486.pdf

Tasca AL, Puccini M (2019) Leather tanning: life cycle assessment of retanning, fatliquoring and dyeing. J Clean Prod 226:720–729. https://doi.org/10.1016/j.jclepro.2019.03.335

Teal SG (1997) Economic analysis of cotton textile finishing processes. Agric Appl Econ

Wang S, Peng Y (2010) Natural zeolites as effective adsorbents in water and wastewater treatment. Chem Eng J 156:11–24. https://doi.org/10.1016/j.cej.2009.10.029

Wang M, Mao M, Zhang M et al (2019) Highly efficient treatment of textile dyeing sludge by $CO_2$ thermal plasma gasification. Waste Manage 90:29–36. https://doi.org/10.1016/j.wasman.2019.04.025

Yuvaraj A, Karmegam N, Ravindran B et al (2020) Recycling of leather industrial sludge through vermitechnology for a cleaner environment—a review. Ind Crops Prod 155:112791. https://doi.org/10.1016/j.indcrop.2020.112791

Zăinescu G, Deselnicu V, Pang X et al (2018) Utilization of sludge from leather tanning residual baths as additive in mortars. Int Conf Adv Mater Syst. https://doi.org/10.24264/icams-2018.x.8

# Effective Sludge Management in the Petroleum and Chemical Sector

Abhishek Kumar, Debabrata Panda, and Krunal M. Gangawane

**Abstract**

Crude oil is the raw substance extracted straight from oil wells. In contrast, petroleum is the purified version of crude oil suitable for fractional distillation to produce various petroleum products. Crude oil is a blend of hydrocarbons found in liquid form in subsurface reservoirs and maintains its liquid state during extraction. Petroleum products are derived from refining crude oil at petroleum refineries and extracting liquid hydrocarbons at natural gas processing facilities. Petroleum is a comprehensive classification that encompasses both crude oil and petroleum products. The phrases "oil" and "petroleum" are sometimes used synonymously.

**Keywords**

Sludge Management · Petroleum Industry · Chemical Industry · Disposal

## 1 Introduction

Petroleum, often known as crude oil, is a naturally occurring flammable liquid composed of a complex combination of hydrocarbons with different molecular weights and other liquid organic compounds. It is found in geological formations under the Earth's surface. Most of the petroleum is classified as a fossil fuel. Fossil fuel is created by burying substantial amounts of deceased animals, often zooplankton and algae, under a sedimentary rock, where they experience significant heat and pressure. Strictly speaking, petroleum denotes crude oil, sometimes used to include all hydrocarbons in solid, liquid, or gaseous form (Robinson and Hsu 2017a; Simanzhenkov and Idem 2003).

Petroleum is extracted chiefly via the process of oil drilling. This topic is addressed after examining structural geology at the reservoir level, sedimentary basin analysis, and reservoir characterization, primarily about porosity and porous structures. It undergoes a process of refinement and separation, primarily based on boiling point, resulting in a wide range of consumer goods. These include petrol (or gasoline), kerosene, asphalt, and chemical reagents used to produce plastics and medicines. Petroleum is used to produce a diverse range of commodities, and it is estimated that global consumption amounts to around 88 million barrels daily.

Petroleum, strictly speaking, encompasses just crude oil. However, everyday use includes all hydrocarbons, whether in liquid, gaseous, or solid form (such as paraffin). Hydrocarbons such as methane, ethane, propane, and butane exist as gaseous substances under the surface at specific pressure and temperature levels. In contrast, pentane and heavier hydrocarbons exist as either liquids or solids. Nevertheless, the composition of gas, drink, and solid in an underground oil reserve is contingent upon subsurface circumstances. The primary output of an oil well is crude oil, which contains a small amount of dissolved natural gas. Due to the disparity in pressure between the surface and underground, a portion of the gas will be released from its dissolved state and subsequently retrieved (or combusted) as related gas or solution gas. A gas well primarily produces natural gas. Due to elevated underground temperature and pressure, the gas may have heavier hydrocarbons, such as pentane, hexane, and heptane, in a gaseous form. Under normal atmospheric circumstances,

A. Kumar (✉) · D. Panda
Department of Chemical Engineering, National Institute of Technology Rourkela, Rourkela 769008, Odisha, India
e-mail: abhisheknitrkl@outlook.com

A. Kumar
Information Management & Dissemination Centre (IMDC), Council of Scientific and Industrial Research–National Metallurgical Laboratory (CSIR–NML), Jamshedpur 831007, Jharkhand, India

K. M. Gangawane
Department of Chemical Engineering, Indian Institute of Technology Jodhpur, Jodhpur 342030, Rajasthan, India

© The Author(s), under exclusive license to Springer Nature Switzerland AG 2025

A. Mandpe and M. P. Shah (eds.), *Wastewater Sludge Nexus*, Sustainable Landscape Planning and Natural Resources Management, https://doi.org/10.1007/978-3-031-45758-6_6

these substances will condense from the gaseous state and transform into natural gas condensate, often called condensate. Condensate resembles petrol's appearance and shares a comparable composition with some volatile light crude oils (Robinson and Hsu 2017b).

The relative abundance of light hydrocarbons in petroleum varies significantly across various oil fields, with more delicate oils containing as much as 97% by weight. In comparison, heavier oils and bitumens may have as low as 50%. The composition of crude oil predominantly consists of alkanes, cycloalkanes, and aromatic hydrocarbons, with other organic molecules that include nitrogen, oxygen, and sulfur. Additionally, trace quantities of metals such as iron, nickel, copper, and vanadium are present. The precise chemical composition exhibits significant variation between different formations.

The composition of greasy Sludge is quite intricate. It consists of an emulsion of oil in water, water in oil, and suspended particles. 1. Oily Sludge has hazardous compounds such as aromatic hydrocarbons, polyaromatic hydrocarbons, and a significant amount of total hydrocarbons. The high viscosity of oily sludge makes it challenging to hydrate. Oily sludge is classified as a dangerous kind of solid waste. Oily sludge comprises approximately 55.13% water, 9.246% sediments, 1.9173% asphaltenes, 10.514% wax, and 23.19% light hydrocarbons. It also contains significant heavy metals, such as vanadium at 204 ppm, iron at 0.6%, and nickel at 506 ppm. This composition renders the oily sludge detrimental to the environment and organisms. Therefore, it is necessary to address this issue to protect the environment (Al-Doury 2019; Al-Mebayedh et al. 2022).

Petroleum is primarily used to produce fuel oil and petrol, which are significant sources of "primary energy" in terms of volume. 84% of the hydrocarbons found in petroleum are transformed into energy-dense fuels, such as petrol, diesel, jet fuel, heating oil, other fuel oils, and liquefied petroleum gas. Oil has emerged as the predominant global energy source since the mid-1950s, owing to its high energy density, ease of transport, and widespread availability. Petroleum is the primary source material for several chemical goods, including medicines, solvents, fertilizers, insecticides, and polymers (Heinberg and Fridley 2016).

Petroleum comprises diverse hydrocarbon compounds, with aromatic hydrocarbons, cycloalkanes, linear or branched alkanes, and asphaltenes being the most frequently encountered molecules. A distinct combination of molecules composes each type of petroleum, dictating its physical and chemical characteristics such as color and viscosity.

The alkanes, alternatively referred to as paraffin, are saturated hydrocarbons characterized by their branched or linear carbon and hydrogen chains. These compounds are denoted by the general formula CnH2n + 2. While the composition typically consists of 5–40 carbon atoms per molecule, minute quantities of shorter or longer molecules might also be detected (Speight and El-Gendy 2018).

Fuel is produced by refining the alkanes from nonane ($C_9H_{20}$) to hexadecane ($C_{16}H_{34}$) and diesel fuel, petroleum, and aviation fuel from pentane ($C_5H_{12}$) to octane ($C_8H_{18}$). Refined alkanes of 16 carbon atoms or more are suitable for lubricating and fuel oil use. At the heavier end of the spectrum, paraffin wax is an alkane composed of around 25 carbon atoms, whereas asphalt contains 35 or more carbon atoms; however, modern refineries typically split asphalt into more valuable byproducts. At ambient temperature, the shortest molecules, consisting of four carbon atoms or less, exist in a gaseous state. These gases are petroleum-based. The disposal method of these gases—flaring off, selling as liquefied petroleum gas under pressure, or utilizing them to fuel the refinery's burners—depends on demand and the expense of recovery. Butane ($C_4H_{10}$) is rapidly merged into the gasoline reservoir during the winter due to its high vapor pressure, which aids in chilly starts. Although best recognized for its use in cigarette lighters, this substance, which is liquefied at a pressure slightly above atmospheric, also serves as a primary fuel source in many developing nations. Propane, which can be melted at moderate pressures, finds application in virtually all petroleum-powered processes, including transportation heating.

Cycloalkanes, referred to as naphthenes, are a class of saturated hydrocarbons characterized by the attachment of hydrogen atoms to one or more carbon rings (CnH2n). Cycloalkanes are comparable to alkanes in that they possess more excellent boiling points. Aromatic hydrocarbons consist of one or more benzene rings and planar six-carbon rings to which hydrogen atoms are covalently bonded (CnHn). Many of them have a fragrant aroma and ignite with a sooty flame. Certain substances are carcinogenic. In an oil refinery, fractional distillation separates these various molecules to produce aviation fuel, petrol, petroleum, and additional hydrocarbons. Illustratively, gasoline frequently contains 2,2,4-trimethylpentane (isooctane), denoted by the chemical formula $C_8H_{18}$, and undergoes an exothermic reaction with oxygen. The combustion of fuels or petroleum in its entirety generates hazardous byproducts. Insufficient oxygen leads to the formation of carbon monoxide. As a result of the elevated temperatures and pressures generated during the combustion of gasoline in automobile engines, exhaust gases frequently contain nitrogen oxides, which are accountable for the formation of photochemical haze (Busca 2007; Kumar and Gangawane 2022a, b; Speight 2011; Tissot and Welte 1978).

## 2 Crude Oil Serves as a Vital Energy Source and Feedstock for Refineries

Diverse types of waste, including chemical and viscous sludge, are produced while refining crude oil. Biosludge, in particular, poses a notable environmental risk due to numerous hazardous constituents. Significant quantities of hazardous waste are produced in petroleum refineries across the globe. Annually, oil refineries in India produce an estimated 20,000 tonnes of oily residue. A significant challenge encountered by oil refineries pertains to the secure dispersal of this viscous residue (Jagaba et al. 2022).

The improper management of these sludges frequently results in ecological contamination and detrimentally impacts their visual appeal. Recycling sludges in an ecologically sustainable fashion represents a viable resolution to the challenge of sludge management. The developed treatment technologies fall into three categories: physical, chemical, and biological remediation.

Due to its oil content and, in most cases, its liquid state, petroleum residue is categorized as hazardous waste and, therefore, cannot be disposed of in a landfill. Oil residue is problematic for two reasons: first, it is typically categorized as hazardous waste (owing to its oil content), and second, it is generally classified as a liquid, which precludes its direct disposal in landfills (Johnson and Affam 2019; Kumar and Gangawane 2022b; Singh and Kumar 2020).

### 2.1 Traditional Approaches to Waste Disposal and Remediation Could Be More Effective When Applied to Oil Residue

Generally, oil residue is unsuitable for incineration due to its excessive oil and water content, rendering it nearly unburnable. Thermal treatment is typically not cost-effective due to excessive water and oil in the waste. Filtration is not feasible for the oil residue due to its extreme particulate content; any attempts to accomplish this will merely obstruct the filtration systems. Due to the high solid and oil content, the oil residue cannot be transported to a wastewater treatment facility, and the waste has an excessively high COD/BOD ratio. Oil residue, as a liquid, is not suitable for disposal in conventional hazardous waste landfills designed to accommodate solid waste only. Due to the scarcity of viable alternatives for managing this particular refuse category, its accumulation has become alarming. The global expansion of oil residue refuse is estimated at 9 billion tonnes, escalating by the minute (Kumar and Gangawane 2023a, b).

Numerous efforts have been made to develop a sustainable, long-lasting solution for the oil residue treatment industry. The industry has made only modest progress thus far. Rather than an absence of viable solutions or resources, the industry's reluctance and lack of motivation to confront this issue directly are the actual causes. The petroleum industry generates hydrocarbon residues during production, refining, transport, and storage. These residues have been accumulated in open earth basins, ditches, and marshes for an extended period. As a result, not only do the storage areas become significantly contaminated, but the surrounding areas do as well (Panda and Gangawane 2023a, b).

Preventing the contamination of surface and groundwater, the surrounding air, fires, explosions, food chain poisoning, and the devastation of verdant spaces are all outcomes that can be achieved through properly treating and disposing of these waste materials. As a result of the ongoing treatment and implemented procedures, substantial areas designated for transient storage are unnecessary. A significant reduction in the likelihood of potential contamination of air and water resources occurs.

The petroleum industry can produce substantial viscous residue throughout its crude oil exploration, production, transportation, storage, and refinement operations. The precipitate produced as a byproduct of the petroleum refining process has garnered growing interest recently. A considerable quantity of petroleum hydrocarbons (PHCs) and other obstinate constituents are present in its composition. Due to its status as a hazardous waste in numerous nations, petroleum sludge inadequately treated or disposed of improperly can present severe environmental and human health risks. Petroleum refineries produce three primary types of sludge: viscous Sludge, biosludge, and chemical Sludge. The majority of crude oils possess the characteristic of separating into lighter and denser hydrocarbons as they are transported and stored. Hydrocarbon residue is typically produced in refineries due to various processes, including pre-export processing, tank farm cleaning, desalt failure, oil emptying from tanks and operation units, pipeline ruptures, and oil processing. These activities include maintaining associated facilities and cleansing crude oil storage tanks. Due to the wide variety in crude oil quality, differences in oil–water separation processes, leakages during industrial processes, and merging with preexisting oily sludge, the composition of oily Sludge can vary. The issue of efficiently remediating oily residue has emerged as a global concern owing to its complex characteristics and the escalating production volume across the globe. The impending prohibition on disposing of surplus petroleum sludge is due to its hazardous properties.

Consequently, there has been a heightened focus on investigating potential technologies for sludge treatment. Refiners and oil production companies are seeking technologies that maximize the extraction of hydrocarbons from sludge while minimizing the quantity of final treated sludge that can be disposed of in a more economical, timely, and environmentally sustainable manner. Over recent years, numerous techniques for treating oily residue have been devised, including biodegradation, solvent extraction, ultrasonic treatment, pyrolysis, photocatalysis, chemical treatment, and landfarming. Utilizing these technologies can reduce or eliminate hazardous constituents, mitigating the adverse environmental and human health effects (Kumar et al. 2022; Ma et al. 2024; Shahbaz et al. 2023; Ting et al. 2024).

## 3 Type of Refinery Sludge

Refinery sludge refers to various byproducts generated during the oil refining process. The types of refinery sludge include (Hasan et al. 2024).

### 3.1 Oily Sludge

Primarily, it consists of three distinct phases—oily, solid, and watery. The substance has many petroleum hydrocarbons (PHCs) and other persistent components. Improper disposal presents significant environmental and health hazards.

### 3.2 Biosludge

Produced as a byproduct of oil exploration, production, transportation, storage, and refining operations. The substance comprises a combination of chemical substances and potentially harmful components. Effective treatment is crucial to avoid environmental pollution.

### 3.3 Chemical Sludge

Occurs during several industrial procedures, such as oil–water separation, tank maintenance, and pipeline ruptures. The composition of crude oil is subject to variation based on factors such as the quality of the oil and the procedures used in the refining process. Efficient remediation techniques are essential because of the dangerous characteristics of the substance. The treatment and disposal of these sludges involve various methods such as incineration, bioremediation, pyrolysis, solvent extraction, and more, aiming to reduce environmental impact and ensure safe handling.

## 4 Crude Oil Sludge Formation

Most oils possess the characteristic of separating into lighter and denser hydrocarbons as they are transported and stored. The denser crude oil fractions settle to the bottom of the storage tank. Paraffin-based crude oil sludge is created when the molecular orbitals of individual hydrocarbons with straight chains come together and generate an induced dipole force that makes it difficult to separate them. When the longer straight-chain hydrocarbon molecules (specifically those with 20 or more carbon atoms) clump together, they separate from a still liquid and settle at the bottom, forming a thick gel-like substance called sludge or wax. It often occurs in storage tanks. The profile undergoes stratification over time as the volatile components in the sludge are released due to variations in temperature and pressure. Removing these unstable elements leads to denser and more viscous fractions in the sludge, resulting in increased density and viscosity and reduced fluidity.

Asphaltic Sludge is created due to the propensity of asphaltenes, resins, and polymeric compounds to precipitate. Neutral resins are aromatic hydrocarbons with a high molecular weight. These resins exhibit insolubility in alkalis and acids while displaying perfect miscibility with petroleum oils, even light fractions. Asphaltenes have similarities with neutral resins. However, they indicate insolubility in light gasoline and petroleum ether. Asphaltenes are precipitated when there is an abundance of petroleum ether. Both asphaltenes and neutral resins exhibit benzene, chloroform, and carbon disulfide solubility (Hasan et al. 2024; Kang et al. 2024; Wan et al. 2024).

### 4.1 Techniques for Treating Sludge

The available methods for recovering and treating oil and redeeming crude sludge include chemical treatments, several distillation processes, cracking, hydro-treatment, solvent treatment, and bioremediation. Here are some traditional approaches to sludge treatment (Hasan et al. 2024; Johnson and Affam 2019):

#### 4.1.1 Manual Cleansing and Combustion

The manual cleaning procedure is cost-effective. The cleaning process involves accessing the tank. The sludge is physically removed from the tank or transferred to pumps inside the tanks. This procedure is time-consuming. Extracting the valuable hydrocarbons from the sludge with this process is challenging (Panda and Gangawane 2022).

#### 4.1.2 Method for Extracting Solvents

This process utilizes a range of solvents capable of decomposing complex molecules in the sludge into their elemental components, namely water, crude oil, and particulate matter. This procedure necessitates blending and stirring equipment (Fig. 1). Sludge contains waxy and non-waxy (asphaltenes) organic compounds, salt, oxides, and other inorganic substances.

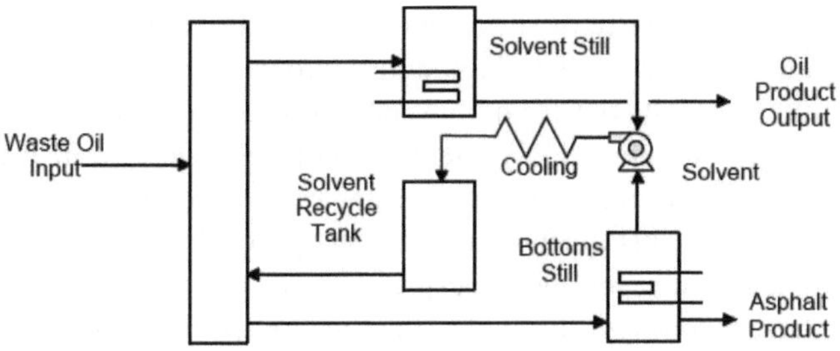


**Fig. 1** Block flow diagram of solvent extraction method (Basnet and Ghimire 2019)

### 4.1.3 These May Be Dissolved by Using a Suitable Solvent

The oily sludge waste is first combined with a solvent in the reactor column. The solvent can specifically dissolve the oil portion of the sludge, while the less soluble contaminants settle at the bottom of the column. Subsequently, the oil solvent solution is conveyed to a solvent distillation apparatus, which extracts the solvent from the oil. The oil that has been separated is classified as oil recovery. In contrast, the solvent vapor that has been divorced may be condensed using a compressor and cooling system and then sent to a solvent recycling tank. The solvent is suitable for performing many extraction cycles. The residual impurities from the reactor column are transferred to a secondary distillation system, where the solvent present in the pollutants is isolated and sent to the solvent recycling tank. However, the waste residues obtained after this separation may need further treatment.

### 4.1.4 Ultra-High-Temperature Gasification

Ultra-high-temperature gasification refers to a process where organic materials, often biomass or waste, are converted into synthetic gas (syngas) at extremely high temperatures, typically exceeding 1,000 °C. This process offers several advantages:

I. *Efficiency*: The high temperatures enhance the efficiency of the gasification process, ensuring thorough conversion of feedstock into syngas.
II. *Syngas Composition*: The resulting syngas consists of carbon monoxide (CO) and hydrogen ($H_2$), which can be utilized in various industrial processes or as a clean fuel.
III. *Waste Management*: Ultra-high-temperature gasification is often employed to dispose of waste materials efficiently, reducing environmental impact.
IV. *Heat Recovery*: The intense heat generated during gasification can be recovered and utilized for cogeneration, improving overall energy efficiency.
V. *Reduced Emissions*: The process can be designed to minimize emissions of pollutants and greenhouse gases, contributing to environmentally friendly energy solutions.

It's crucial to note that the success of ultra-high-temperature gasification depends on proper engineering, control systems, and feedstock selection.

### 4.1.5 Oil Sludge Separation Using Cyclone

Using cyclones for oil sludge separation is a procedure that effectively employs cyclonic separation technology to distinguish the constituents of oil sludge, which generally consists of solid particles, water, and oil. As part of this procedure, the oil residue is introduced into a cyclone separator, which generates high-velocity rotation. The generation of centrifugal force due to this rotation results in the heavier solid particles spiraling toward the outer wall of the cyclone. As a result, a slurry cake is formed due to the accumulation of these heavier solids at the cyclone's base. The lighter oil and water phases are concurrently directed toward the cyclones center and discharge through distinct apertures. By effectively removing solid particles from the oil sludge, this separation method enhances environmental outcomes associated with sludge disposal and facilitates the recovery of valuable oil. As a result of its cost-effectiveness and efficacy in segregating components according to their densities, cyclonic separation is utilized in a vast array of industries. If additional purification is required, the oil and water phases that have been separated may be subjected to other treatments (Hasan et al. 2024).

### 4.1.6 Surfactant

Surfactants are paramount in the petroleum and chemical sectors of sludge management as they significantly contribute to the efficient treatment and remediation of viscous sludge. The application of surfactants in sediment management encompasses various critical components (Urum 2004):

I. *Emulsification and Dispersal*: Surfactants break down and spread hydrophobic substances in oily sludge. Surfactants aid in the creation of stable emulsions by decreasing the tension between oil and water at their interface, hence improving the separation of oil from the sludge matrix.
II. *Enhanced Oil Recovery (EOR)*: Surfactants sometimes enhance oil extraction from sludge or polluted soil. Surfactants modify the capacity of surfaces to be wetted, improving the liberation of contained oil and easing its removal.

III. *Washing and Solubilization*: Surfactants facilitate the cleansing and dissolving of water-repellent impurities in sludge, allowing for their elimination. It is especially beneficial in managing sludge that contains intricate combinations of hydrocarbons.
IV. *Foam Generation*: Surfactants may be used to produce foams to improve the treatment of polluted areas. Foams facilitate the transportation of surfactants and other treatment agents into the sludge, enhancing overall efficiency.
V. *Biodegradation Enhancement*: Surfactants can improve the availability of pollutants by enhancing their ability to dissolve in a liquid, stimulating the growth of microorganisms involved in breaking down these contaminants. It is particularly advantageous in the biological treatment of sludge.
VI. *Soil Washing*: Surfactants are used in soil washing procedures to mobilize and eliminate pollutants from soil particles. This strategy is often used in combination with other remedial methods. When selecting surfactants for sludge management, environmental impact, and the sludge's properties are crucial factors to consider. By strategically utilizing surfactants, wastewater treatment processes can be considerably more effective, promoting environmentally conscious and sustainable practices within the petroleum and chemical sectors.

## 4.2 Microwave Irradiation

The disclosed procedure includes using microwave frequencies ranging from 900 to 2,450 MHz on an industrial scale. Microwave energy treats greasy sludge by causing molecular interaction with the electromagnetic field. This methodology results in rapid and effective heating, exceeding traditional approaches. Increased temperature induces demulsification of the oil and water emulsion, resulting in decreased viscosity and accelerated sedimentation of water droplets.

The rapid temperature rise also causes the decomposition of dense hydrocarbons into lighter molecules. When dealing with oily sludge, when water is the main component, the absorption of microwave radiation is more noticeable in water than in oil. The water volume is expanded due to the absorption in the inner phase, which decreases the interfacial layer between water and oil. This technique enables the act of separating. The optimal microwave irradiation parameters were identified as 420 W for power and 12 s for treatment duration. During a targeted experiment, microwave radiation on 188 containers of water–oil mixture led to the division of 146 and 42 oil containers. This shows superior efficacy in comparison to traditional approaches.

Nevertheless, the effectiveness of microwave irradiation may be limited by many criteria, such as microwave power, duration, presence of surfactants, pH, salt level, and specific features of the sludge. Although microwave irradiation has limits, it stands out from conventional heating methods due to its ability to quickly increase molecular energy in the substance, resulting in much faster reaction rates within a short time. This attribute makes the approach very energy-efficient. However, the widespread use of this technology in industrial settings is limited by the need for specific equipment and high operational expenses (Chen et al. 2019).

## 4.3 Electrokinetic Method

The electrokinetic approach utilizes a low-intensity direct current (DC) applied to an absorbent material via a pair of electrodes. This leads to the migration of electrons from an area of lower concentration to a region of greater concentration via a porous medium in the liquid phase. In addition, the electrodes enable the transfer of charged ions and electrophoresis in a colloidal system. The electrokinetic approach often reveals a three-stage process during separating petroleum sludge. In the first phase, colloidal aggregates are disintegrated in an electric field. This results in the movement of colloidal particles from the sludge and solid phase toward the anode region, a process known as electrophoresis. Simultaneously, the separated liquid phase migrates toward the cathode area, a phenomenon called electro-osmosis.

The application of electrokinetics in the treatment of oily residue yields a substantial reduction of 43% in light hydrocarbons and 63% in water content. Fifty percent of the light hydrocarbons were removed through the combination of electrokinetics and a surfactant. The efficacy of the electrokinetic approach may be affected by several parameters, including pH, electrical voltage, resistance, and electrode spacing. The electrokinetic technique for oil recovery from oily sludge consumes less energy than other recovery processes, such as centrifugation and pyrolysis. It is important to emphasize that research on electrokinetic techniques for recovering oily sludge has mainly been conducted in laboratory settings. Additional examination is required to evaluate the effectiveness and financial consequences on a broader scope (Fellah Jahromi and Elektorowicz 2018).

## 4.4 Ultrasonic Irradiation

The ultrasonic therapy procedure utilizes ultrasonic waves to generate compression and rarefaction inside a treatment chamber. This technique is a very efficient approach for separating solid and liquid constituents in concentrated suspensions by inducing the destabilization of water–oil mixtures.

Ultrasonic irradiation causes ultrasonic waves to travel through the treatment medium, resulting in alternating cycles of compression and rarefaction. The compression cycle applies positive pressure, causing molecules to come closer together. In contrast, the rarefaction cycle applies negative pressure, causing molecules to move farther apart. Microbubbles are formed as a result of this adverse pressure. When these little bubbles reach a precarious size, they implode forcefully, generating shockwaves that increase warmth and pressure. Cavitation, characterized by the formation and collapse of vapor-filled cavities, causes the emulsion system to experience elevated temperature, decreased viscosity, and improved liquid phase mass transfer. Consequently, this leads to the instability of the water–oil combination [85]. Ultrasonic irradiation is advantageous compared to other approaches because it can effectively clean the surfaces of solid particles and reach different areas within a system consisting of many phases.

Ultrasonic therapy, while efficient and environmentally friendly, has several limits. Various factors might affect its efficiency, including the length and frequency of treatment, water content, temperature, size of solid particles, starting concentration of petroleum hydrocarbons (PHCs), power and intensity of sonication, salinity, and surfactants. Moreover, the approach is linked to substantial expenses (Check and Mowla 2013; Herrera Urbina 2003; Xu et al. 2019; Yi et al. 2017).

### 4.5 Froth Flotation

This method utilizes air bubbles in an aqueous slurry to capture oil droplets and tiny particles. Afterward, the oil droplets are suspended in the liquid, and small solid particles are gathered in a foam layer. To start the procedure, a sludge slurry is created by introducing a certain quantity of water to the oily sludge. Aeration produces small air bubbles inside the sludge mixture, interacting with the oil droplets. The decrease and separation of the water layer between the oil droplets and air bubbles facilitate the movement of oil droplets toward the air bubbles. The resultant amalgamation of air bubbles and oil droplets ascends to the water surface, enabling the extraction of the accumulated oil for further refinement (Wang et al. 2022).

## 5 Petroleum Sludge Disposal Methods

Several disposal options are used once petroleum sludge has been extracted from all valuable hydrocarbons and oil. Biodegradation, solidification/stabilization (S/S), oxidation, and incineration are some of these processes (Johnson and Affam 2019; Kumar and Gangawane 2023b; Murshid et al. 2018, 2021).

### 5.1 Incineration of Petroleum Sludge: Process and Efficiency

Petroleum industry waste sludge is incinerated by allowing it to burn completely with plenty of air and fuel. For this reason, rotary kilns and fluidized bed incinerators are the most common. The usual combustion temperature range for rotary kiln incinerators is 980–1,200 °C, and the residence duration is around 30 min. Conversely, fluidized bed incinerators have a more extended residence period, measured in days, and operate at lower temperatures, ranging from 730 to 760 °C.

The fluidized bed incinerator's excellent combustion efficiency, low pollutant emissions, fuel flexibility, and excellent mixing efficiency make it ideal for treating poor-quality sludge. When tested on three kinds of oily sludge, a fluidized bed incinerator could treat them with a 98–99% combustion efficiency without needing supplementary fuels. It should be mentioned that sludge that contains a large amount of water must undergo pre-treatment to decrease its viscosity before it can be fed into the incinerator.

### 5.2 Enhanced Solidification/Stabilization (S/S) Techniques for Petroleum Sludge Treatment

Solidification/Stabilization (S/S) is a waste management technique that uses a binder to encapsulate or seal garbage, preventing harmful substances' release into the environment. The objective is to transform the items into environmentally acceptable building materials or non-toxic trash for landfill disposal. The process of treating hazardous waste with cement involves three primary stages:

i. *Remediation of chemical pollutants*: This entails chemical reactions between the hydration products of the cement and the impurities.
ii. *Physically absorbing pollutants*: This process entails absorbing contaminants already on the surface of hydrated cement products.
iii. *Containment of polluted waste or soil*: Accomplished using hardened pastes with minimal permeability.

The composition of hydration products and pollutants is vital in the first two stages, influencing the outcomes. In contrast, the paste's hydration products and pore structure properties affect the third stage. S/S is extensively used worldwide to treat hazardous waste, low-level radioactive and mixed wastes, and the remediation of polluted areas. The US Environmental Protection Agency (USEPA) states that S/S is the most effective technology for managing 57 hazardous wastes. Hydraulic types of cement are often used as binding agents, and the resulting material may be utilized for building purposes based on its specific qualities.

Although S/S is a rapid and economical method, it might release highly concentrated organic contaminants when exposed to environmental leachants. Elevating the cement content in the solidification/stabilization process of refinery oily sludge might result in amplified levels of polycyclic aromatic hydrocarbons (PAHs) and total petroleum hydrocarbons (TPH) in the liquid that is produced when the sludge is leached.

In order to enhance the efficacy of S/S in fully immobilizing dangerous pollutants, additional substances are included in the mixture. The combination of activated carbon and Portland cement has shown substantial efficacy in the remediation of organic pollutants. High carbon power plant fly ash (HCFA), a substance with pozzolanic properties, offers a cost-effective solution by combining both sorption and binding qualities. The use of HCFA as a sorbent in the S/S treatment of waste generated from petroleum exploration and production has been seen to reduce the leaching of petroleum hydrocarbons (PHCs).

The oily sludge is subjected to a S/S treatment process. Concrete molds are used to prepare the samples, which are then covered in plastic bags to avoid carbonation. The samples are then allowed to cure before undergoing testing. The use of rice husk ash (RHA) as a replacement for cement has been shown to enhance the compressive strength of the resulting product, offering a viable option for minimizing toxicity before ultimate disposal.

## 5.3 Enhancing Oily Sludge Treatment Through Advanced Oxidation Processes

Chemical and other oxidation agents are used in oxidation treatment, a degrading process, to break down organic pollutants. The organic molecules in oily sludge are oxidized to water, carbon dioxide, or non-hazardous materials when reactive agents are added. Several oxidation reagents have been used to treat oily sludge. These include hypochlorite, ozone, permanganate, persulfate, hypochlorite, and ultrasonic irradiation. Hydrogen (H*), hydroxyl (OH*), hydroperoxyl ($HO_2$*), and hydrogen peroxide ($H_2O_2$) are some of the intermediate radicals that may be effectively generated by ultrasonic irradiation during the oxidation process. These radicals facilitate the reduction of complicated hydrocarbons to their more straightforward, more soluble counterparts.

It found that phenols, PAHs, and other pollutants may be reduced in soil polluted with oily sludge using a Fenton-type reagent, even at a pH as low as 3.0. Applying oxidation treatment to oily sludge results in biodegradable byproducts that are environmentally benign and take a concise amount of time. Nevertheless, there are several potential obstacles that large-scale applications may encounter, including the need for vast quantities of chemical reagents, much equipment, higher energy usage, and constantly high operating expenses.

## 5.4 Microbial Biodegradation in Environmental Pollution Remediation

Microorganisms are essential for breaking down and eliminating environmental contaminants. They are often used in land treatment, biopile/composting, and slurry procedures.

### 5.4.1 Efficiency and Parameters of Oily Sludge Land Farming

The land farming treatment method entails the degradation of oily sludge pollutants by combining them with soil, using biological, chemical, and physical processes. This technique is favored over alternative approaches because of its cost-effectiveness, low energy use, capacity to manage substantial sludge quantities, and uncomplicated operational protocols. Nevertheless, this process is characterized by its time-consuming nature, need for vast amounts of land, and potential inefficiency in colder climates.

### 5.4.2 Biopile and Composting: Oily Sludge Treatment Efficiency

The biopile treatment technique entails converting petroleum waste into heaps, where indigenous or external microorganisms are used for degradation. Adding organic materials to improve efficiency is called composting, which may serve as an alternative to land treatment. The study showed that using a bulking material such as cotton stalks greatly enhances the metabolic microbial activity during the composting process of oily sludge. It was discovered that adding manure throughout the composting process improves microbial activity and diversity.

Adding crude manure and straw in bioaugmentation resulted in a significant decrease in total petroleum hydrocarbon (TPH) content by 46–53% within 56 days in the composting heaps, as opposed to a 31% drop in control piles. It was discovered that the organic material with the most significant capacity to reduce TPH (total petroleum hydrocarbons) in viscous residue is kitchen refuse compost. The sand amendment, mature oil compost, and pulverized refuse wood are all less effective than it. Biopile/composting treatment, despite being ecologically sustainable and requiring less land area than conventional cultivation, remains time-consuming and land-intensive.

### 5.4.3 Bio-Slurry Treatment for Oily Sludge: Efficiency and Costs

This treatment technique combines sludge-related solids and water, with a concentration ranging from 5 to 50% by weight per volume. The objective is to dissolve contaminants into the water phase, resulting in the extraction of solubilized pollutants. Microbial degradation reduces toxicity or converts the final product into carbon dioxide and water. It has noted a reduction in total petroleum hydrocarbons (TPH) ranging from 40.7 to 53.2% after two weeks and from 63.7 to 85.5% after six weeks using bio-slurry treatment on oily sludge. Adding bio-surfactants might lead to a substantial reduction of 80–99% in total petroleum hydrocarbons (TPH) during 10–20 days during the biodegradation process of oily sludge. While bio-slurry treatment is very effective and fast, requiring little land use, it is rather costly compared to other disposal methods.

## 6 Conclusion

Petroleum, an essential worldwide energy resource, undergoes complex procedures to produce valuable goods but also creates problematic waste, namely oily sludge. Due to the constraints of traditional disposal techniques, there is an increasing need for sophisticated solutions. It is essential to address the treatment of various refinery sludges to preserve the environment. The industry's dedication to innovation and sustainability is vital in tackling waste management problems for a cleaner and more sustainable future. Crude oil sludge is formed when hydrocarbons separate, resulting in the production of sludges that are based on paraffin and asphalt. Multiple therapeutic approaches have been devised to tackle this problem, each with merits and constraints. The techniques included in this list are manual cleaning, solvent extraction, ultra-high-temperature gasification, cyclonic separation, surfactant application, microwave irradiation, electrokinetic methods, ultrasonic irradiation, froth flotation, and microbial biodegradation. The choice of a particular technique is contingent upon many criteria, including the properties of the sludge, environmental issues, and objectives related to waste management. The disposal techniques for petroleum sludge include burning, solidification/stabilization (S/S), advanced oxidation processes, and microbial biodegradation. Nevertheless, there are ongoing obstacles, such as excessive operating costs and the need for specialized equipment. The industry must navigate these hurdles to choose the most appropriate solution, considering the distinct properties of the sludge, environmental consequences, and waste management goals.

## References


Al-Doury MMI (2019) Treatment of oily sludge using solvent extraction. Petrol Sci Technol 37(2):190–196

Al-Mebayedh H, Niu A, Lin C (2022) Petroleum hydrocarbon composition of oily sludge and contaminated soils in a decommissioned oilfield waste pit under desert conditions. Appl Sci 12(3):1355

Basnet R, Ghimire RM (2019) Study on re-refining of used lubricating oil in Bagmati Zone, Nepal. Study on re-refining of used lubricating oil in Bagmati Zone by Guru Prasad Chaudhari Rabin Basnet Rudra Mani Ghimire

Busca G (2007) Acid catalysts in industrial hydrocarbon chemistry. Chem Rev 107(11):5366–5410

Check GR, Mowla D (2013) Theoretical and experimental investigation of desalting and dehydration of crude oil by assistance of ultrasonic irradiation. Ultrason Sonochem 20(1):378–385

Chen G et al (2019) Synergistic effect of surfactant and alkali on the treatment of oil sludge. J Petrol Sci Eng 183

Fellah Jahromi A, Elektorowicz M (2018) Electrokinetically assisted oil-water phase separation in oily sludge with implementing novel controller system. J Hazard Mater 358:434–440

Hasan AMA, Kamal RS, Farag RK, Abdel-raouf ME (2024) Petroleum sludge formation and its treatment methodologies: a review. Environ Sci Pollut Res 1–18

Heinberg R, Fridley D (2016) Our renewable future: laying the path for one hundred percent clean energy. Island Press

Herrera Urbina R (2003) Recent developments and advances in formulations and applications of chemical reagents used in froth flotation. Miner Process Extr Metall Rev 24(2):139–182

Jagaba AH et al (2022) Diverse sustainable materials for the treatment of petroleum sludge and remediation of contaminated sites: a review. Cleaner Waste Syst 2:100010

Johnson OA, Affam AC (2019) Petroleum sludge treatment and disposal: a review. Environ Eng Res 24(2):191–201

Kang C et al (2024) Sustainable treatment and recycling of oily sludge: incorporation into gel-system to improve oil recovery. J Cleaner Prod 140640

Kumar A, Gangawane KM (2022b) Synthesis and effect on the surface morphology & magnetic properties of ferrimagnetic nanoparticles by different wet chemical synthesis methods. Powder Technol 410:117867. https://linkinghub.elsevier.com/retrieve/pii/S0032591022007483

Kumar A, Gangawane KM, Ramanjaneyulu B (2022) Ferrofluids for waste-water treatment. In: International conference on chemical, bio and environmental engineering. Springer, pp 723–744. https://doi.org/10.1007/978-3-030-96554-9_48

Kumar A, Gangawane KM (2022a) Effect of precipitating agents on the magnetic and structural properties of the synthesized ferrimagnetic nanoparticles by co-precipitation method. Powder Technol 401:117298. https://doi.org/10.1016/j.powtec.2022.117298

Kumar A, Gangawane KM (2023a) Advanced computational approaches for water treatment. Advanced computational approaches for water treatment. CRC Press, New York, pp 1–12

Kumar A, Gangawane KM (2023b) Molecular dynamics for the membrane process. Advanced computational approaches for water treatment. CRC Press, New York, pp 13–38

Ma C et al (2024) Recent advancements in pyrolysis of halogen-containing plastics for resource recovery and halogen upcycling: a state-of-the-art review. Environ Sci Technol

Murshid NB et al (2021) Alternative remediation for raw petroleum sludge by using solidification/stabilization method. J Solid Waste Technol Manage 47(2):216–224

Murshid N, Kamil N, Kadir AA (2018) Treatment of petroleum sludge by using solidification/stabilization (S/S) method: effect of hydration days to heavy metals leaching and strength. In: IOP conference series: earth and environmental science. IOP Publishing, p 12054

Panda D, Gangawane KM (2023a) Development of superhydrophobic hybrid silica-cellulose aerogel as promising thermal insulation and sound absorption. Int J Mater Res 114(4–5):405–417. https://doi.org/10.1515/ijmr-2022-0066/html

Panda D, Gangawane KM (2023b) Recycled cellulose–silica hybrid aerogel for effective oil adsorption: optimization and kinetics study. Sadhana Acad Proc Eng Sci 48(3). https://doi.org/10.1007/s12046-023-02161-9

Panda D, Gangawane KM (2022) Superhydrophobic hybrid silica-cellulose aerogel for enhanced thermal, acoustic, and oil absorption characteristics. J Mater Sci. https://doi.org/10.1007/s10853-022-07506-z

Robinson PR, Hsu CS (2017a) Introduction to petroleum technology. Springer handbook of petroleum technology, pp 1–83

Robinson PR, Hsu CS (2017b) Petroleum and its products. Handbook of industrial chemistry and biotechnology, pp 13–106

Shahbaz M et al (2023) A review of waste management approaches to maximise sustainable value of waste from the oil and gas industry and potential for the State of Qatar. Fuel 332(P2):126220

Simanzhenkov V, Idem R (2003) Crude oil chemistry. CRC Press

Singh B, Kumar P (2020) Physicochemical characteristics of hazardous sludge from effluent treatment plant of petroleum refinery as feedstock for thermochemical processes. J Environ Chem Eng 8(4):103817

Speight JG (2011) Handbook of industrial hydrocarbon processes. Elsevier

Speight JG, El-Gendy NS (2018) Introduction to petroleum biotechnology. Elsevier

Ting EGY et al (2024) Examining the synergistic effects through machine learning prediction and optimisation in the anaerobic co-digestion (ACoD) of palm oil mill effluent (POME) and decanter cake (DC) with economic analysis. J Cleaner Prod 140666

Tissot BP, Welte DH (1978) Composition of crude oils. Petroleum Formation and Occurrence, pp 333–368

Urum K (2004) Biosurfactant enhanced treatment of petroleum oil contaminated soils

Wan G, Sun L, Xu L, Lin L (2024) Emission of nitrogen/sulfur pollutants and migration of heavy metals during combustion of oily sludge from the oil refining process in fluidized bed. J Energy Inst 112:101476

Wang C et al (2022) Separation of emulsified crude oil from produced water by gas flotation: a review. Sci Total Environ 845:157304

Xu X et al (2019) Research on ultrasound-assisted demulsification/dehydration for crude oil. Ultrason Sonochem 57:185–192

Yi M, Huang J, Wang L (2017) Research on crude oil demulsification using the combined method of ultrasound and chemical demulsifier. J Chem

# Streamlining Sludge Management in Petroleum and Chemical Industries

Sunita Saklani and Nitesh Kumar

**Abstract**

Petroleum and chemical industries are considered the backbone of a nation whether it is developed or developing and act as a major contributor in revenue generation of the economy. As these industries fulfil the greater human need, the waste generated in these poses a huge problem of environmental pollution and human health. Sludge from petroleum is termed hazardous waste due to the presence of compounds like aliphatic hydrocarbons and polycyclic aromatic hydrocarbons (PAHs), PCBs (polychlorinated biphenyls), asphaltene, and nitrogen sulphur-oxygen components and heavy metals, lead, mercury, zinc barium, chromium, arsenic, etc. Petroleum sludge is generated during different processes, i.e. extraction, transportation, and cleansing. Similarly, sludge from chemical industries may be an explosive, oxidizing waste, organic peroxides waste, poisonous infectious substance, toxic and ecotoxic and also categorized as hazardous waste. There are various methods for treatment to minimize the toxicity. However, sludge accumulation in landfills or the environment is not suitable so there is a requirement to look forward to the reuse potential of sludge in a sustainable manner and environmental impact can be prevented. Therefore, in this chapter, streamlining of petroleum and chemical industries sludge is reviewed to utilize it in an environment-friendly manner.

S. Saklani (✉)
Department of Zoology, Sidharth Govt. Utkrisht College Nadaun, Nadaun, Himachal Pradesh, India
e-mail: sunsaki786@gmail.com

N. Kumar
Department of Biosciences, Himachal Pradesh University, Shimla, India

**Keywords**

Petroleum · Chemicals · Sludge · Management · Hazardous waste · Environment pollution

## 1 Introduction

The petroleum industry covers various steps starting initially from discovery, and extraction and up to its consumption which involves land and underwater exploration, digging, extraction of crude oil and natural gas, transportation, refining, and delivery to the end users. All these phases lead to the production of various types of waste (Jagaba et al. 2022a; Borgeihpour et al. 2018). Waste generated in the petroleum industry is divided into two groups: Toxic, corrosive, inflammable, abrasive, and dangerous to the environment and to human beings is considered hazardous waste, whereas non-dangerous, nontoxic but which can impact the environment is considered non-hazardous waste (Al Qallaf et al. 2016). As per the RCRA Resource Conservation and Recovery Act, sludge is considered hazardous waste besides other hazardous waste (USEA 2008). The sludge is generally referred to the remains accumulated at the base of storage facilities and tanks. The elements present in petroleum sludge are Phosphorous, Nitrogen, Iron, Potassium, Calcium, Copper, Cadmium, Magnesium, Phosphate, Zinc, Chromium, Sodium, and Lead (Asia 2006). Petroleum industry sludge is mainly contained remains of petroleum and wax base. Accumulation of sludge is aided by density, viscosity, and distribution of particles of oil components. Petroleum sludge is mainly contributed by two major contents: One is constituted by inorganic remains like sand, clay, scale, and sediments, and the other by storage tanks and pipelines. The asphaltene adsorption is the main activity at oil–water intact.

Sludge is mainly accumulated at the base of tanks and other various storage vessels and facilities and in the refining

© The Author(s), under exclusive license to Springer Nature Switzerland AG 2025

A. Mandpe and M. P. Shah (eds.), *Wastewater Sludge Nexus*, Sustainable Landscape Planning and Natural Resources Management, https://doi.org/10.1007/978-3-031-45758-6_7

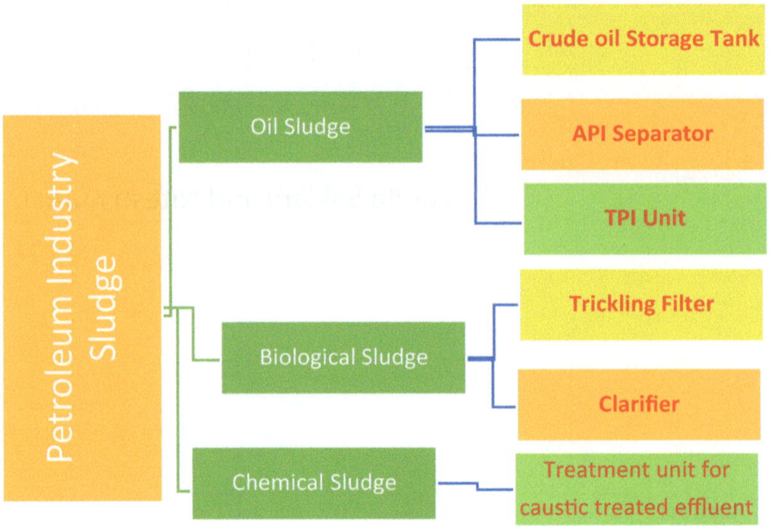


**Fig. 1** Sludge sources in oil refinery (modified after Jagaba et al. 2022a)

process. In the storage vessels of crude oil, the base of the vessel contained sludge which is composed of hydrocarbons, paraffin, asphaltene, water, and inorganic material like sulphides and oxides of iron. In the petroleum refining process, large amounts of sludge is produced which mainly comprises polycyclic aromatic hydrocarbon (PAH) and homologs of benzene which are hazardous and not easily absorbed. These are generally produced because of hydrocarbon's accumulation across sedimentary fields over a long time period (Liu et al. 2017; Jagaba et al. 2022a). Petroleum industry sludges production has also been supplemented by the different refinery activities (Fig. 1). The types of sludge commonly found in petroleum sector include oil drilling sludge, landing sludge, petroleum wastewater activated sludge, tank bottom sludge, refining sludge, oil water separator sludge, etc., are some important sludges produced (Fig. 2) Petroleum industry sludge is generated from the recovery of crude oil from the oil well and it includes crude oil, fractions of gas, and condensates. After the extraction of gas from crude oil, crude is left behind, that comprises condensates and oil. When the oil is also separated, the remaining material is termed PIS. A significant amount of sludge from the petroleum industry is produced which needs to be disposed of. Approximately 50 tons of sludge which is oily is produced in one year by petroleum refineries having production ability of 105,000 drums in a day (Ling and Isa 2006). A number of other researchers give different amounts of sludge production annually for refineries having different daily output capacities. Therefore, it is expected that 0.3–0.5% sludge is produced by one ton of crude oil (Jagaba et al. 2022a, b) However, this range is different from the range's estimation on the basis of capabilities of refining of petroleum facilities, i.e. 0.1% to 0.5%. Moreover, three different views about PIS generation are optimistic, realistic, and pessimistic for 0.1%, 0.5%, and 1.5% of the refinery. By considering all the different studies conducted earlier, it was resolved that crude oil production is responsible for production of sludge ranges between 0.1% and 0.5% annually. The average production of petroleum industry sludge is 0.3% from the crude oil data (U.S. Energy information Administration) of output of seven regions.

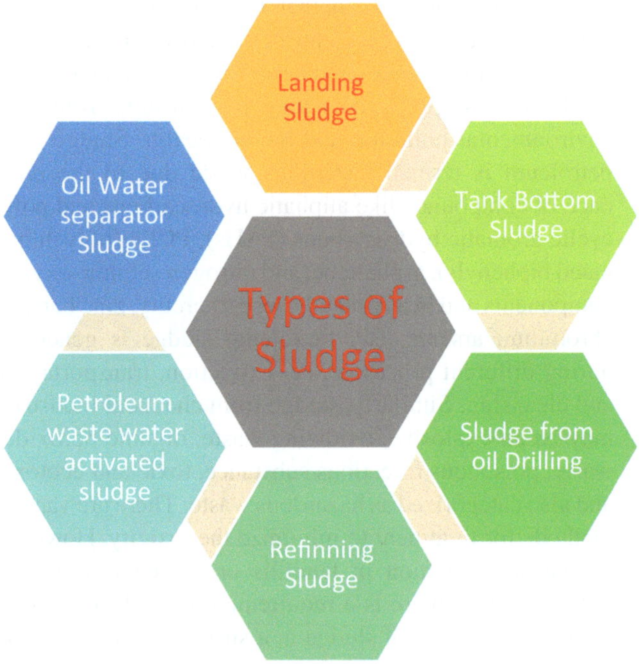


**Fig. 2** Some main sludge types

## 2 Features of Petroleum Industry Sludge and Impact on Ecosystem

The characteristic features of petroleum industry sludge are highly variable and this variation depends on the crude oil supply, condition of storage, wastewater treatment methods, processing plant, and technology employed for oil processing and it is also influenced by sample time and location (Jagaba et al. 2022b). Therefore, the petroleum industry sludge is highly variable in heating value, density, viscosity and in composition (Aldhawi et al. 2022). The petroleum industry sludge comprises mainly of grease and oil, poisonous metals' variety, organic compounds such as phenolic acid, benzene, toluene, xylene, and non-acid species in small amount, i.e. ketones, esters, and amides (Birniwa et al. 2022). It also contains the TENORM material technologically enhanced naturally occurring radioactive materials like U-238 series, Th-232 series, and K-40 and produced in equipment of oil and gas, because of alkaline earth metal's precipitation, i.e. silicates, carbonates, and sulphates (Attallah et al. 2020). The amount of these present varies with sample location and production age of oil and gas (Attahlah et al. 2020). There are a number of metal oxides present in petroleum sludge whose percentage composition varies in different sludge types produced in the petroleum industry and includes Cao, $SiO_2$, $SO_3$, $Fe_2O_3$, $Al_2O_3$, MgO, and $P_2O_5$. These all-metal oxides are found in different percent compositions in petroleum sludge, petroleum effluent treatment plant sludge, petroleum refinery sludge, dried oil sludge, crude oil sludge, oil sludge ash, petroleum oily sludge, petroleum waste sludge, petroleum sludge ash, petrochemical sludge, and tank oil sludge. Among all metal oxides, $Fe_2O_3$ found in ample amounts, i.e. 51wt% in petroleum sludge, 10.14 wt% in dried oily sludge, and 33.99 wt% in petroleum refinery sludge. Presence of $Fe_2O_3$ in the material changes the volatile properties and minimizes absorption of water properties; therefore, it is utilized as suitable material for roadbed and production of syngas that can be used as fuel (Jagaba et al. 2022a). Silica ($SiO_2$) is present in petroleum oil sludge, crude oil sludge, petroleum sludge ash, and oil sludge ash in appreciable amounts. Therefore, this material can be used after pyrolysis of oil sludge for making bricks, tiles and can also be used as cement (Pakpahan et al. 2016a). Due to higher percentage of BaO (39.08 wt%) in petroleum waste sludge as compared to silica ($SiO_2$), it can also be used for production of ceramic with the addition of some biomass enhancing material (Jagaba et al. 2022a).

It is evident from the past available information in the general public that the petroleum industry has been infamous for environmental-related issues especially with the issues of oil spilling which causes vast environmental issues around the world. So far as the sludge is concerned, it is a hazardous and most prominent waste in petroleum industry. Due to increasing population and increase in development around the world, consumption of oil is also increasing and so the amount of petroleum sludges. Sludge of petroleum industry is among the most hazardous components released into the environment which is having a great degree of impact on the biotic and abiotic constituents which ultimately affect human health and happen in the different ecosystems. PIS obtained from hydrocarbons of petroleum oil which constitutes nitrogen compound, sulphur compound, bacteria, wax, PHC's, colloid, asphaltene, heavy metals, and organic pollutants like PAHs which is considered as unpleasant, toxic, hazardous (Huang et al. 2014). These constituents are responsible for various ailment like CNS disorders, renal and respiratory problems among humans (Jasmine and Mukherji 2015). So, it is very important to treat these components to prevent land, water, and air from getting polluted. Although storage of sludge in landfills can be a good option but with the increase in production of crude oil, it is costly and less effective, so composting can be done but it can lead to generation of oily compost leachate which in long run have polluting impact on abiotic factors (Choromański Karwowska and lebkowska 2016; Sayed et al. 2021). Sulphur produce $SO_2$ on burning and it also damage catalysts used for extraction and hydrocarbons remain can leach into groundwater, rivers, and various water bodies which severely damage the fish eggs and result in biodiversity loss (Almajidi 2019; Bhattacharyya and Sekdar 2003). The intrusion of PIS in soil specially hydrocarbons can result in decreasing fertility of soil, obstruction in germination of seed, variation in microbial flora and reduction of crop growth by affecting the microbial activity, permeability of soil, hydraulic conductivity, and soil enzymes (Varjani et al. 2020).

In India, the Central Pollution Control Board is primarily responsible for regulation and controlling of industrial activities. The specific parameters and guidelines about discharge of industrial wastes/sludge vary on the type of industry and its location. Establishing large-scale projects, like in the petroleum sector, is subjected to environmental impact assessment (EIA) studies to detect their potential to harm the environment. CPCB in assistance with state pollution control board are responsible for enforcement of parameters and guidelines about limit of pH level of sludge, limit of heavy metals and toxic substances, treatment of sludge, and disposal of hazardous waste. India being a fastest growing economy became the third largest consumer of the petroleum in the world, and it accounts for 4.6% of total petroleum consumption of world which is 97,103,871 barrels/per day in 2016) (https://www.worldometers.info/oil/india-oil/), and in FY 2021-22, India consumed 204.23 million metric tons petroleum product with 5.1% growth over previous year. Indian demand for petroleum is expected to enhance remarkably up to 40%, i.e. 6.7 million barrels/ day by 2030 (https://www.investindia.gov.in/sector/oil-gas).

Similarly, around the world, the consumption of petroleum is also enhanced in various developing economies present in different continents and subcontinents and so the problem of petroleum waste sludge. To combat the problem of PIS, there is a need to adopt a suitable streamlining technique of reutilization of petroleum sludges so that harmful effects can be reduced and the goal of sustainable environment and economy can be achieved. Lack of efficient management technique of petroleum sludge can cause soil and water pollution, adverse effect on human health, economic loss, and extra cost to area cleans up and unpleasant site (Asia 2006; Al Qallaf et al. 2016). Therefore, it is essential to review the different techniques of streamlining and utilization of this PIS to minimize environmental hazards and produce the product that can be utilized for sustainable development.

**Waste Sludge Streamlining of Petroleum Industry**. The best and most important use of PIS is in construction which through proper monitoring and regulation we can ensure compliance, safety, and security of the environment as this sludge may contain toxic and hazardous material. So, it is used as kilned bricks and blocks, stabilization of roads, concrete adhesive, thermal insulation, etc. (Shperber et al. 2011a; Jagaba et al. 2021).

**Production of Cement Clinker**: In the growing economy around the world, the demand for cement in the construction industry is increasing at tremendous pace and the cement industry is facing raw material shortage, greenhouse gas emitting issues, and energy crisis. Simultaneously the voice for conservation of resources and protection of the environment is also on its top these days. So, to resonate the concerns and issues, there is a need to inculcate material with properties similar to cement so as to reduce the cost as well as decrease the dependency on the natural resources. As the PIS contains a high amount of anhydrate calcium sulphate ($CaSO_4$) so it can be a good replacement for gypsum which acts as setting retarder during strength development. Only 5% PIS could be used as it contains large amounts of $SO_3$ (Aeslina and Ali 2017). Although PIS use does not give results as expected, these are enough to meet the required standard. This addition is tagged as PS5% and in comparison with PS0, composition of PS5 for $SiO_2$ and CaO is 19% and 70%, whereas for PS0 is 20.4 and 71%, respectively (Kankia et al. 2021). Surprisingly, 100% replacement of gypsum by PIS showed an increase in UCS value by 28.8% after curing for 28 days (Benlamoudi et al. 2017). Though PIS can be used as alternative fuel in cement plants but due to the presence of carbon, sulphur, and ash contents, emission issue of toxic compounds may rise (Al-Futaisi et al. 2007). So, blend of sludge with biofuels (obtained from biowaste of agro industry) is used in cement plant in a customized way so as to reduce emission of greenhouse gases, pollutants into the atmosphere, and decrease in the formation of undesirable clinker (Tsiligiannis and Tsiliyannis 2020).

**Production of Mortar**: Mortar is a mixture which is used in construction as a bonding material where Portland cement is among the main components responsible for binding properties as it provides strength and durability required to hold construction material. Researchers have investigated PIS potential as a cement alternative after thermal treatment (to produce PSA) as its particle size, mineral phase, chemical composition, and attached functional groups suit the desirability as that of cement. As replacement of less than 20% of cement with PIS enhances the strength appreciably (Jagaba et al. 2021; Pakpahan et al. 2016b). Moreover, PIS is experimentally used in the manufacture of geopolymers by Kanika et al. (2021). These days geopolymer materials are used as an alternate Cementitious material due to its lower carbon footprints, rapid settings and resistance to corrosion. When 10% PSA (petroleum sludge ash) is used to make geopolymer mortar it enhances compressive strength up to 31 Mpa, whereas 20% PSA increases hardened density but reduces its ease to mix, flow, and plasticity.

**Use of Petroleum Sludge as Block for Buildings** Petroleum industry sludge can be used in construction of blocks that is feasible if it is encapsulated with plasters or tiles for prevention of toxic wastes percolating into soil (Hassan et al. 2005; Hago et al. 2007). For adhesion requirement, treatment of blocks is done with 8–13% mixture of petroleum sludge and vegetable oil. After assembling the product with the basic requirement of blocks, the resultant compressive strength is found to be enhanced in comparison with conventional construction blocks. Moreover, the level of C.O.D is not of that much concern. Therefore, blocks from petroleum sludge are environmentally supportive (Johnson et al. 2015).

**Ceramic Production**. Ceramic industry is one of the areas identified by the researcher for incorporation of treated PIS. It has been reported in several studies treated PIS with addition of $CaCO_3$ and $NaHCO_3$ at varying composition used to produce light weighted ceramic product with foam like cavities, bricks, concrete formulation, and aggregate. The ceramic synthesis process starts with clay and PIS incorporated with suitable amount of $NaHCO_3$ would yield least density value ceramic when less than 45% sample without additive (1.223 g/cm$^3$) is used. (Martinez et al. 2018). The red ceramic is manufactured from PIS and bentonite clay as this mixture contains most of the heavy metal within permissible range so it could be using as low-cost methods in civic works (Pinheiro and Holanda 2010). Red ceramic tiles (with 30% PIS) have weak mechanical strength but can be used satisfactorily in making roofing tiles and clay bricks in an industrial setting and meeting all the necessary criteria required for their applications (Yaro et al. 2022). To produce refractory ceramic, 60%

of bauxite ore is mixed with 40% PIS containing silicon and barium oxide, when heated at high temperature oxides react to form mullite minerals and barium aluminate which enhance its refractory properties so it is used as lining of furnaces in industries. The changes in properties like decrease/increase in compressive strength, bulk density, porosity, etc., are determined by the firing temperature and content of PIS (Khalil et al. 2018). Sintered ceramic is a good way to reduce waste such as PIS. As it unveiled that 5% PIS sintered ceramic (850–1000 °C) when crushed at 25 MPa, it was found that there is no major alteration in the structure although mechanical strength increase and density remain unchanged (Rodrigues et al. 2005).

**Fabrication of Bricks and Tiles**: Use of 5% PIS is most common and reliable addition in clay bodies to produce bricks and tiles without any major negative alteration in properties as the hazardous metals turned dormant as the mixture is fired although its workability enhanced (Abdullahi et al. 2022). Similarly, 5% solid PIS is burnt (1240C) with porcelain stoneware tile composition (as a replacement for natural kaolin). This method shows satisfactory result in consumption of waste in porcelain stoneware tile making although it decreases density, flexural strength, linear shrinkage, and increasing water absorption properties. The heavy metals (Ba, Hg, Cd, Ag, etc.) in these tiles are also within limits (Pinheiro and Holanda 2013). The use of 5% PIS in manufacturing of vitrified floor tiles at optimum firing heat in place kaolin which result in remarkable changes in composition and morphology during firing process. Although densitification is negatively affected but can be reduced using greater than 2.5% PIS. So far as vitrified floor tiles technical parameters are taken into account, these are quite comparable to porcelain tiles (Pinheiro and Holanda 2013).

**For Construction of Material for Highway**. PIS along with soil is also used as structural bed for road surface, waterproofing layers of roadbed, and shield for filtration prevention. The possibility of sludge for solidifying the other hazardous waste in controlling release of toxic metal during the process named stabilization/solidification of material. 0.5 found suitable optimum ratio for controlling heavy metal release during the process. It is concluded that stabilization/solidification of heavy metal was in control and parameters (physicochemical) are in agreement with bitumen (Shperber et al. 2011b; g point (Li et al. 2015)). Likewise, PIS along with phosphogypsum and 20% binder, i.e. formed by combination of cement (Portland) fly ash and silica fumes during stabilization solidification process is utilized for construction of roadbed material. It was found that this phosphogypsum-based binder upgrades the compressive strength decrease release of heavy metals and clarifies the structure of pore and showed microstructure improvement (Xiao et al. 2019). The possibility PIS addition in bitumen has bee also investigated to decrease the dependency on petroleum-based bitumen. It was found that strength of bitumen is decreased by using PSMB, though this stiffness found to be increased by increasing temperature and mixing speed. This improvement in stiffness is attributed to the increased ratio of asphaltenes to maltenes fractions present in PSMB under high conditions of mixing. Moreover, increased flexibility is also seen in bitumen having addition of PIS and stability of storage and plasticity is like that bitumen (Memon et al. 2020). Therefore, it could be better option to add PIS to bitumen.

The use of petroleum oil sludge has also been evaluated to manufacture the asphalt along with bitumen and aggregates. The best result was obtained by the use of 50% of oil sludge and mixing of 50% oil sludge with 2% polymer. It is found that there is decrease in resistance of bitumen due to diminishing effect of oil sludge on bitumen. Because oil sludge blocks the polysulphide formation between bitumen and sulphur molecule. However, it is upgraded by adding polymer of styrene–butadiene–styrene (Karami and Behbahani 2018). Polymer bitumen binder having best properties extensibility, penetration, brittleness, adhesion, and softening temperature can be prepared by processing of lignite with PIS containing bitumen 85%, resin 11% and styrene–butadiene–styrene polymer. The binder produced is names as bitumen-polymer-plasticizer formed after heating at 160 °C for duration five hours and give favouring properties after using it as an asphalt concrete binder. Therefore, PIS can be used in processing of resin and provide economic means of disposal of waste (Noor et al. 2021).

**Bioremediation and production of Biofertilizers** Effective microorganism technology is used for bioremediation of processed sludge having high content of P, N, K, and organic matter. Therefore, these can be added to soil (1:1) to prepare the EM biofertilizers and used for onion cultivation. The onion cultivated contains the content of heavy metals within acceptable quantity and indeed yield was improved (Ahmad 2017). Similarly, use of oil sludge in land farming can be used as carbon source and give soil condition which are acceptable for growing of jack beans (*Canavalia ensiformis*) and found suitable for decreasing the phytotoxicity of jack bean. The total TPH found to be lowered by 57.4% after growing beans for 120 days and aliphatic hydrocarbons left in root and aromatic hydrocarbons found in stem (Jagaba et al. 2022a).

**Waste Water Treatment by Using PIS**: It includes use of PIS as an absorbent material and recovered bacteria from PIS in wastewater treatment. Use of PIS as an absorbent material: Based on available written material, it can be easily concluded that PIS has a significant importance in environment cleaning efforts as it shows remarkable property of adsorption is utilized to prepare carbonaceous adsorbent and chemically treated PS used in wastewater treatment and for removal of oil spills, respectively. The property of adsorption mainly contributed due to the presence of asphaltene. The PIS after de-oiling is converted into activated carbon, as its asphaltene amount is higher so, it increases the surface area from 1103 to 3292 $m^2/g$, due to decrease in pore size of activated carbon which is confirmed through methylene blue adsorption from 17.8 mL/0.1 g to 64.6 mL/0.1 g which is comparable to activated carbon prepared from commercial asphalt. This activated carbon has a remarkable feature of leaching dangerous chemicals in comparison with heavy metal adsorption which is a significant characteristic for activated carbon potential application. Moreover, activated carbon made from PIS and rice husk shows more methylene blue adsorption with 758 mg/g (in comparison with 588 mg/g with activated carbon of PIS) which indicates that additional mesopores are formed due to formation of carbon skeleton of asphaltene with PIS, rice husk wrapped inside (Wang et al. 2017).

Porous carbonaceous adsorbent (CA) prepared from pyrolysis of PIS through its activation by KOH and without it. These carbonaceous adsorbents are used in mine to remove Cd from its effluent. Cd adsorption by activated CA and non-activated CA found to be 98% and 78%, respectively. This variation is due to formation of more micro- and mesopores in activated CA improved pore (Mohammadi and Mirghaffari 2015).

The studies have shown that chemically treated PIS shows higher efficiency in adsorbing crude oil than untreated PIS which is 2 g/g and 1.388 g/g, respectively. This property is used to resolve oil spillage issues in saline waters and can help us to resolve environmental hazards (Sayed and Zayed 2006). Moreover, crude oil adsorbent can also be prepared by using saw dust, oil sludge, and a chemical activator ($ZnCl_2$). S-AB-ZnCl2 is an adsorbent, prepared from oil sludge that shows good adsorption at certain specific conditions and has 85% regeneration efficiency after use (three times) although the loss rate subsequently increases.

Bio remedial capacities of PIS are manifested in its inhabited bacterial presence. Bacteria isolated from PIS show significant property of degrading various pollutant and complex aromatic organic compound specially *Bacillus cereus* (MA1) and *Pseudomonas guariconensis* (AT18) belonging to different genera where MA1 strain shows their remarkable calibre to break down some specific pollutant like prometryn, fluometuron, and sulfamethoxazole (Alhefeiti et al. 2021), and AT 18 strain shows their affinity for $Cr^{3+}$, $Cu^{2+}$, $Zn^{2+}$, and $Mn^{2+}$ metallic ions in their solution, with maximum 100% biosorption of $Cr^{3+}$ ion at 7–7.72 Ph and minimum biosorption 20% of $Mn^{2+}$ ion (Silva et al. 2009). Metal adsorption by P. aeruginosa AT18 can be calculated by equation given below.

$Q = V(C0 - Ce)/X$, where Q = metal intake in mg metal/g biosorbent; V = metal solution volume in mL; $C_0$ = metal initial concentration in the solution in mg metal/L, Ce = metal final concentration in the solution in mg/L; X = dry weight of biomass in g (Silva et al. 2009).

A similar study is done to see capability of Pseudomonas aeruginosa PSA5 and Rhodococcus sp. NJ2 to decomposed benzo(a)pyrene [B(a)P]. After an incubation period of 25 days, breakdown of 50 ppm of substrate by 88% and 47% is observed, respectively. Two bacteria with the presence of enzymes 2-carboxybenzaldehyde dehydrogenase (774 nmol/mg protein) and salicylate hydroxylase (840 nmol/mg protein), respectively, produce biosurfactants. These biosurfactants reduce surface tension and increase cell surface hydrophobicity which leads to higher emulsification indexes that help in breaking down pollutants by making them more water soluble and easy to break (Mishra and Singh 2014). Similarly another strain of Pseudomonas spp(BB) used to break down pyrene which shows 82% breakdown of 100 ppm substrate in 28-day incubation period (Singh et al. 2013).

**Generation of Gas through Anaerobic Digestion**: For enormous problems look towards microorganism for solution, anaerobic digestion by microorganism is an emerging field for energy and environment-related issues. It involves multi-microbial fermentation of the substrate which result in considerable amount of biogas production. This process undergoes in three phases which involves: generation of methane, hydrolysis and acid, and acetate and hydrogen. Study of PIS anaerobic digestion is taken in bioreactors with semicontinuous flow where oily biological sludge acts as a substrate and as an inoculum and its efficiency depends upon types and output of gas produced. In the bioreactor, volatile fatty acids are broken down after retention period of 40 days with increase in PH and maximum production of 864 ml of methane gas, which constitute 60.4% of the whole composition (Zhidong 2011). To enhance the biogas, yield pre-processed PIS (substrate) and sugarcane bagasse(cosubstrate) mixture under thermochemical condition gives highest biogas production of 9,268 mL, with PIS–sugarcane ratio as 2:193, co-substrate–inoculum ratio 0.18 and C/N ratio 30 (Ghaleb et al. 2021). As the carbon–nitrogen ratio is low in PIS so, we need to add some organic waste (sugarcane bagasse) with higher carbon nitrogen ratio for smooth running and to achieve reasonable yield from anaerobic digestion, as we increase sugarcane bagasse component biogas yield enhanced, it is due to increase in volatile solid content of the mixture. Moreover, it also reduces the toxic impact of PIS. Thus, sugarcane bagasse act as cosubstrate in mesophilic conditions under codigestion of PIS (Ghaleb et al. 2021).

For an anaerobic digestion process with more than 90% conversion of PIS, temperature above 1300 °C is needed to yield 1.8 $Nm^3$ $H_2$/kg oil sludge when total of 2.3 $Nm^3$/Kg oil sludge gas is produced. However, by using mixture of PIS and sugarcane, the yield of hydrogen can be enhanced to 3.5 $Nm^3$ $H_2$/kg oil sludge. This hydrogen can be used in integrated system where hydrotreatment process and PIS/biomass cogasification replace significant amount of hydrogen requirement for hydrodesulphurization of diesel which ultimately reduces considerable amount of greenhouse gases (Gonzalez et al. 2018).

**Other Application** Besides the above given bioremedial useful ways of PIS with the help of microorganism, there is a vast scope to tap various other useful applications of this petroleum waste sludge with the use of microorganisms. In this context, Bacillus licheniformis NS032 potential to yields levan, used in cosmetics as a glossy material and its amount as a product can be enhanced by increasing sucrose content (Birniwa and Abdullahi 2019). The potential of Acinetobacter radioresistance strain KA2 is observed in mineral-based media to degrade total hydrocarbon present in petroleum of oil refinery with 72–80% removal of TPH in 120 days (Poorsoleiman et al. 2020). Ralstonia pickettii bacteria degrading phenol is isolated from PIS which consume phenol as its carbon source (Al-Zuhair and El-Naas 2012).

Thermally treated PIS shows properties of significant metallic catalyst, especially as of iron catalyst. It increases catalytic behaviour and the thermal energetic capacity which is comparable with catalyst commercially available. (Castro-Leon et al. 2020). The pyrolysis method is used to produce biochar from PIS. Its redox capacity is determined by carbon hydrogen mol ratio, at 700 °C it shows good condensation and product of high aromicity obtained as a result of great capacity of electron transfer. When this material is used as cathod electrode material, it shows remarkable performance for lithium-ion battery (Abdullahi et al. 2021).

The potential of PIS in rubber industry has also been identified as the addition of 8–10% (by mass) oil sludge organic waste in rubber material designed to be used as strip fillers shows remarkable scope. The plasticization effect mechanism shows that, the organic content oil sludge addition increases elastic properties of rubber, whereas there is a negligible decrease in tensile strength of the compounds. Minerals of oil sludge in rubber material are marginally substituted among fillers to produce filler strip. (Turebekova et al. 2019) (Fig. 3).

## 3 Chemical Sludge

Chemical sludge refers to a semi-solid or solid waste material generated as a by-product of various chemical processes or industrial activities (Demirbas et al. 2017). It is typically composed of a mixture of chemicals, water, and other materials, and it can vary in composition and properties depending on the specific industrial process that produced it (Van der Bruggen et al. 2003). Chemical sludge is often considered hazardous waste due to its potential to contain harmful substances, such as heavy metals, organic chemicals, or toxic compounds.

The generation of chemical sludge can occur in a variety of industries, including:

**Wastewater Treatment**: Chemical sludge can be produced during the treatment of industrial or municipal wastewater, where chemicals are used to precipitate or flocculate contaminants for removal (Amuda and Amoo 2007).

**Chemical Manufacturing**: In the production of various chemicals, such as pharmaceuticals, petrochemicals, and specialty chemicals, by-products in the form of sludge may be generated (Awaleh and Soubaneh 2014).

**Metal Processing**: Sludge can result from processes like electroplating, where metals are deposited onto surfaces using chemical baths (Demirbas et al. 2017).

**Paper and Pulp Industry**: Chemical sludge can be created as a result of the pulping and bleaching processes in paper manufacturing (Van der Bruggen et al. 2003).

**Textile Industry**: Dyeing and finishing operations in textiles can generate chemical sludge (Awaleh and Soubaneh 2014).

Managing and disposing of chemical sludge is a significant environmental and regulatory challenge. Due to its potential toxicity and harmful effects on ecosystems and human health, it must be treated and disposed of in accordance with strict environmental regulations and guidelines (Chang et al. 2001). Common methods for managing chemical sludge include dewatering, solidification, and incineration, with the goal of reducing its volume and hazardous characteristics before disposal (Zhang et al. 2017). In many cases, industries are encouraged to reduce the generation of chemical sludge through process optimization and cleaner production practices to minimize their environmental impact and the costs associated with sludge management (Chang et al. 2001).

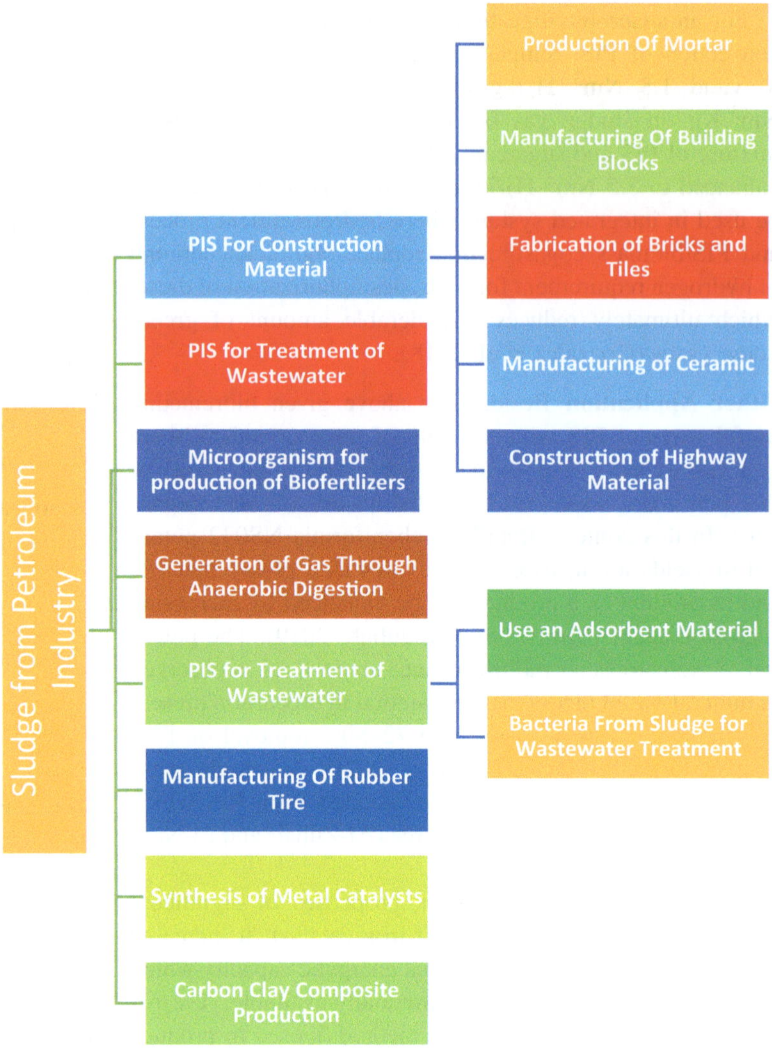


**Fig. 3** Application or utilization of sludge in different field (modified after Jagaba et al. 2022a)

## 4 Types of Chemical Sludges

Chemical sludges are residues or by-products formed during various chemical processes and wastewater treatment. They can vary in composition and characteristics depending on the specific industry, chemicals involved, and treatment methods. Here are some common types of chemical sludges:

**Industrial Process Sludge**: This type of sludge is generated in various manufacturing and industrial processes. Industrial process sludge is a by-product of various manufacturing and industrial processes (Tyagi and Lo 2013). It typically contains a mixture of solid materials, chemicals, and waste residues from these processes. The composition and characteristics of industrial process sludge can vary widely depending on the specific industry and the chemicals involved (Hu et al. 2013). Proper disposal and treatment of this sludge are essential to prevent environmental pollution and health risks. Many industries have regulations in place to manage and dispose of industrial process sludge safely (Cieślik et al. 2015). Examples include metal-working sludges, paint sludges, and chemical manufacturing residues.

**Wastewater Treatment Sludge**: Wastewater treatment plants produce sludge as a result of removing solids and contaminants from sewage or industrial effluents (Cai et al. 2007). There are two main types of wastewater sludges:

- **Primary Sludge**: This sludge is settled out in the primary clarifier and consists of solids that settle naturally (Andreoli et al. 2007).
- **Secondary Sludge (Activated Sludge)**: Secondary sludge results from biological treatment processes, such as the activated sludge process, which involves microorganisms breaking down organic matter (Wilén 1995).

**Hazardous Waste Sludge**: Hazardous waste sludge is a type of sludge generated during the treatment or disposal of hazardous wastes. It contains a range of toxic and harmful substances, including chemicals, heavy metals, and other pollutants (Mantis et al. 2005). Handling and disposal of hazardous waste sludge require strict adherence to environmental regulations to prevent contamination and health risks (Misra and Pandey 2005). Specialized processes, such as secure landfill disposal or incineration, are often used to manage this type of sludge. Proper management is essential to protect both the environment and public health (Mantis et al. 2005).

**Mining and Mineral Processing Sludge**: Mining and mineral processing sludge is a by-product of extracting and processing minerals from ore. It typically contains a mixture of mineral particles, chemicals, and water (Whitworth et al. 2022). The composition of this sludge can vary widely depending on the type of ore and processing methods used. Proper disposal and containment of mining and mineral processing sludge are crucial to prevent environmental contamination and the release of heavy metals or other harmful substances (Meng et al. 2022). Various methods, such as tailings ponds or sludge dewatering, are employed to manage this type of waste.

**Petrochemical Sludge**: Sludge generated in the petroleum and petrochemical industries can contain hydrocarbons, heavy metals, and other contaminants. Petrochemical sludge is a residue generated in the petroleum and petrochemical industry (Ayol and Yurdakoş 2019). It consists of a complex mixture of hydrocarbons, chemical compounds, and solid particles. This sludge can be a result of refining, storage tank cleaning, or spill clean-up. Proper handling and disposal are essential due to the potential presence of hazardous materials (Teng et al. 2021). The management of petrochemical sludge often involves specialized treatment and disposal methods to minimize environmental and health risks (Ayol and Yurdakoş 2019).

**Textile Industry Sludge**: The textile industry produces sludge from dyeing and finishing processes, which may contain dyes, chemicals, and organic materials. Textile Industry Sludge is a by-product of textile manufacturing processes (Karthik and Rathinamoorthy 2015). It contains a combination of dyes, chemicals, organic matter, and solid residues. This sludge is often produced during dyeing and finishing operations in the textile industry (Madhav et al. 2018). Proper disposal and treatment are crucial to prevent the release of harmful chemicals and contaminants into the environment. Regulatory compliance and responsible waste management practices are essential in handling textile industry sludge (Karthik and Rathinamoorthy 2015).

**Food Processing Sludge**: Food processing sludge is a residue produced in the food processing industry, comprising organic matter, food residues, and cleaning chemicals. It can result from activities such as washing, cooking, and packaging food products (Morales et al. 2016). Proper handling and disposal of this sludge are essential to prevent the release of organic contaminants and potential pathogens into the environment (Vico et al. 2018). Compliance with food safety and environmental regulations is crucial for the responsible management of food processing sludge. Sustainable disposal options, such as composting or anaerobic digestion, may be considered to reduce its environmental impact (Morales et al. 2016).

**Pharmaceutical Sludge**: Pharmaceutical manufacturing processes can produce sludge containing various chemicals and pharmaceutical residues (Ihsanullah et al. 2022). Pharmaceutical sludge is a waste product generated during the manufacturing of pharmaceutical drugs. It may contain a variety of pharmaceutical residues, chemicals, and organic materials (Buchberger 2007). The composition of this sludge can be complex and variable, depending on the specific pharmaceutical processes involved. Proper disposal and treatment are necessary to ensure the safe handling of potentially hazardous substances and to prevent environmental contamination (Ihsanullah et al. 2022). Compliance with pharmaceutical industry regulations and environmental standards is essential in managing pharmaceutical sludge.

These are just some examples of the types of chemical sludges encountered in various industries. Proper management and disposal of chemical sludges are crucial to prevent environmental contamination and health hazards, and regulations vary depending on the type and location of the sludge generation.

## 5 Composition of Chemical Sludges

Chemical sludges, which are generated in various industrial processes and wastewater treatment, can have diverse compositions depending on the industry and specific chemicals involved (Sommers 1977). Chemical sludges, generated across a range of industrial processes and wastewater treatment, exhibit diverse compositions contingent on the industry and the particular chemicals involved (Sommers et al. 1976). These sludges primarily consist of chemical residues, encompassing unreacted or partially reacted chemicals employed in the industrial process, which can encompass acids, bases, solvents, and various reactants (Gambus and Gorlach 1998). Heavy metals, such as lead, mercury, cadmium, chromium, and arsenic, may be present, often introduced by the chemical processes. Inorganic solids, like sand, clay, and other particulate matter, are commonly found in chemical sludges. Additionally, a substantial water content can be attributed to the use of water in industrial processes (Sommers 1977). Furthermore, organic compounds, including oils, greases, and hydrocarbons, may be part of the sludge. Some chemical sludges also harbour toxic or hazardous chemicals, like cyanide or chlorinated compounds, especially in industries dealing with hazardous materials (Gambus and Gorlach 1998). pH adjusters and other additives are added as necessary for process control. Due to the potential hazards associated with chemical sludges, strict adherence to environmental regulations is essential for their management and disposal (Sommers 1977). Treatments such as stabilization, solidification, or incineration are often employed to mitigate the risks (Gambus and Gorlach 1998). Accurate characterization and appropriate handling are imperative to ensure the safe and compliant disposal of chemical sludges, with the exact composition contingent on the origin of the sludge and necessitating comprehensive analysis for precise determination (Sommers 1977). While the composition can vary significantly, here is a general overview of what chemical sludges may contain:

**Chemical Residues**: These are the primary components of chemical sludges, and they can include unreacted or partially reacted chemicals used in the industrial process. The specific chemicals will depend on the industry (Speight 2002). Examples include acids, bases, solvents, and various chemical reactants.

**Heavy Metals**: Some chemical processes can introduce heavy metals into the sludge. Common heavy metals found in chemical sludges may include lead, mercury, cadmium, chromium, and arsenic.

**Inorganic Solids**: Chemical sludges often contain inorganic solids such as sand, clay, and other particulate matter (Islam 2015). These solids can come from raw materials, catalysts, or other substances in the process.

**Water**: Many chemical sludges have a significant water content, as water is often used in industrial processes and may become part of the sludge (Ahmad et al. 2016).

**Organic Compounds**: Depending on the nature of the chemical process, organic compounds, including oils, greases, and other hydrocarbons, may be present in the sludge (Bharathiraja et al. 2014).

**Toxic or Hazardous Chemicals**: Some chemical sludges may contain toxic or hazardous chemicals, especially in industries where hazardous materials are used (Sommers et al. 1976). These can include substances like cyanide, chlorinated compounds, and highly reactive chemicals.

**pH Adjusters**: In some cases, chemicals used to adjust the pH of the process may become part of the sludge, contributing to its composition (Sommers et al. 1976; Bharathiraja et al. 2014).

**Other Additives**: Depending on the specific process, there may be other additives, stabilizers, or conditioners that find their way into the sludge (Sommers et al. 1976).

Due to the potentially hazardous nature of chemical sludges, their management and disposal typically require adherence to strict environmental regulations. Treatment processes like stabilization, solidification, or incineration may be used to reduce the risk associated with these sludges. Proper characterization and handling are crucial to ensure the safe and compliant disposal of chemical sludges. The specific composition of chemical sludges will depend on the industrial process from which they originate, and an in-depth analysis is often necessary to determine the exact constituents and their concentrations.

## 6 Source of Chemical Sludges

Chemical sludges originate from a variety of industrial processes and wastewater treatment facilities. One common source of chemical sludges is municipal or industrial wastewater treatment plants, where precipitation and flocculation lead to sludge formation (Islam 2015). The petrochemical industry generates sludges as by-products, containing oil, grease, and heavy metals. In mining and mineral processing, sludges form during the extraction and processing of ores, often containing metals and minerals (Samer 2015). Textile manufacturing produces chemical sludges as a result of dyeing and finishing processes, which involve various dyes and chemicals. Electroplating operations contribute to sludge production with metal residues and plating chemicals (Velusamy et al. 2021). In the paper and pulp industry, sludges are generated through pulping and bleaching processes, containing wood fibres and chemicals. Additionally, food processing facilities create organic sludges during operations such as meat rendering, dairy processing, and fruit and vegetable processing (Islam 2015). These diverse sources of chemical sludges necessitate proper management and disposal to prevent environmental contamination (Velusamy et al. 2021). Here are some common sources of chemical sludges:

**Wastewater Treatment Plants**: Many chemical sludges are generated in municipal or industrial wastewater treatment plants. These sludges often result from the precipitation or flocculation of solids during the treatment process (Canziani and Spinosa 2019). Wastewater treatment plants, whether serving municipalities or industries, are significant sources of chemical sludges. These sludges typically form as a consequence of the precipitation or flocculation of solid particles during the wastewater treatment process (Czerwionka et al. 2020). As water is purified and contaminants are removed, the solids that settle out create these chemical-laden sludges (Canziani and Spinosa 2019). Proper management and disposal of these sludges are crucial to ensure that they do not pose environmental hazards. This underscores the importance of adhering to regulations and guidelines governing the handling of chemical sludges to safeguard the environment (Czerwionka et al. 2020).

**Petrochemical Industry**: Refineries and petrochemical plants produce chemical sludges as by-products of their operations. These sludges can include oil and grease residues, heavy metals, and other contaminants (Hu et al. 2013). The petrochemical industry, encompassing refineries and petrochemical plants, generates chemical sludges as inherent by-products of their operational processes. These sludges often contain a mixture of oil and grease residues, heavy metals, and various contaminants (Jagaba et al. 2022a). Due to the nature of petrochemical operations, these sludges can be particularly environmentally hazardous. Safeguarding against their release and ensuring proper disposal is a vital aspect of environmental responsibility in this industry (Hu et al. 2013). Adherence to environmental regulations and guidelines is crucial in managing these chemical sludges.

**Mining and Mineral Processing**: The mining and mineral processing industry generates sludges as a result of extracting and processing ores. These sludges can contain heavy metals, minerals, and other contaminants (Babel and del Mundo Dacera 2006). The mining and mineral processing industry is a significant source of chemical sludges, which emerge during the extraction and processing of ores (Samer 2015). These sludges are often laden with heavy metals, minerals, and a host of other contaminants. In mining operations, ore-containing rock is often crushed and chemically treated to extract valuable minerals, leading to the formation of such sludges (Babel and del Mundo Dacera 2006). The presence of heavy metals, such as lead, arsenic, and cadmium, in these sludges poses serious environmental concerns, as they can leach into soil and water, causing contamination and harm to ecosystems (Velusamy et al. 2021). Proper disposal and responsible management of these sludges are imperative to mitigate environmental impacts and ensure regulatory compliance within the mining and mineral processing sector (Babel and del Mundo Dacera 2006).

**Textile Industry**: In the textile industry, chemical sludges are a common by-product, primarily arising from the dyeing and finishing processes. These sludges consist of a mix of dyes, chemicals, and colorants used in the textile manufacturing and colouring procedures (Karthik and Rathinamoorthy 2015). Dyeing involves the application of various chemicals to fabrics to achieve the desired colour and texture, resulting in the generation of these sludges. The presence of these chemicals in the sludges can pose environmental concerns if not managed appropriately (Velusamy et al. 2021). Proper disposal and treatment of textile industry chemical sludges are essential to prevent contamination and adhere to environmental regulations. Sustainable practices, such as the recycling and responsible disposal of these sludges, are increasingly important within the textile sector to reduce its ecological footprint (Karthik and Rathinamoorthy 2015).

**Pharmaceutical Industry**: The pharmaceutical industry is a source of chemical sludges, particularly during the production of pharmaceuticals. These sludges often consist of a mixture of active pharmaceutical ingredients, solvents, and various other chemicals used in drug manufacturing processes (Velusamy et al. 2021). The presence of these compounds in the sludges can raise concerns about their environmental impact. Proper disposal and treatment of pharmaceutical industry chemical sludges are essential to prevent potential contamination and adhere to strict regulations and guidelines, given the potential risks associated with pharmaceutical compounds (Islam 2015). Sustainable practices, such as efficient recycling and responsible management of these sludges, are increasingly vital in the pharmaceutical sector to minimize ecological and health-related implications and ensure compliance with regulatory standards (Velusamy et al. 2021).

It's important to note that the composition and properties of chemical sludges can vary widely, and their proper management, treatment, and disposal are crucial to prevent environmental contamination. Regulations and guidelines for handling chemical sludges also exist in many countries to ensure they are managed safely and responsibly (Fig. 4).

## 7 Treatment for Chemical Sludges

The treatment of chemical sludges is a multifaceted process that depends on various factors, including the sludge's composition, contaminants, and regulatory requirements (Raheem et al. 2018). Common treatment methods include dewatering to reduce water content, chemical conditioning

**Fig. 4** Main sources of chemical sludge

to aid in solid–liquid separation, and stabilization to limit the mobility of hazardous components (Zhang et al. 2017). For sludges with organic content, biological treatments like composting or anaerobic digestion may be employed to break down organic compounds (Zhang et al. 2014). Additionally, thermal treatment methods such as incineration can thermally destroy organic contaminants, while chemical treatments may be used to transform or remove specific contaminants (Johnson and Sumpter 2001). The choice of treatment method must align with local regulations and guidelines to ensure the responsible management of chemical sludges, mitigating environmental and health risks (Zhang et al. 2017). Ultimately, the selection of treatment processes should be tailored to the specific characteristics of the sludge to achieve effective and environmentally responsible results.

## 8 Resources Utilization of Sludges

Resource utilization of chemical sludges is an essential aspect of sustainable industrial practices. Chemical sludges are typically generated as a result of various industrial processes, often containing valuable components and potentially hazardous materials (Tyagi and Lo 2013). To maximize the utilization of these resources while minimizing their environmental impact, several strategies can be employed. One common approach is the recovery of precious metals. Chemical sludges may contain trace amounts of valuable metals such as gold, silver, or platinum. Utilizing advanced extraction techniques, these precious metals can be separated from the sludge and reintroduced into the production cycle (Kroiss 2004). This not only reduces the need for environmentally damaging mining operations but also provides a financial incentive for resource recovery (Tyagi and Lo 2013). Another vital resource within chemical sludges is phosphorus. Phosphorus is an essential nutrient for agriculture, and chemical sludges often contain significant concentrations of this element. By extracting and processing the phosphorus, it can be converted into phosphorus-based fertilizers, reducing the demand for phosphate rock mining and ensuring a sustainable source of nutrients for crop production. Additionally, chemical sludges can be a source of energy (Kroiss 2004). Through incineration or gasification, these sludges can be converted into heat or electricity, contributing to a circular economy approach by reducing waste and harnessing the energy content of the sludges. Water reuse is another critical aspect of resource utilization for chemical sludges (Syed-Hassan et al. 2017). By treating and purifying the water contained in the sludges, industries can recycle this resource for various non-potable purposes, reducing the strain on freshwater supplies and minimizing wastewater discharge (Tyagi and Lo 2013).. Furthermore, innovative technologies are emerging that enable the conversion of chemical sludges into useful chemicals or fuels. These processes transform what was once considered waste into valuable products, contributing to sustainability goals and reducing environmental impacts (Syed-Hassan et al. 2017). Certain chemical sludges can also serve as soil conditioners. These materials can be applied to agricultural land or used in land reclamation projects to improve soil fertility and structure. This not only reduces the need for chemical fertilizers but also helps restore degraded land. In summary, the resource utilization of chemical sludges is a multifaceted process that can bring about environmental and economic benefits (Mulchandani and Westerhoff 2016). It allows industries to reduce their environmental footprint, comply with waste management regulations, and contribute to a more sustainable and circular economy by recovering valuable resources from what was once considered waste (Tyagi and Lo 2013). Proper management and the use of advanced technologies are essential in realizing the full potential of chemical sludge utilization Table 1.

**Table 1** Resource potential of chemical sludge

| Serial no | Methods use | Extraction | Application | References |
|---|---|---|---|---|
| 1 | Advanced extraction methods | Gold, silver, and platinum extraction | Reintroduce in production cycle | Kroiss (2004) |
| 2 | Incineration and gasification | Generation of heat and electricity | Circular economy approach | Syed-Hassan et al. (2017) |
| 3 | Extraction and processing of phosphorus | Phosphorus-based fertilizers | Sustainable source of nutrient for crop production | Kroiss (2004) |
| 4 | Purification of water | Treatment of water | Nonpotable use of water | Tyagi and Lo (2013) |
| 5 | Production of fuel | – | Contribution in sustainable goal and reducing environmental impact | Syed-Hassan et al. (2017) |

## 9 Conclusion

It is clear that petroleum and chemical industry is growing continuously to meet out the increased energy and others products demand around the globe due to increasing populations. Therefore, the amount of petroleum and chemical sludge from petroleum and chemical industry is increasing in recent times and projected to increase in near future, since the sludge contained hazardous components having detrimental effect on environment. Although there are different waste treatment methods/treatment are there to reduce the risk. But still, the problem of leftover sludge management is there. Sludge landfills is also an option, but is not effective in long run approach. So, there is need to look forward for potential use of sludge to bring it in cyclic without impact the environment harnessing it for economical purpose. Different researchers have studied different methods or techniques for resource utilization of sludge, some of these are really effective; however, for others, there is need to improve these to make effective. So that, harmful effect of sludge on environment can be minimized along with streamlining it fully to achieve the goal of sustainable development. Furthermore, the genetic engineering technique can be investigated and employed for the microorganism isolated from the sludge to make them efficient in degrading the waste present in sludge and useful product can obtained for energy conservation.

## References


Abdullahi SS, Musa H, Habibu S, Birniwa AH, Mohammad REA (2021) Facile synthesis and dyeing performance of some disperse monomeric and polymeric dyes on nylon and polyester fabrics. Bull Chem Soc Ethiop 35(3):485–497. https://doi.org/10.4314/bcse.v35i3.2

Abdullahi SS, Musa H, Habibu S, Birniwa AH, Mohammad REA (2022) Comparative study and dyeing performance of as-synthesized azo heterocyclic monomeric, polymeric, and commercial disperse dyes. Turk J Chem 46(6):1841–1852. https://doi.org/10.55730/1300-0527.3484

Aeslina AK, Ali B (2017) Petroleum sludge treatment and reuse for cement production as setting retarder. IOP Conf Ser: Mater Sci Eng 203:012010. https://doi.org/10.1088/1757-899X/203/1/012010

Ahmad J (2017) Bioremediation of petroleum sludge using effective microorganism (EM) technology. Pet Sci Technol 35(14):1515–1522. https://doi.org/10.1080/10916466.2017.1356850

Ahmad T, Ahmad K, Alam M (2016) Sustainable management of water treatment sludge through 3 'R'concept. J Clean Prod 124:1–13

Al-dhawi B, Mohamed K, Rahman S, Baloo L, Almahbashi N, Ghaleb A, Jagaba AH, Kumar V, Saeed A (2022) Treatment of synthetic wastewater by using submerged attached growth media in continuous activated sludge reactor system. Int J Sustain Build Technol Urban Develop 13:2–10. https://doi.org/10.22712/susb.20220002

Al-Futaisi A, Jamrah A, Yaghi B, Taha R (2007) Assessment of alternative management techniques of tank bottom petroleum sludge in Oman. J Hazard Mater 141(3):557–564. https://doi.org/10.1016/j.jhazmat.2006.07.023

Alhefeiti MA, Athamneh K, Vijayan R, Ashraf SS (2021) Bioremediation of various aromatic and emerging pollutants by *Bacillus cereus* sp. isolated from petroleum sludge. Water Sci Technol J Int Assoc Water Pollut Res 83(7):1535–1547. https://doi.org/10.2166/wst.2021.065

Almajidi M (2019) New coulometric method for determination of the chemical compositions of the refinery waste. Egypt J Chem 62(11):2109–2117

Al-Zuhair S, El-Naas MH (2012) Phenol biodegradation by ralstonia pickettii extracted from petroleum refinery oil sludge. Chem Eng Commun 199(9):1194–1204

Amuda O, Amoo I (2007) Coagulation/flocculation process and sludge conditioning in beverage industrial wastewater treatment. J Hazard Mater 141(3):778–783

Andreoli CV, Von Sperling M, Fernandes F (2007) Sludge treatment and disposal. IWA publishing

Asia IO, Enweani IB, Eguavoen I (2006) Characterization and treatment of sludge from the petroleum industry. Afr J Biotechnol 5:461–466

Awaleh MO, Soubaneh YD (2014) Waste water treatment in chemical industries: the concept and current technologies. Hydrol Curr Res 5(1):1–12

Ayol A, Yurdakoş ÖT (2019) Chemical and thermal characteristics of petrochemical industrial sludge. Desalin Water Treat 172:29–36

Babel S, del Mundo Dacera D (2006) Heavy metal removal from contaminated sludge for land application: a review. Waste Manage 26(9):988–1004

Benlamoudi A, Aeslina AK, Khodja M (2017) Petroleum sludge as gypsum replacement in cement plants: its impact on cement strength. IOP Conf Ser Mater Sci Eng 226:012059. https://doi.org/10.1088/1757-899X/226/1/012059

Bharathiraja B, Yogendran D, Ranjith Kumar R, Chakravarthy M, Palani S (2014) Biofuels from sewage sludge—a review. Int J ChemTech Res 6(9):4417–4427

Bhattacharyya JK, Shekdar AV (2003) Treatment and disposal of refinery sludges: Indian scenario. Waste Manage Res 21(3):249–261. https://doi.org/10.1177/0734242X0302100309

Birniwa A, Abdullahi S (2019) Study on physico-mechanical behaviour of acacia nilotica (gum tree) and glass fiber blend reinforced epoxy resin composite. Chem Search J 10(2):46–53

Birniwa AH et al (2022) Application of agricultural wastes for cationic dyes removal from wastewater. In: Muthu SS, Khadir A (eds) Textile wastewater treatment. sustainable textiles: production, processing, manufacturing & chemistry. Springer, Singapore. https://doi.org/10.1007/978-981-19-2832-1_9

Borgheipour H et al (2018) Application of DEA technique in SWOT analysis of oily sludge management using fuzzy data. Glob J Environ Sci Manage Gjesm 4(2):183–194

Buchberger WW (2007) Novel analytical procedures for screening of drug residues in water, waste water, sediment and sludge. Anal Chim Acta 593(2):129–139

Cai Q-Y, Mo C-H, Wu Q-T, Zeng Q-Y, Katsoyiannis A (2007) Occurrence of organic contaminants in sewage sludges from eleven wastewater treatment plants, China. Chemosphere 68(9):1751–1762

Canziani R, Spinosa L (2019) Sludge from wastewater treatment plants. In: Industrial and municipal sludge. Elsevier, pp 3–30

Castro-León G, Baquero-Quinteros E, Loor BG, Alvear J, Montesdeoca Espín DE, De La Rosa A, Montero-Calderón C (2020). Waste to catalyst: synthesis of catalysts from sewage sludge of the mining, steel, and petroleum industries. Sustainability 12(23)

Chang G, Liu J, Lee D (2001) Co-conditioning and dewatering of chemical sludge and waste activated sludge. Water Res 35(3):786–794

Choromański P, Karwowska E, Łebkowska M (2016) The influence of petroleum products on the methane fermentation process. J Hazard Mater 301:327–331. https://doi.org/10.1016/j.jhazmat.2015.09.011

Cieślik BM, Namieśnik J, Konieczka P (2015) Review of sewage sludge management: standards, regulations and analytical methods. J Clean Prod 90:1–15

Czerwionka K, Wilinska A, Tuszynska A (2020) The use of organic coagulants in the primary precipitation process at wastewater treatment plants. Water 12(6):1650

Demirbas A, Edris G, Alalayah WM (2017) Sludge production from municipal wastewater treatment in sewage treatment plant. Energy Sourc Part a: Recov Util Environ Effects 39(10):999–1006

Gambus F, Gorlach E (1998) Chemical composition of sludges from sewage treatment plants of the Cracow province as a criterion of their usage. Acta Agraria et Silvestria. Series Agraria (36)

Ghaleb A, Kutty SRM, Hayder TSG, Jagaba AH, Noor A, Kumar V, Saeed AA, Al-dhawi BN (2021) Sugarcane bagasse as a Co-substrate with oil-refinery biological sludge for biogas production using batch mesophilic anaerobic co-digestion technology: effect of carbon/nitrogen ratio. Water 13(5):590

González A, Lora EES, Palácio JCE, Olmo OAA (2018) Hydrogen production from oil sludge gasification/biomass mixtures and potential use in hydrotreatment processes. Int J Hydrogen Energy 43(16):7808–7822. https://doi.org/10.1016/j.ijhydene.2018.03.025

Hago AW, Hassan HF, Rawas AA, Taha RA, Al-Hadidi S (2007) Characterization of concrete blocks containing petroleum-contaminated soils. Constr Build Mater 21:952–957

Hassan HF, Taha RA, Rawas AA, Shandoudi BA, Gheithi KA, Barami AM (2005) Potential uses of petroleum-contaminated soil in highway construction. Constr Build Mater 19:646–652

Hu G, Li J, Zeng G (2013) Recent development in the treatment of oily sludge from petroleum industry: a review. J Hazard Mater 261:470–490

Huang Q, Mao F, Han X, Yan J, Chi Y (2014) Characterization of emulsified water in petroleum sludge. Fuel 118:214–219

Ihsanullah I, Khan MT, Zubair M, Bilal M, Sajid M (2022) Removal of pharmaceuticals from water using sewage sludge-derived biochar: a review. Chemosphere 289:133196

Islam B (2015) Petroleum sludge, its treatment and disposal: a review. Int J Chem Sci 13(4):1584–1602

Jagaba AH, Kutty SRM, Hayder G, Baloo L, Noor A, Yaro NSA, Saeed AAH, Lawal IM, Birniwa AH, Usman AK (2021) A systematic literature review on waste-to-resource potential of palm oil clinker for sustainable engineering and environmental applications. Materials (Basel, Switzerland) 14(16):4456. https://doi.org/10.3390/ma14164456

Jagaba AH, Kutty M, Rahman S, Lawal I, Haruna B, Abdullahi A, Augustine AY, Shehu N, Usman A, Kilaco U, Ibrahim A, Sule N, Azmatullah, B, Soja U, Yakubu A (2022a) Circular economy potential and contributions of petroleum industry sludge utilization to environmental sustainability through engineered processes—a review. Cleaner Circ Bioecon 3:100029. https://doi.org/10.1016/j.clcb.2022.100029

Jagaba AH et al (2022b) Effect of environmental and operational parameters on sequential batch reactor systems in dye degradation. In: Muthu SS, Khadir A (eds) Dye biodegradation, mechanisms and techniques. sustainable textiles: production, processing, manufacturing & chemistry. Springer, Singapore. https://doi.org/10.1007/978-981-16-5932-4_8

Jasmine J, Mukherji S (2015) Characterization of oily sludge from a refinery and biodegradability assessment using various hydrocarbon degrading strains and reconstituted consortia. J Environ Manage 149:118–125. https://doi.org/10.1016/j.jenvman.2014.10.007

Johnson AC, Sumpter JP (2001) Removal of endocrine-disrupting chemicals in activated sludge treatment works. Environ Sci Technol 35(24):4697–4703

Johnson OA, Madzlan N, Kamaruddin I (2015) Encapsulation of petroleum sludge in building blocks. Constr Build Mater 78:281–288. https://doi.org/10.1016/j.conbuildmat.2014.12.122

Kankia MU, Baloo L, Danlami N, Samahani WN, Mohammed BS, Haruna S, Jagaba AH, Abubakar M, Ishak EA, Sayed K, Amila Bt Wan Zawawi N (2021) Optimization of cement-based mortar containing oily sludge ash by response surface methodology. Materials (Basel, Switzerland) 14(21):6308. https://doi.org/10.3390/ma14216308

Karami E, Jafari Behbahani T (2018) Upgrading Iranian petroleum sludge using polymers. J Petrol Explor Prod Technol 8:1319–1324

Karthik T, Rathinamoorthy R (2015) Recycling and reuse of textile effluent sludge. Environmental implications of recycling and recycled products, pp 213–258

Khalil NM, Algamal YM, Saleem QM (2018) Exploitation of petroleum waste sludge with local bauxite raw material for producing high-quality refractory ceramics. Ceram Int 44(15):18516–18527. https://doi.org/10.1016/j.ceramint.2018.07.072

Kroiss H (2004) What is the potential for utilizing the resources in sludge? Water Sci Technol 49(10):1–10

Li G, Guo SH, Ye HF (2015) Thermal treatment of heavy oily sludge: resource recovery and potential utilization of residual asphalt-like emulsion as a stabilization/solidification material. RSC Adv 5(127):105299–105306

Ling CC, Isa MH (2006) Bioremediation of oily sludge contaminated soil by co-composting with sewage sludge. J Sci Indus Res 65:364–369

Liu TL et al (2017) Contamination of fresh water by petroleum sludge. Petrol Sci Technol 35(4):413–418

Madhav S, Ahamad A, Singh P, Mishra PK (2018) A review of textile industry: wet processing, environmental impacts, and effluent treatment methods. Environ Qual Manage 27(3):31–41

Mantis I, Voutsa D, Samara C (2005) Assessment of the environmental hazard from municipal and industrial wastewater treatment sludge by employing chemical and biological methods. Ecotoxicol Environ Saf 62(3):397–407

Martinez JD et al (2018) Ceramic light-weight aggregates production from petrochemical wastes and carbonates ($NaHCO_3$ and $CaCO_3$) as expansion agents. Constr Build Mater 180:124–133

Memon AM et al (2020) Modeling and optimization of mixing conditions for petroleum sludge modified bitumen using response surface methodology. Constr Build Mater 291:123288

Meng S, Wen S, Han G, Wang X, Feng Q (2022) Wastewater treatment in mineral processing of non-ferrous metal resources: a review. Water 14(5):726

Mishra S, Singh SN (2014) Biodegradation of benzo(a)pyrene mediated by catabolic enzymes of bacteria. Int J Environ Sci Technol 11(6):1571–1580

Misra V, Pandey S (2005) Hazardous waste, impact on health and environment for development of better waste management strategies in future in India. Environ Int 31(3):417–431

Mohammadi S, Mirghaffari N (2015) Optimization and comparison of Cd removal from aqueous solutions using activated and non-activated carbonaceous adsorbents prepared by pyrolysis of oily sludge. Water Air Soil Pollut 226:1–11

Morales AB, Bustamante MA, Marhuenda-Egea FC, Moral R, Ros M, Pascual JA (2016) Agri-food sludge management using different co-composting strategies: study of the added value of the composts obtained. J Clean Prod 121:186–197

Mulchandani A, Westerhoff P (2016) Recovery opportunities for metals and energy from sewage sludges. Biores Technol 215:215–226

Noor A, Kutty SRM, Jagaba AH, Yusuf M, Akram MW, Adil MR, Ahmad N, Jamal M (2021) Kinetic modelling of nutrient removal of petroleum industry wastewater remediation. In: Third international sustainability and resilience conference: climate change, Sakheer, Bahrain, pp 216–220. https://doi.org/10.1109/IEEECONF53624.2021.9667961

Pakpahan EN, Shafiq N, Isa MH, Mohamed K, Shamsul R, Mustafa M (2016a). In: Zawawi N (ed) Engineering challenges for sustainable future. CRC press, pp 251–256

Pakpahan E, Shafiq N, Isa MH, Mohamed K, Shamsul R, Mustafa M (2016b) Petroleum sludge thermal treatment and use in cement replacement—a solution towards sustainability. In: Proceedings of the 3rd international conference on civil, offshore and environmental engineering (ICCOEE 2016, Malaysia, 15–17 Aug 2016). https://doi.org/10.1201/b21942-50

Pinheiro BC, Holanda JN (2013) Obtainment of porcelain floor tiles added with petroleum oily sludge. Ceram Int 39:57–63

Poorsoleiman MS, Hosseini SA, Etminan A, Abtahi H, Koolivand A (2020) Bioremediation of petroleum hydrocarbons by using a two-step inoculation composting process scaled-up from a mineral-based medium: effect of biostimulation of an indigenous bacterial strain. Waste Biomass Valorization 12:2089–2096

Qallaf YA, Al-kandari S, Yousef K, Mutairi SE, Fouzy H (2016) Analysis and improvement possibilities of waste management at Kuwait Oil Company (KOC). Waste Manage 202:73–84

Raheem A, Sikarwar VS, He J, Dastyar W, Dionysiou DD, Wang W, Zhao M (2018) Opportunities and challenges in sustainable treatment and resource reuse of sewage sludge: a review. Chem Eng J 337:616–641

Rodrigues MF, Holanda JNF, Sthel MS, Vargas H (2005) Thermal characterization of sintered clay-petroleum waste mix by the photoacoustic technique. In: Salgado L, Filho FA (eds) Advanced powder technology Iv. Switzerland, pp 512–516

Samer M (2015) Biological and chemical wastewater treatment processes. Wastewater Treatment Eng 150:212

Sayed SA, Zayed AM (2006) Investigation of the effectiveness of some adsorbent materials in oil spill clean-ups. Desalination 1–3:90–100

Sayed K, Baloo L, Yekeen ST, Kankia MU, Jagaba AH (2021) Determination of total petroleum hydrocarbons concentration in coastal seawater of teluk batik beach, perak, malaysia. Key Eng Mater 888:119–128

Shperber ER, Bokovikova TN, Shperber DR (2011a) Methods for processing petroleum wastes. Chem Technol Fuels Oils 47:237–242. https://doi.org/10.1007/s10553-011-0288-x

Shperber ER, Bokovikova TN, Shperber DR (2011b) Methods for processing petroleum wastes. Chem Technol Fuels Oils 47:237–242. https://doi.org/10.1007/s10553-011-0288-x

Silva RMP, Abalos Rodríguez A, Gómez Montes De Oca JM, Cantero Moreno D (2009) Biosorption of chromium, copper, manganese and zinc by *Pseudomonas aeruginosa* AT18 isolated from a site contaminated with petroleum. Bioresourc Technol 100(4):1533–1538. https://doi.org/10.1016/j.biortech.2008.06.057

Singh R, Das AJ, Pokhrel S, Gautam P (2013) Biodegradation of pyrene by bacterial strain isolated from petroleum sludge collected from barauni oil refinery, Begusari, Bihar (India). J Pure Appl Microbiol 7(2):1441–1445

Sommers L (1977) Chemical composition of sewage sludges and analysis of their potential use as fertilizers (0047–2425)

Sommers L, Nelson D, Yost K (1976) Variable nature of chemical composition of sewage sludges (0047–2425)

Speight JG (2002) Chemical process and design handbook. McGraw-Hill Education

Syed-Hassan SSA, Wang Y, Hu S, Su S, Xiang J (2017) Thermochemical processing of sewage sludge to energy and fuel: fundamentals, challenges and considerations. Renew Sustain Energy Rev 80:888–913

Teng Q, Zhang D, Yang C (2021) A review of the application of different treatment processes for oily sludge. ESPR 28:121–132

Tsiligiannis A, Tsiliyannis C (2020) Oil refinery sludge and renewable fuel blends as energy sources for the cement industry. Renew Energy 157:5–70

Turebekova GZ, Shapalov SK, Alpamysova GB, Issayev GI, Bitemirova A, Lyubchyk S, Yergubekova ZS, Pusurmanova GZ (2019) Recycling of wastes of petroleum refining industry by using them in the manufacture of tire rubbers. Sci News Natl Acad Sci Republic Kazakhstan-Ser Chem Technol 1:16–20

Tyagi VK, Lo S-L (2013) Sludge: a waste or renewable source for energy and resources recovery? Renew Sustain Energy Rev 25:708–728

USEA (2008) Hazardous waste listings—a user-friendly reference document draft [Internet]. [cited 21 Oct 2017]. Available from: http://www.epa.gov/osw/hazard/wastetypes/pdfs/listingref.pdf

Van der Bruggen B, Vandecasteele C, Van Gestel T, Doyen W, Leysen R (2003) A review of pressure-driven membrane processes in wastewater treatment and drinking water production. Environ Prog 22(1):46–56

Varjani S, Pandey A, Upasani VN (2020) Oilfield waste treatment using novel hydrocarbon utilizing bacterial consortium—a microcosm approach. Sci Total Environ 745:141043. https://doi.org/10.1016/j.scitotenv.2020.141043

Velusamy S, Roy A, Sundaram S, Kumar Mallick T (2021) A review on heavy metal ions and containing dyes removal through graphene oxide-based adsorption strategies for textile wastewater treatment. Chem Rec 21(7):1570–1610

Vico A, Pérez-Murcia MD, Bustamante MA, Agulló E, Marhuenda-Egea FC, Sáez J, Paredes C, Pérez-Espinosa A, Moral R (2018) Valorization of date palm (*Phoenix dactylifera* L.) pruning biomass by co-composting with urban and agri-food sludge. J Environ Manage 226:408–415

Wang J, Sun C, Lin B-C, Huang Q-X, Ma Z-Y, Chi Y, Yan J-H (2017) Micro- and mesoporous-enriched carbon materials prepared from a mixture of petroleum-derived oily sludge and biomass. Fuel Process Technol 171:140–147. https://doi.org/10.1016/j.fuproc.2017.11.013

Whitworth AJ, Forbes E, Verster I, Jokovic V, Awatey B, Parbhakar-Fox A (2022) Review on advances in mineral processing technologies suitable for critical metal recovery from mining and processing wastes. Cleaner Eng Technol 7:100451

Wilén B-M (1995) Effect of different parameters on settling properties of activated sludge

Xiao W, Yao X, Zhang FY (2019) Recycling of oily sludge as a roadbed material utilizing phosphogypsum-based cementitious materials. Adv Civil Eng 1–10. https://doi.org/10.1155/2019/6280715

Yaro NSA, Napiah M, Sutanto MH, Usman A, Jagaba AH, Mani AU, Ahmad A (2022) Geopolymer utilization in the pavement industry-an overview. In: IOP conference series: earth and environmental science, vol 1022, no. 1. IOP Publishing, p 012025

Zhang J, Lü F, Shao L, He P (2014) The use of biochar-amended composting to improve the humification and degradation of sewage sludge. Bioresour Technol 168:252–258

Zhang Q, Hu J, Lee D-J, Chang Y, Lee Y-J (2017) Sludge treatment: current research trends. Biores Technol 243:1159–1172

Zhidong L (2011) A study of anaerobic digestion of excess petrochemical plant sludge. Energy Sourc Part A-Recov Util Environ Effects 33(10):899–907

# Sludge Management in Petroleum and Chemical Industry

Debjani Bhakta, Swapnadipa Patra, and Subhasis Roy

**Abstract**

Hazardous waste is produced substantially by chemical, cosmetic, pharmaceutical, and oil refinery industrial processes. Globally, cost-effective treatment and proper disposal pose serious challenges because of the complex and mutable composition of sludge generated by the chemical and petroleum sectors. Refineries and other chemical businesses produce enormous amounts of oily waste, biological sludge, and chemical sludge waste, which has several detrimental consequences on the environment. Industrial wastewaters are incredibly diverse and specific to each industrial operation. The concentrations of pharmaceuticals in sludge are not regulated due to their micro size. There is no question regarding the detrimental effects of these micropollutants on the environment, as pharmaceutical wastes can significantly affect the health of living organisms. Therefore, appropriate procedures should be used to remove pharmaceutical waste from sludge. Low oil content sludges are hazardous from refineries with less than 40% recoverable oil in them. Before being disposed of, these sludges must be processed and rendered harmless. Encouragement of sustainable resource exploration and exploitation, which include both the treatment and management of these waste items, is crucial today. Waste can be processed in biological, physical, and chemical ways to reduce its environmental pollution to tolerable levels. Industrial wastewater treatment uses activated sludge treatment, encapsulation, and biodegradation in land farming, bioreactors, landfilling, incineration, thermal plasma, low-temperature conversion, etc., in a step-by-step approach to treating industrial wastewater. Generally, the sludge management technique to be explicitly applied depends on the nature and composition of the industrial wastewater and varies from industry to industry. In this chapter, the typical sludge management techniques used in the petroleum, chemical, and pharmaceutical sectors have been highlighted from the angles of sustainability, safety, and sound economics.

**Keywords**

Petroleum · Pharmaceuticals · Chemicals · Sludge management · Environment · Remediation · Health

D. Bhakta · S. Patra · S. Roy (✉)
Department of Chemical Engineering, University of Calcutta, Kolkata, India
e-mail: subhasis1093@gmail.com; srchemengg@caluniv.ac.in

## 1 Introduction

Wastewater treatment facilities are designed and run with the goal of minimizing the harm that human activity causes to the environment and public health. As an outcome of the process of treatment of wastewater, sewage sludge is produced by the main sedimentation tank, secondary sedimentation tank, and various other related operations. Wastewater sludge must be treated and disposed of using a number of complex processes, including concentration, drainage, thermal drying, aerobic and anaerobic digestion, incineration, and disposal in landfills, to prevent environmental harm. Negligent handling of organic waste can lead to significant environmental contamination, including discoloration, the spread of illnesses, and global warming. The difficulty of sludge processing and disposal is getting worse because sewage treatment plants cannot accommodate the amount of sludge that has to be treated (Singh et al. 2011). This problem is made worse by the fast growth of cities and industry that has happened over the last 30–40 years, particularly in developing nations. The goal of treating biosolids, or sludge, is to reduce its bulk and volume as much as possible to reduce disposal costs and limit any potential health risks (Barry et al. 2019). Organic waste has the potential to become among the most readily accessible

© The Author(s), under exclusive license to Springer Nature Switzerland AG 2025

A. Mandpe and M. P. Shah (eds.), *Wastewater Sludge Nexus*, Sustainable Landscape Planning and Natural Resources Management, https://doi.org/10.1007/978-3-031-45758-6_8

resources due to the rising need for organics and renewable energy on a worldwide scale. Because it may be used for recuperating energy or providing nutrients, sludge is an environmentally friendly resource with commercial potential that might be used to create value-added goods (Spinosa et al. 2011). It also includes a significant quantity of renewable organics. For instance, methane from the AD procedure or bio-oil coming from the thermochemical method helps attain energy self-sufficiency; is frequently employed to provide a supplemental source of carbon for natural nitrogen (N) and P elimination procedure in wastewater; and bio-hydrogen from sludge in the fermentation process can be utilized as a pure fuel (Pradel et al. 2016).

Due to its numerous links to other economic sectors, the oil and gas industry is a significant part of the production sector. Any of a broad range of compounds produced from natural gas and crude oil, either directly or indirectly, as outputs that are employed in commerce, are known as petroleum distillates, as well as petrochemicals. The petrochemical industries' effluents are rich in metal derivatives, phenols, polycyclic and aromatic hydrocarbons, surface-active chemicals, sulfides, naphthylenic acids, and other compounds. Water bodies become contaminated with harmful compounds found in effluents as a result of ineffective purification systems, endangering marine and human lives. While petrochemicals provide us with a plethora of beneficial items, they may also pose a threat to human health and the environment. When these compounds are discharged into the atmosphere, they may cause contamination of the soil, water, and air. Important environmental contaminants such as the aromatic compounds found in petrochemicals can enter the ecosystem through a variety of means, including natural oil seeps, production waste byproducts, and releases, storage facility wastes, unintentional spills from pipelines, waste products from the preparation of coal tar, petrochemical wastewater and emissions, etc. One of the main sources of greenhouse gases that cause global warming is the petrochemical sector. Air pollution, acid rain, ozone layer depletion, and other environmental effects are also present. The petrochemical sector releases potentially hazardous materials that are flammable, offensive-smelling, or toxic (Sharma et al. 2017).

Chemical classes used to improve daily living and avoid or treat illnesses are referred to as pharmaceuticals and personal care products (PPCPs). These classes cover a wide range of goods, including over-the-counter drugs like ibuprofen, prescription drugs like metformin, and personal care items like triclosan (TCS) and triclocarban (TCC). Beauty products, toothpaste, detergents, shampoos, and conditioners have all contained the final two TCS and TCC molecules. In addition to being broken down by environmental microorganisms, PPCPs can also be broken down by microorganisms and abiotic chemical reactions in the wastewater treatment process. Transformation products are the end results of the process. There have been reports on the makeup of PPCPs and a few transformation products that compose the influent that enters wastewater treatment facilities worldwide. Waste streams are complex mixtures of trace chemicals that reflect the industry and people they serve. There are difficulties when analyzing the body of scientific literature that is currently available when PPCPs are usually introduced into wastewater systems and treatment plants. This is because individual studies often only involve one or a small number of treatment facilities, and it is difficult to make straightforward comparisons between studies due to differences in the assortment of target analytes (De Boer et al. 2020). The current chapter has examined common sludge management approaches from the perspectives of reliability, security, and good economics in the petroleum, chemical, and pharmaceutical industries. The hazards related to both the industries and their effluents have also been discussed.

## 2 Petrochemical Refinery

As the world's population and economy expand, so does the need for petroleum and other energy resources. By 2035, it is predicted that the world's energy needs will have increased by 37% from their present level, with petroleum products providing a significant percentage of that energy. However, the petroleum sector faces significant hurdles due to a lack of water supplies and worries over environmental damage. It is now well-acknowledged that the production of petroleum energy and water are related. To support the petroleum sector and safeguard the environment, innovative approaches to wastewater treatment and water management are desperately needed. Four broad segments comprise the petroleum industry: extraction, production, and exploration; hydrocarbon processing (petrochemicals and refineries); preservation, transport, and supply; and retail or sales. These four domains are sometimes referred to as downstream, midstream, and upstream processes. While all industry sectors have an influence on the environment, petroleum refineries and petrochemicals plants (PRPPs) utilize the most water and produce the most effluent when processing hydrocarbons. In addition to polluting the environment, releasing these effluents without using efficient treatment methods to separate the oil and water lowers product recovery and throws away the chance of recycling the water to be employed in other PRPP processes (Gong et al. 2017).

Several procedures and activities in PRPP generate wastewater. Certain types of wastewater are produced immediately during manufacturing, including process water, wasted caustic in crackers, vapor condensation, and aromatic

**Fig. 1** Sources of petroleum sludge

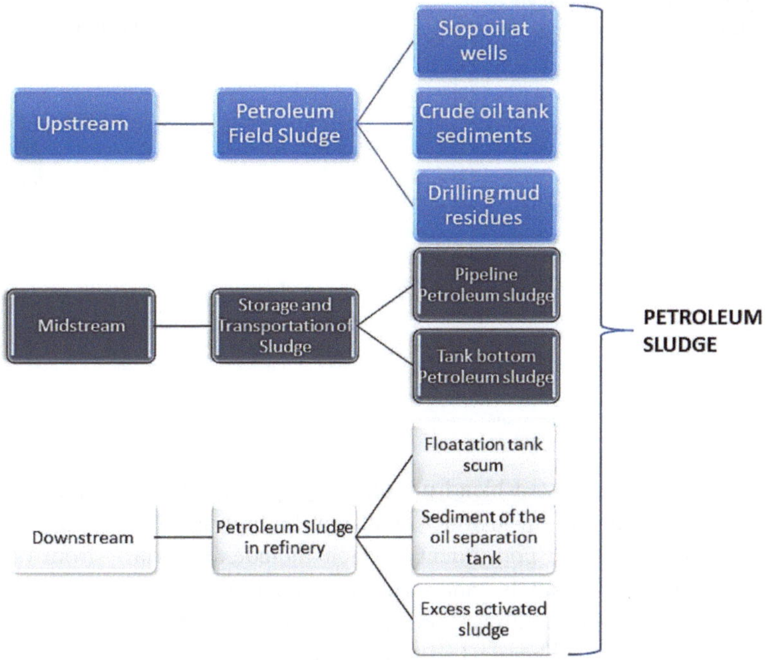


plants. Others come from compressor, pump, and cooling tower blowdown. Another important source of environmental contaminants from PRPP is surface runoff from paved utility area drains that collect rainwater tainted by spilled petroleum products, chemical solvents, and crude oil. Sizable PRPPs can produce a significant amount of sanitary wastewater. These many wastewater kinds can be separated through treatment procedures, but they can also occasionally be mixed together. It is evident that efficient methods for wastewater treatment and reuse are essential to the management of wastewater from the PRPP. These technologies help the petroleum industry by increasing crude oil recovery, lowering and mitigating environmental effects, and offering an extra water supply through water recycling.

## 2.1 Source of Refinery Sludge

Many wastes are produced during the drilling and extraction of petroleum that are considered hazardous waste and cannot be disposed of without treatment. The sections of these waste oils that can no longer be utilized are treated at the refineries' Effluent Treatment Plants (ETPs), while the remaining waste oils are recycled back to the refining facilities. An ETP's main goals are to lessen the amount of oil and grease in these wastes, remove suspended pollutants, and break down dissolved contaminants to bring their concentrations down to allowable discharge limits. The biological, chemical, and physical processes are combined to form the ETPs. Some wastes are produced throughout each step that are referred to as "petroleum refinery sludge," "bottom tank sludge," or "effluent treatment plant sludge." Petroleum sludge from various upstream, midstream, as well as downstream sources, is depicted in (Fig. 1) (Thong et al. 2021).

***Petroleum Refinery Sludge***—Oily sludge, sometimes referred to as petroleum refinery sludge (PRS), is produced when effluent from petroleum refineries is mechanically treated. This process uses physical procedures to separate immiscible fluids (free and emulsified oil) from dense solids in separating tanks such as dissolved air floatation units. Typically, oily sludge is produced when cleaning crude oil storage tanks.

***Chemical Sludge***—Chemical sludge is produced during treatments employing physicochemical procedures like floatation or clariflocculation to remove dissolved oil and lighter solids.

***Biosludge***—Therapies involving biological procedures, such as the activated sludge process, which removes dissolved organic contaminants, produce biosludge.

Due to its complex makeup, which includes a mixture of heavy metals, asphaltenes, long-chain paraffinic wax, sediments, and petroleum hydrocarbons, the sludge produced by petroleum refineries has attracted more attention recently (Bhattacharyya and Shekdar 2003).

## 2.2 Characteristics of Petroleum Sludge

According to the characteristics of the refinery wastewater and the applicable treatment techniques, the volume and attributes of petrochemical sludge might differ from plant to plant. This

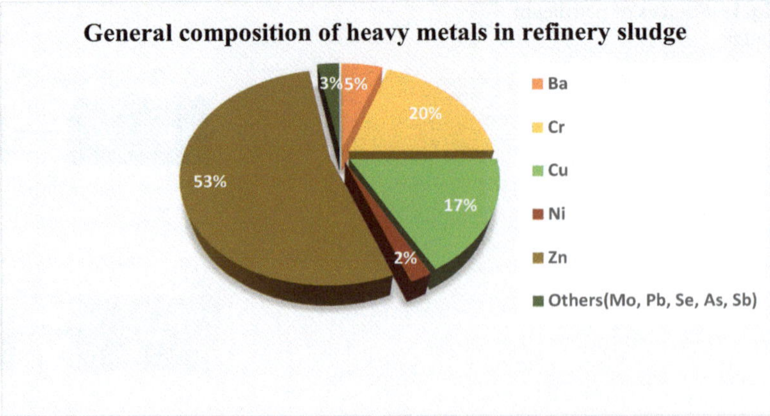


**Fig. 2** General mass percentage of heavy metals in refinery sludge

sludge is a complicated blend of solid components, water-in-oil emulsion, and oil-in-water emulsion. Its high level of hazardous and toxic constituents, which include bacteria, toxic and heavy metals, and organic contaminants like aromatic polycyclic hydrocarbons, petroleum hydrocarbons (PHCs), and certain chemical additives, make it a potentially dangerous solid waste material even though it appears to be a straightforward two-phase system. Huge amounts of sludge have been created in petrochemical companies throughout various stages of petroleum refining and wastewater treatment. Zn, Cr, Cu, Ba, Ni, Mo, and Pb are the amounts of heavy metals from various petrochemical industry sources (https://doi.org/10.5004/dwt.2019.24778) (chart). The amount of sludge produced during petroleum refining processes relies on several variables, including the qualities of crude oil (such as density and viscosity), the refinery's processing strategy, the way oil is stored, equipment and reagents used in the entire process and—most importantly—the capacity of the refinery (Ayol and Yurdakoş 2019).

The general chemical properties of petroleum sludge are as follows-

***pH***—Oily sludges typically have a pH of 6.5 to 7.5.

***Total Nitrogen (TN)***—As the majority of them were identified in the distillate residue, the TN content in oily or petroleum-based sludges is 3% or less.

***Proximate Analysis***—The sludge's composition is generally low in fixed carbon (about 5 wt%), moderate in ash (about 26 wt%), and abundant in volatile matter (about 69 wt%). The amount of carbon contained in biomass that would remain as biochar after pyrolysis is indicated by the fixed carbon content. The sample's volatile organic to fixed carbon ratio (i.e., 69/5) suggests that gasification and pyrolysis might be used to treat the sludge. A greater heating value corresponds to a higher volatile content.

***Ultimate Analysis***—The petroleum sludge generally consists of about 64% C, 6% H, 0.8% N, 0.96% S, and 27.99% O.

***Heavy Metal Composition***—Fig. 2 shows the amounts of heavy metals present in the petrochemical sector that come from various sources. Zn, Cr, Cu, Ba, Ni, and trace amounts of other metals such as Mo, Se, Sb, and Pb are found to be the most prevalent heavy metals.

***Low Heating Value (LHV)***—A petroleum sludge sample's low heating value (LHV) is typically 27 MJ/kg of dissolved sample. With C and H, LHV is proportionate; whereas with N and S, it is the opposite.

***High Heating Value (HHV)***—A petroleum sludge sample's low heating value (LHV) is typically 28 or 29 MJ/kg of dissolved sample with a slight deviation from LHV (Ayol and Yurdakoş 2019).

## 2.3 Classification of Refinery Sludge

The waste produced in refineries can be roughly categorized into the following groups based on source and chemical composition.

- ***Hydrocarbon waste***:—It consists of waste oils/solvents, dissolved air floatation float, tank bottoms, FFU sludge, desalter bottoms, slop oil emission solids, and API separator sludge.
- ***Spent catalyst***:—It consists of a hydroprocessing catalyst, fluid cracking catalyst, and other used inorganic clays.
- ***Inorganic waste***:—It consists of used caustic, used acids, and used amines.
- ***Contaminated soil and solids***:—Sludge used to clean heat exchanger bundles, waste coke, carbon, charcoal, waste sulfur, and various contaminated soils are all included.
- ***Aqueous waste***:—It contains wasted sulfuride solutions, high/low pH water, oil-contaminated water (not effluent), and biomass.

## 2.4 Sludge Treatment Methods

Chemical treatments, different distillation techniques, cracking, hydro-treating, solvent treatment, and bioremediation are only a few of the different strategies used to recover oil from crude sludge and redeem it. The following are a few examples of traditional sludge treatment techniques:

- ***Manual cleaning and incineration***:

The cheapest cleaning approach is manual cleaning. By stepping inside the tank, the cleaning is completed. Sludge is removed from the tank manually or using tanks that have pumps. This process requires a lot of time. This approach makes it challenging to extract the useable hydrocarbons from the sludge.

- ***Solvent extraction method***:

This process makes use of a variety of solvents that can break down the complex molecules in the sludge into its three fundamental components: water, crude oil, and particles. For this procedure, mixing and agitation equipment is needed. Sludge is non-oily and waxy.

- ***Ultra-high-temperature gasification***:

Sludge is thermally oxidized using this technique. Using an oxygen-free plasma arc, the sludge is heated to a very high temperature (1000 °C). This process turns the sludge into pyrogas.

- ***Microwave heating method***:

Through convection, conduction, and heat radiation from the material's surfaces, heat is transported to the substance during traditional thermal heating. While microwave energy works through chemical interaction with the electromagnetic field to reach the materials directly. Thermal gradients cause energy to travel during heat transfer, while microwave heating is an energy conversion process that converts electromagnetic energy to thermal energy. As a result, there are numerous benefits to employing microwaves to process materials. Since microwaves can deposit energy inside of materials, heat can be produced throughout the material's volume. It is feasible to quickly and uniformly heat objects without relying on heat diffusion from the surface because the transfer of heat energy does not depend on the thickness of materials.

- ***Centrifugation method***:

This method uses centrifugal force to separate components according to their densities (such as water, solids, oil, and pasty mixes in oily sludge). The viscosity of oily sludge is reduced using a particularly high-speed spinning apparatus, demulsifying agents, tensioactive chemicals, the injection of steam, and direct heating and steam. After that, heated steam is used to treat the liquid in a pre-treatment tank to lessen its viscosity. Water is combined with this less viscous petroleum for high-speed centrifugation. After centrifugation, the water that contained a high concentration of petrochemical hydrocarbon was drained for additional wastewater treatment. The water and solid-containing separated oil is then transported to a gravimetric separator for additional separation in order to recover the oil. Water from the separator that has been separated is transported for sewage treatment. Centrifugation and separator sediments are collected as a solid residue for further processing. For the treatment of oily sludge, centrifugation is a relatively clean and well-established technology, and it is successful at separating the oil from the sludge.

- ***Electrokinetic method***:

In this procedure, an electrode pair is placed on either side of a porous medium, and a small amount of direct current is conducted through the medium, causing the liquid phase to move electro-osmotically, ions to move, and charged particles to move electrophoretically in a colloidal system. Specific electrodes: Water, oil, and solids can be separated from oily sludge using based on three mechanisms and performed using the electrokinetic approach to the electric field, colloidal aggregates in oily sludge can be disrupted, which results in the As a result, greasy sludge solid and colloidal particles flow toward the anode using electrophoresis. Electro-osmosis causes water and oil to flow in the direction of the cathode. Near the anode, the separated solid phase electrocoagulates, increasing the concentration of the solid phase and the sediments.

- ***Froth flotation method***:

Oily sludge and effluent from refineries can be treated using the froth flotation process. This process creates an oily sludge slurry by combining water with oily sludge. Air bubbles are created in the water-sludge combination when air is circulated through the sludge slurry. These air bubbles in the slurry mixture approach the oil droplets. Oil spreads into the air bubbles when the water coating separating the oil from it becomes so thin that it ruptures. In a water-oil mixture, an accumulation of oil droplets with air bubbles can rise swiftly to the top, where it can be skimmed off and collected for further filtration.

- ***Ultrasonic irradiation***:

Compressions and rarefactions are caused by ultrasonic waves in the medium they travel through. By forcing molecules closer together, the compression cycle applies positive pressure to the medium. In an emulsion, smaller droplets move more quickly than bigger ones when exposed to ultrasonic radiation. This may enhance the frequency of their collisions to form aggregates and droplet coalescence then encourage the separation of the water and oil phases.

## 2.5 Refinery Sludge Utilization Techniques

Multiple alternatives have been proposed for beneficiation, recycling, and advancement of petroleum sludge as well as sludge management techniques. Some of the common techniques have been enlisted in Table 1 below.

## 2.6 Sludge Disposal Method

After oil has been recovered, oily sludge should be disposed of using a variety of techniques, including combustion, stabilization/solidification, oxidation, and biodegradation. The safe disposal of oily sludge in the environment is one of the main issues that oil refineries encounter.

- ***Incineration***:

  This approach involves completely burning oily wastes while using extra air and auxiliary fuels. Rotary kilns and fluidized bed incinerators are the most frequently utilized incinerators. For rotary Kelvin, the combustion temperature range is 980–1200 °C, and the residence time is about 30 min. The combustion temperature in a fluidized bed incinerator can range from 730 to 760 °C, and the residence period can be measured in days. Due to its fuel flexibility, high mixing efficiency, high combustion efficiency, and comparatively low pollutant emissions, the fluidized bed incinerator is more efficient for low-quality sludge.
- ***Stabilization/solidification***:

  By changing the pollutants into a less harmful or soluble state (stabilization), contaminants are immobilized using this technique. Solidification can be used to encapsulate the pollutants. This procedure makes it simple to dispose of inorganic trash. Organic wastes don't work as well with this technique.
- ***Oxidation method***:

  Adding reactive chemicals to oily wastes causes chemical oxidation, which converts organic components into non-hazardous products like inorganic salts or carbon dioxide and water.
- ***Bioremediation***:

  This sludge is less biodegradable. Utilizing bacteria, fungi, certain algae, and plants, this method reduces or gets rid of hazardous contaminants. These microorganisms could exist naturally or could be created in a lab. They either consume the contaminants (organic compounds) or absorb all dangerous substances (heavy metals) into themselves from the environment. Utilizing fertilizers, compost, bulking agents, and some chemicals, such as oil dispersants, can improve bioremediation. In several regions of the world, bioremediation of waste oil in soil (land farming) or even land spreading has been practiced.
- ***Land farming method***:

  Farming on land is more cost-effective and environmentally friendly. Inland spreading, the sludge is equally spread across a plot of land, where it can be broken down over the course of months or years by the local microbial flora. Utilizing tilling tools, the sludge is incorporated into the soil, and fertilizers speed up the rate of soil deterioration in farming. In land spreading and land farming, biodegradation, vaporization, oxidation, and, to a lesser extent, degradation by sunshine and leaching, are the main processes by which hydrocarbons disappear. The area of land might be used once more for additional sludge treatment once the sludge has been significantly degraded (Islam 2015).

# 3 Pharmaceutical and Personal Care Products

The potential for sludge agricultural reuse to transmit developing organic pollutants, including pharmaceuticals, to the soil system has drawn attention to the existence of sewage sludge containing these contaminants. The fate of pharmaceuticals in soil has been shown in Fig. 3. Therefore, the presence of pathogens, unstable organic materials, and organic and inorganic contaminants may restrict the direct agricultural reuse of sewage sludge.

Large amounts of organic sludge are produced during industrial fermentations used to produce a variety of helpful

**Table 1** Various utilization techniques of refinery sludge

| Utilization technique | Brief explanation |
|---|---|
| Anaerobic digestion (AD) of refinery waste sludge (Choudhury et al. 2022) | AD may result in the reduction of solids and the recovery of energy for advantageous reuse from industrial sludges. This system works well for treating a variety of agricultural and animal wastes in addition to municipal wastewater treatment sludge. Numerous studies have documented the effective generation and improvement of biogas from oil refinery discharges using different AD technology procedures. It has been discovered that AD technology without pretreatment has an extremely lengthy hydraulic retention time (HRT) and produces relatively little biogas. This may be due to the difficulty in producing methane in the absence of an appropriate inoculum, which is made up of prospective bacteria capable of breaking down the complicated structure of sludge, which is made up of cyclic and polycyclic aromatic hydrocarbons, alkanes, alkenes, and branched alkanes. The benefit of using pre-treatment procedures in refinery sludge is that they can increase solubilization and separate the solid, water, and oil fractions |
| Anaerobic co-digestion with food waste and swine manure (Lee et al. 2022) | Anaerobic microorganisms can be impeded by hazardous compounds such as heavy metals along with oil components in oil sludges, and the oil sludges themselves have a low anaerobic degradability. One potential solution to address the aforementioned issues is to co-digest the oil sludges along with organic wastes like food scraps and swine dung. Positive benefits like moisture management, balancing of macro-or trace-nutrients, and dilution of harmful compounds can be obtained via co-digestion, or combining two or more substrates for AD. Co-digestion can also result in synergistic effects, which suggest greater efficiency than that of digestion of a single substrate |
| Microbial electrochemical technologies (METs) for refinery wastewater treatment (Priyadarshini et al. 2022) | Petrochemical Wastewater often contains organic materials with chemical bonds that capture more than 7.6 kJ/L of chemical energy. Consequently, it is important to capture and apply this chemical energy for the treatment of wastewater. As a result, a class of cutting-edge technologies officially known as microbial electrochemical technologies (METs) has emerged. In addition to treating wastewater, the METs may also generate energy or other high-value products like methane and hydrogen. The PW often contains aliphatic and aromatic hydrocarbons, which METs can efficiently break down in their various chambers while also converting the chemical energy contained in these hazardous substances into bioelectricity or other assets. Furthermore, by using the appropriate cathode catalysts, the cathodic chamber of METs may also be used to create chemicals and fuels, while the anodic chamber can be used to remove organic pollutants from PW that are resistant to removal. Additionally, METs directly convert chemical energies into bioelectricity by using electrogenic bacteria as inexpensive catalysts for separating away the bonds in chemical substances |
| Construction materials (Mohd Pauzi and Ahmad Shah 2022) | By lowering the fire temperature and reducing the quantity of water, clay, and fuel needed, the utilization of sludge in brickmaking enables a decrease in the quantity of ingredients and energy used in the process. Additionally, Gorbaev et al. investigated the viability of utilizing waste oil sludge as a combustible addition in the brick-making process. They found that the bulk density of the bricks would drop by up to 25% due to the impurities and organic materials present in oil sludge. This would provide thinning and combustible qualities, resulting in the creation of lightweight bricks. The creation of building materials for roads is an additional means of recycling petroleum waste |
| Technology advancement (Jafarinejad and Jiang 2019) | New methods to boost the Petrochemical treatment plant's efficiency and discharge water's quality that is suitable for reuse are required due to the rise in petroleum-based products, water shortages, and harsher environmental laws. Membrane separation is one of the many technologies under investigation that has promise because of its comparatively low cost and notable effluent water quality improvement. Finally, the economic, environmental, technical, and regulatory restrictions should all be taken into consideration when choosing a Petrochemical wastewater treatment system |

medications; this sludge is typically disposed of via landfilling and cremation (Fytili and Zabaniotou 2008). Actually, there is still no data to support their appropriateness for this usage in terms of macronutrient composition, phytotoxicity, and organic matter stability, which limits their agricultural reuse. Composting is thought to be the most effective method for raising sludge quality. The easily accessible organic materials are broken down or changed into stable humic-like molecules throughout the composting process, which increases the final compost's maturity and stability. Nearly all infections are rendered inactive by the elevated temperatures (55–70 °C) attained throughout the procedure and their maintenance over several days. Furthermore, during the composting of sewage sludge, a drop in the concentration of organic contaminants and a decline in the bioavailability of metal traces have also been documented (Bin et al. 2013). Due to its growing volume and the presence of contaminants including heavy metals and pharmaceutical and personal care products (PPCPs), treating sewage sludge is becoming an issue in the world. Currently,

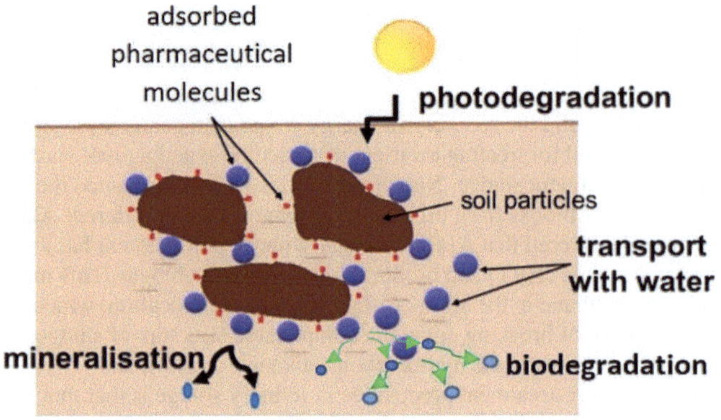


**Fig. 3** Phases a pharmaceutical waste undergoes in soil (Reprinted with permission from Bolesta et al. (2022))

incineration or the use of sludge as fertilizer in agriculture accounts for more than 70% of sludge treatment. New thermochemical techniques are now being developed, and they should soon be put into use (Tarpani et al. 2020). Along with raw sewage, pharmaceuticals enter the treatment facility (Sellier et al. 2022). In the 1960s and 1970s, shortly after several synthetic chemicals had been widely released into the environment, adverse impacts on the ecosystem and human health began to emerge. Since then, those chemicals that are harmful and/or persistent in the environment have been the main focus of the assessment of the effects of chemical pollution. Examples include organic pollutants (POPs), which include polychlorinated biphenyls (PCBs), polycyclic aromatic hydrocarbons (PAHs), polychlorinated dibenzo-p-dioxins (PCDDs), and polychlorinated dibenzofurans (PCDFs), as well as organochlorine pesticides like DDT (Cucina et al. 2018).

## 3.1 Source of Pharmaceutical Sludge

***Human and Veterinary***—Agricultural practices and livestock are two main sources of pharmaceutical poisoning. Antibiotics used in veterinary medicine and aquaculture may leave residues in meat and fish products. Numerous investigations discovered that the concentration of antibiotics in animals was higher than the allowable maximum residual limit. This suggests that pharmaceutical and cosmetic products are frequently misused and that the necessary withdrawal period was not given (Vishnuraj et al. 2016).

***Hospitals***—They discovered that the overall antibiotic burden was borne more by general consumers, with just 13 and 28% of the total coming from 16 hospitals in the summer and winter, respectively. Another analysis of the waste waters of four healthcare facilities in Portugal with wastewater treatment plant influents demonstrates that hospitals contribute less to the overall tetracycline concentration in the towns, with the exception of minocycline. The range of TC in Iran's hospital effluent was 516–1240 ng $L^{-1}$. It is evident that hospitals have a major role in the environmental pollution caused by pharmaceuticals (Scaria et al. 2021).

***Pharmaceutical industries***—Significant research has been performed over the last year in the rapidly expanding area of pharmaceuticals and personal care products (PPCP) in sludge and biosolids. A study by McClellan and Halden (2010) examined 110 composite specimens for 72 PPCPs from the National Sewage Sludge Survey performed in 2001. Of the 72 analytes, they discovered that at least one was present in 54% of the samples, with triclocarban and triclosan being the most prevalent. Additionally, they calculated that 210–250 metric tons of PPCPs are loaded annually into American soils via the reprocessing of biosolids. When Snyder et al. (2010) looked into triclocarban in biosolids, they discovered that 23 samples had a mean content of 19 ± 11 mg/kg. The analytical techniques for measuring PPCPs have also advanced (Brisolara and Qi 2011).

## 3.2 Classification of Pharmaceutical Sludge

***Class A***—According to the US Environmental Protection Agency's (EPA) standards 40 CFR Part 503, Class A sewage sludge is referred to as sewage sludge that has gone through treatment to eliminate infectious agents, comprising pathogenic microbes, enteric viruses, and viable helminthes ova, to detectable levels. Class A sewage sludge does not require limits on location or harvesting in order to be disseminated for use as a soil supplement. The Part 503 Biosolids Rule Class A pathogen reduction standards can be satisfied, for example, by using treatment techniques including the process of composting thermal drying, as well as extremely high-temperature aerobic digestion.

***Class B***—Sewage sludge classified as class B has fewer pathogens per unit of material. Class B wastewater sludge can continue to contain certain pathogens, thus for a while after usage, there must be site limitations limiting public access, agricultural harvests, and animal grazing. Class B waste sludge is unable to be sold or distributed for land application at public contact places in bags or other containers, in contrast to Class A sewage sludge. If Class B sewage sludge complies with 40 CFR Part 503's restrictions on metallic substances, vector attraction diminution, and other management criteria, it can be used in bulk at suitable land application locations, such as woods, reclamation sites, and fields for agriculture (Jones-Lepp and Stevens 2007). Figure 4 represents the influent and effluent concentrations of different pharmaceutical components from a sewage treatment plant in Tokyo, Japan.

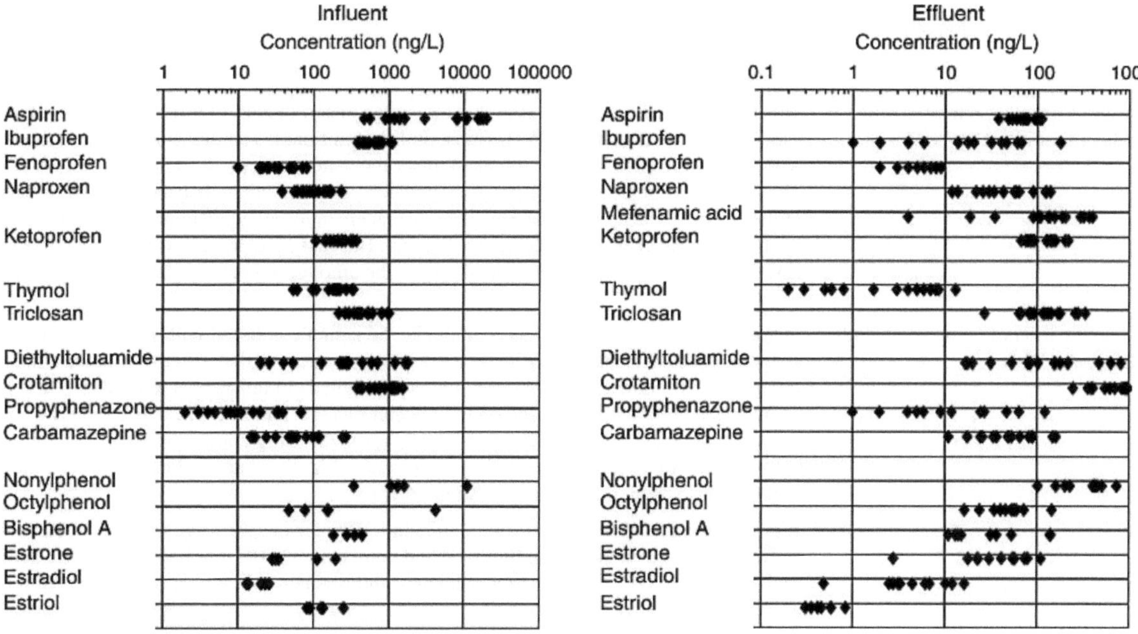


**Fig. 4** Pre-treatment and post-treatment pharmaceutical concentrations of wastewater from a sewage treatment plant in Tokyo (Reprinted with permission from Nakada et al. (2006))

### 3.3 Treatment and Disposal of Pharmaceutical Sludge

According to estimates, wastewater treatment facilities effectively dispose of more than 70% of relatively hydrophobic contaminants from classes including heavy metals, persistent organic pollutants (POPs), brominated flame retardants, and some personal care products (PCP), surfactants, plastic additives, hormones, some PCP, some medicines, and household chemicals. The physicochemical characteristics of micropollutants, such as molecular weight, octanol–water partition coefficient ($K_{ow}$), acid dissociation constant (pKa), solubility, or biodegradability, determine how they behave in a typical municipal sewage treatment facility (Bolesta et al. 2022). Table 2 illustrates some of the ways to treat the pharmaceutical components in wastewater sludge.

## 4 Hazards Associated with Petrochemical Industry and Personal Care Products

### 4.1 Petrochemical Industry

The petrochemical industry is essential to several manufacturing sectors. This industry discharges a lot of harmful and poisonous compounds as effluents. Solid waste that is challenging to handle and dispose of is produced by releasing pollutants into the atmosphere. When released, the majority of the petrochemical compounds may have negative effects on the environment, including soil, water, and air pollution. The Importance of aromatic compounds found in petrochemicals environmental toxins that could be brought into environmental effects of natural oil leaks and industrial waste products and pollutants, waste from oil storage, and unintentional spills from coal tar processing wastes, petrochemical waste, and oil tankers industrial emissions and effluents, etc. the petrochemical sector is a significant generator of the main greenhouse gases. In charge of causing global warming. Additional environmental effects include the destruction of the ozone layer, acid rain, air pollution, etc. Chemicals that could be dangerous from the petrochemical industry discharge anything flammable, offensive-smelling, or poisonous. The petrochemical sector may also result in environmental degradation and biodiversity (Yu et al. 2006). Polycyclic and aromatic hydrocarbons, phenols, metal derivatives, surface-active compounds, sulfides, naphthylenic acids, and other chemicals are widely present in petrochemical industry effluents (Israel et al. 2008). Water bodies become contaminated by toxic compounds that are present in effluents due to ineffective purification systems, which is harmful to both aquatic and human life (López-Navarro et al. 2013).

Petrochemical exposure can happen in a variety of ways, including by ingestion or skin absorption. Additionally, they can damage the brain, nerves, and liver, cause birth abnormalities, cancer, asthma, and hormone imbalances, and have an impact on human existence by building up in tissues and organs. The long-term repercussions of acne include

**Table 2** Schematic presentation of pharmaceutical components of wastewater sludge

| Treatment technique | Process description |
|---|---|
| Agricultural application of anaerobically digested sludge (Yoshida et al. 2013) | This technique involves digesting thicker sludge anaerobically to produce biogas, which is then used to keep the digester at 35 °C (mesophilic conditions). The digested sludge is then combined with a dewatering polymer. In order to lower the water content, the agent is routed to a belt filter. The product, which has a 24% dry matter (DM) content, is then distributed to be utilized as a replacement for artificial fertilizers. The displacement of an equivalent volume of synthetic NPK 15-15-15 fertilizer was ascribed to the system. Based on the processed sludge's 16 kg/1000 kg DM phosphorus and nitrogen concentration, the amount of synthetic fertilizer that was required was calculated |
| Agricultural application of composted sludge (Sablayrolles et al. 2010) | In this approach, sludge that has been thickened is first combined with a bulking agent, like wood chips or sawdust. These weren't taken into account because they usually constitute waste. The mixture is subsequently decomposed under controlled circumstances to produce compost with the required composition. Based on the phosphorus level, the system was credited with producing an equivalent amount of synthetic fertilizer using compost instead. The biosolids are seen as being of good quality, similar to the anaerobic digestate application, with a minimal runoff contribution to the eutrophication potential |
| Incineration | A polymer is added to thicken sludge to help it dewater in centrifuges and boost its calorific value. The sludge is burned in fluidized bed combustors (FBCs) at 850 °C with the addition of fuel and lime to boost combustion efficiency and regulate acid gases, respectively, after achieving the DM concentration of 35% in the centrifuges. Both energy and heat are produced through incineration, and the system was credited with producing both. Bottom ash is dumped in a sanitary landfill, and fly ash is disposed of as hazardous trash |
| Pyrolysis | This method calls for the first filter pressing to get the thickened sludge's water content down to 30% DM. Depending on the pyrolysis temperature (300–900 °C), the sludge is subsequently dried and pyrolyzed to create syngas, tar, and char. The produced syngas are utilized to facilitate the pyrolytic process, and any surplus can be exported. Here, it was supposed to be utilized as a natural gas substitute for heating purposes, for which the system was given credit. The comparable amounts of heavy fuel oil and charcoal were also attributed to the generation of tar and char |
| Wet air oxidation (Foglar 2003) | The thickened sludge is pumped to the reactor following the addition of a dewatering polymer, where it will be combined with oxygen at a high temperature and pressure (in this example, 235 °C and 40 bar). A carbon-rich effluent that can be used for denitrification is the process' output. Methanol substitutes include wastewater treatment facilities. The inclusion of a catalyst due to a paucity of data, (copper sulfate) was eliminated. The volume of based on the projected amount of methanol that was prevented by utilizing the wastewater, Methanol's energy content (35 MJ/kg), and the energy recovered by the (7.5 GJ) procedure |
| Vermicomposting of pharmaceutical waste with cattle dung using *Eisenia fetida* (IndraKumar Singh et al. 2022) | Cattle dung makes a great addition to allopathic pharmaceutical business waste since it is a great source of earthworm feed. The earthworms fared extremely well to the AS75 feed combination that contained cattle manure. There are additional nutrients (N, P, and K) in the final vermicompost. The final vermicompost was found to be less dangerous and stable based on the results of the genotoxicity and phytotoxicity tests. The vermicompost was stable and mature, as shown by SEM and FTIR measurements, suggesting that the earthworms and their gut microorganisms mineralized the meal components during the vermicomposting process. Ultimately, the best combination of all the feed mixes was found in AS50. Therefore, it can be concluded that an economical and environmentally beneficial process (vermicomposting) using the right amount of cattle manure and *Eisenia fetida* could convert allopathic wastewater into a nutritious, detoxified, robust, and full-grown vermicompost for use in agriculture. Additionally, it could act as a springboard for future sludge management approaches used by the allopathic pharmaceutical sector |
| Membrane processes (Lan et al. 2019) | A variety of membrane technologies, including as reverse osmosis, nanofiltration, and ultrafiltration, are extensively used as a reliable and affordable mature method for eliminating impurities from water. It was discovered that the traditional biological method in conjunction with membrane technologies (reverse osmosis and nanofiltration) effectively reduced the number of tetracycline-resistant genes in swine effluent. Osmotic pressure is used in forward osmosis, an energy-efficient natural process that transfers water molecules from a diluted Tetracycline-containing mixture to a concentrated brine, in contrast to pressure-driven membrane technologies like RO and NF. This approach has the potential to be used for separating and recovering pharmaceuticals from wastewater due to its minimal fouling of membranes and high material recovery capabilities |

(continued)

**Table 2** (continued)

| Treatment technique | Process description |
|---|---|
| Biochar in sludge management (Mamera et al. 2021) | Because biochar may bind many inorganic and organic contaminants, it is a good solution for managing fecal sludge. The quality of water resources may increase if biochar is used as a cheap adsorbent in pit latrine sludge treatment. Additionally, sludge that has been altered with biochar may be used again as an affordable by-product |
| Carbon as an adsorbent for pharmaceutical components (Bouhcain et al. 2022) | As a type of adsorbent, carbon extracted from argon nutshells was tested for its elimination of two newly discovered pollutants that were used as model pollutants: coffee and diclofenac, a pain-relieving medication. For diclofenac and caffeine, the maximum adsorption capacities were around 126 and 210 mg per gram of activated charcoal, respectively. A pseudo-second-order kinetic model was used to explain the adsorption process, and thermodynamic variables demonstrated that the procedure was endothermic in the case of caffeine and spontaneous and exothermic in the case of diclofenac |
| Advanced oxidation process (Deng and Zhao 2015) | The widely researched method used for the degradative elimination of pharmaceuticals along with other persistent organic contaminants is the advanced oxidation procedures (AOPs), which are regulated by exceptionally reactive oxidants, primarily hydroxyl radicals. The remarkable benefit of using these processes, despite their high cost, is the potential for harmful pollutants to degrade and mineralize into less toxic or even non-toxic compounds. There are several AOPs that may be selected based on their remedy aim and the properties of raw or wastewater based on the technique of the radical production pathway. The most often used AOPs for the decomposition of pharmaceuticals are UV/$H_2O_2$, the ozonation Fenton/photo-Fenton, oncolysis, and heterogeneous photocatalysis |
| Date palms (Bhakta et al. 2023) | Samir et al. used date stem as a precursor for calcination in order to create activated carbon (AC). A hydroxylation technique was used to change the AC and produce AC-OH by increasing the number of hydroxyl groups on the surface. Additionally, AC was injected into a $TiO_2$ solution to create AC-$TiO_2$, ensuring that photodegradation occurred. On the basis of atenolol (AT) and propranolol, the resulting materials were assessed for the removal of pharmaceutical contaminants (PR). The elimination of AT and PR was found to be 92% achieved by adsorption and 94% through photodegradation, according to the results. When the two procedures were compared, adsorption turned out to be more effective in eliminating pollutants from water due to its low energy consumption. This suggested that AC-OH was an inexpensive and ecologically benign material that could be used for wastewater treatment in industries |

skin inflammation, ulceration, and allergic dermatitis (Sharma et al. 2017).

A complex blend of several organic components makes up petroleum. These substances include monocyclic and polycyclic aromatic hydrocarbons as well as straight-chained, branched, and cyclic aliphatic hydrocarbons. Typically, oils with longer carbon chains and more benzene rings have higher levels of toxicity. Besides, lead and sulfur mixture oil is quite dangerous as well. The most poisonous petroleum product is benzene, which is also a dangerous carcinogen. Other extremely poisonous chemicals, like benzene, ortho- and para-xylenes, as well as toluene (methylbenzene). Despite differences in the toxicity of various forms of Oil, all petroleum-derived goods are bad for people's bodies, and the environment. A significant source of hazardous and toxic air pollutants like benzene, toluene, xylene, and ethylbenzene is petroleum refineries. They are also significant emitters of air pollutants such as sulfur dioxide, nitrogen oxides, and carbon monoxide (CO). Volatile aliphatic hydrocarbons, sulfur dioxide ($SO_2$), and hydrogen sulfide ($H_2S$). Additionally, the combination of volatile hydrocarbons and nitrogen oxides aids in the creation of ozone (VAH), as well as particles (PM). The variety of sources that wastewater in refineries can encounter (including equipment leaks, spills, and desalination of crude oil), may be extremely contaminated. This polluted water may be stormwater or process wastewater from desalination, distillation, cooling tower water, or cracking. It might contain harmful wastes like lingering oil (Mostafa 2021).

### 4.2 Pharmaceutical and Personal Care Products (PPCPs)

The values of acceptable daily intake (ADI) were computed for every plant tissue reported PPCP. The quantity of PPCP that a person may ingest on a daily basis throughout their lifetime without experiencing negative effects is known as the ADI value. The smallest possible adult daily therapeutic dosage (mg/day) was divided by a factor of safety of 1000 and a body weight of 70 kg to produce the acceptable daily intake (ADI) values for medicines used in human treatment. According to DEFRA (2007), the National Health and Medical Research Council of Australia (2008), and WHO (2011), the safety factor of 1000 is made up of three factors of ten that are applied to account for variations in human reaction, the possible sensitivity of demographic subgroups (such as kids and infants), and the smallest daily therapeutic dose that is not a dose that reflects no effect (Prosser and Sibley 2015).

Compounds that show endocrine system action (e.g., progesterone, testosterone) were given an extra safety factor of 10. A factor of safety of 300 was applied to a no visible negative impact level of 25 mg/kg weight of the person (bw)/day in order to determine the ADI levels for triclocarban and triclosan. The 90-day oral toxicology research with rats as well as the 2-year oral toxicity study with rats that produced no visible negative impact for triclosan and triclocarban, respectively, were conducted by the European Commission's Health and Consumer Protection Directorate-General in 2005 and by Health Canada and Environment Canada in 2012. According to the European Commission Health and Consumer Protection Directorate-General (2005) and Health Canada and Environment Canada (2012), the safety value of 300 corresponds to a value of 3 for a restricted set of studies and a value of 10 for each intraspecies and interspecies heterogeneity. The Food and Agriculture Organization of the United Nations (2011) gave the values that served as the basis for the ADI standards for veterinary medicines (Andersson et al. 2018). The ADI values for the synthetic perfumes galaxolide and tonalide have been determined by applying a factor of safety of 300 to no observable impact limitation values of 20 and 5 mg/kg/day, respectively. On the basis of a value supplied by the US Food and Drug Administration (USFDA 2010), the ADI value for the veterinary antibacterial flunixin is determined. The dietary developmental research and the 90-day oral toxicology test conducted on rats yielded the no effect limitation values for galaxolide and tonalide, accordingly (European Chemicals Agency, 2008a, b). For the synthetic scent ambrettolide, in which a no-impact limitation value was not possible to determine, the identical ADI value utilized for tonalide was likewise used. Since tonalide had a value that was lower than galaxolide and hence indicated the more cautious option, it was chosen to adopt its ADI value. The European Food Safety Authority's (2008) level of concern served as the basis for the ADI estimate of caffeine (European Food Safety Authority's 2009). The estimated daily intake (EDI) amount can be calculated as:

$$EDI = \frac{Cfood * IRveg * \beta\frac{\mathrm{g}}{\mathrm{cup}} * \beta\frac{\mathrm{ww}}{\mathrm{dw}}}{m}$$

The PPCP content in plant tissue is represented by $C_{food}$ (ng per g of dry weight). Dry weight is represented by dw. $IR_{veg}$ stands for 2.80 cup equivalents of vegetables per day, or the 95th percentile of vegetable consumption rates (dry beans and peas not included). The USEPA Risk Considerations Handbook was consulted to determine the appropriate consumption rates for rice and wheat grain, which were 24 and 186 g/day, correspondingly. It was believed that the water content of both these grains was 14% and 11.6%, respectively. The weight of a cup of freshly harvested tissue diced to less than 0.5 $cm^3$ was measured in the lab and is represented by the symbol βg/cup. The majority of a cup of fresh collard was substituted with the weight of a cup of cabbage. The USEPA uses βww/dw, or mean wet-to-dry conversion factor, for plant tissues for determining soil screening values (i.e., 0.085). The symbol m stands for the mean weight of an adult (20 to < 65 years), or 76.6 kg, and a child (1 to < 4 years), or 15.4 kg (Miralbés 2004). One method that people might become exposed to PPCPs is via consuming plant tissue that has been infected with them. In light of the possibility that people may be exposed by other routes, such as ingestion of other matrices, cutaneous contact, or inhalation, a hazard quotient of $\geq 0.1$ was designated as a possible risk to human health for the purposes of this study (Pan et al. 2014).

## 5 Conclusion

A prominent source of pollution is petrochemical processing plants, which often house polluting chemical companies alongside oil refineries near industrial poles. The quality of life is compromised by living close to these locations, which worsens the ecological circumstances of their immediate surroundings. Contamination exposure may lead to uncertainty, anguish, and psychological anxiety, all of which have a variety of negative consequences on health and well-being. Researchers launched a concentrated effort to comprehend the existence and effects of PPCPs being discharged into the atmosphere ten years prior to the beginning of the twenty-first century. There are a few documented cases of endocrine disruption in both aquatic and terrestrial creatures, but there aren't many that directly link to health issues for humans. Primary PPCP accumulating in crops doesn't seem to pose a significant risk to human health. On the other hand, the discovery of bacteria and genes resistant to antibiotics, together with the existence of traces of antibiotics, raises the possibility of unintentional harm to human health. In particular, people who encounter this sort of industrial complex attempt to make reasonable assessments of the hazards they face, but their knowledge of the matter is frequently inadequate or lacking. This chapter has thrown light on the major sources of industrial sludge, management strategies of effluents, and the potential hazards related to the petroleum and PPCP industries.

## References


Andersson N, Arena M, Auteri D et al (2018) Guidance for the identification of endocrine disruptors in the context of regulations (EU) No 528/2012 and (EC) No 1107/2009. EFSA J 16. https://doi.org/10.2903/j.efsa.2018.5311

Ayol A, Yurdakoş ÖT (2019) Chemical and thermal characteristics of petrochemical industrial sludge. Desalination Water Treat 172:29–36. https://doi.org/10.5004/dwt.2019.24778

Barry D, Barbiero C, Briens C, Berruti F (2019) Pyrolysis as an economical and ecological treatment option for municipal sewage sludge. Biomass Bioenergy 122:472–480. https://doi.org/10.1016/j.biombioe.2019.01.041

Bhakta D, Gayen D, Bhattacharjee J, Chatterjee R, Dastidar P, Roy S (2024) Repurposing industrial and agricultural wastes as sustainable alternatives in photovoltaic cells and dye degradation. ChemistrySelect 9(38):e202403145

Bhattacharyya JK, Shekdar AV (2003) Treatment and disposal of refinery sludges: Indian scenario. Waste Manage Res 21:249–261. https://doi.org/10.1177/0734242X0302100309

Bin HY, Zakaria MP, Latif PA, Saari N (2013) Degradation of veterinary antibiotics and hormone during broiler manure composting. Bioresour Technol 131:476–484. https://doi.org/10.1016/j.biortech.2012.12.194

De Boer J, Garrigues P, Gu J-D et al (2020) The handbook of environmental chemistry

Bolesta W, Głodniok M, Styszko K (2022) From sewage sludge to the soil—transfer of pharmaceuticals: a review. Int J Environ Res Public Health 19

Bouhcain B, Carrillo-Peña D, El Mansouri F et al (2022) Removal of emerging contaminants as diclofenac and caffeine using activated carbon obtained from Argan fruit shells. Appl Sci (Switzerland) 12. https://doi.org/10.3390/app12062922

Brisolara KF, Qi Y (2011) Biosolids and sludge management. Water Environ Res 83:1351–1375. https://doi.org/10.2175/106143011x13075599869416

Choudhury SP, Saha B, Haq I, Kalamdhad AS (2022) Use of petroleum refinery sludge for the production of biogas as an alternative energy source: a review. In: Advanced organic waste management: sustainable practices and approaches. Elsevier, pp 277–297

Cucina M, Tacconi C, Sordi S et al (2018) Valorization of a pharmaceutical organic sludge through different composting treatments. Waste Manage 74:203–212. https://doi.org/10.1016/j.wasman.2017.12.017

Deng Y, Zhao R (2015) Advanced oxidation processes (AOPs) in wastewater treatment. Curr Pollut Rep 1:167–176

European Food Safety Authority (EFSA) (2009) Flavouring group evaluation 49, (FGE.49): Xanthin alkaloids from the Priority list from chemical group 30. EFSA J 7. https://doi.org/10.2903/j.efsa.2009.741

Foglar A (2003) Wastewater denitrification process-the influence of methanol and kinetic analysis

Fytili D, Zabaniotou A (2008) Utilization of sewage sludge in EU application of old and new methods—a review. Renew Sustain Energy Rev 12:116–140

Gong Z, Du A, Wang Z et al (2017) Experimental study on pyrolysis characteristics of oil sludge with a tube furnace reactor. Energy Fuels 31:8102–8108. https://doi.org/10.1021/acs.energyfuels.7b01363

IndraKumar Singh S, Singh WR, Bhat SA et al (2022) Vermiremediation of allopathic pharmaceutical industry sludge amended with cattle dung employing Eisenia fetida. Environ Res 214. https://doi.org/10.1016/j.envres.2022.113766

Islam B (2015) Petroleum sludge, its treatment and disposal : a review

Israel AU, Obot IB, Umoren SA et al (2008) Effluents and solid waste analysis in a petrochemical company—a case study of Eleme petrochemical Company Ltd, Port Harcourt, Nigeria. E-J Chem 5(1):74–80. ISSN: 0973-4945

Jafarinejad S, Jiang SC (2019) Current technologies and future directions for treating petroleum refineries and petrochemical plants (PRPP) wastewaters. J Environ Chem Eng 7. https://doi.org/10.1016/j.jece.2019.103326

Jones-Lepp TL, Stevens R (2007) Pharmaceuticals and personal care products in biosolids/sewage sludge: the interface between analytical chemistry and regulation. In: Analytical and bioanalytical chemistry, pp 1173–1183

Lan L, Kong X, Sun H et al (2019) High removal efficiency of antibiotic resistance genes in swine wastewater via nanofiltration and reverse osmosis processes. J Environ Manage 231:439–445. https://doi.org/10.1016/j.jenvman.2018.10.073

Lee C, Kim S, Park MH et al (2022) Valorization of petroleum refinery oil sludges via anaerobic co-digestion with food waste and swine manure. J Environ Manage 307. https://doi.org/10.1016/j.jenvman.2022.114562

López-Navarro MÁ, Llorens-Monzonís J, Tortosa-Edo V (2013) The effect of social trust on citizens' health risk perception in the context of a petrochemical industrial complex. Int J Environ Res Public Health 10:399–416. https://doi.org/10.3390/ijerph10010399

Mamera M, van Tol JJ, Aghoghovwia MP et al (2021) Potential use of biochar in pit latrines as a faecal sludge management strategy to reduce water resource contaminations: a review. Appl Sci (Switzerland) 11

McClellan K, Halden R (2010) Pharmaceuticals and personal care products in archived U.S. biosolids from the 2001 EPA national sewage sludge survey. Water Res 44:658–668

Miralbés C (2004) Quality control in the milling industry using near infrared transmittance spectroscopy. Food Chem 88:621–628. https://doi.org/10.1016/j.foodchem.2004.05.004

Mohd Pauzi NN, Ahmad Shah MS (2022) A review on the treatment of petroleum sludge and use as construction materials. Civil Sustain Urban Eng 2:12–20. https://doi.org/10.53623/csue.v2i1.74

Mostafa MMG (2021) Environmental hazards of petroleum refinery in Bangladesh: a review. Petro Chem Indus Int 4:15–21

Nakada N, Tanishima T, Shinohara H et al (2006) Pharmaceutical chemicals and endocrine disrupters in municipal wastewater in Tokyo and their removal during activated sludge treatment. Water Res 40:3297–3303. https://doi.org/10.1016/j.watres.2006.06.039

Pan M, Wong CKC, Chu LM (2014) Distribution of antibiotics in wastewater-irrigated soils and their accumulation in vegetable crops in the Pearl River Delta, Southern China. J Agric Food Chem 62:11062–11069. https://doi.org/10.1021/jf503850v

Pradel M, Aissani L, Villot J et al (2016) From waste to added value product: towards a paradigm shift in life cycle assessment applied to wastewater sludge—a review. J Clean Prod 131:60–75

Priyadarshini M, Ahmad A, Das S, Madhao Ghangrekar M (2022) Application of microbial electrochemical technologies for the treatment of petrochemical wastewater with concomitant valuable recovery: a review. https://doi.org/10.1007/s11356-021-14944-w/Published

Prosser RS, Sibley PK (2015) Human health risk assessment of pharmaceuticals and personal care products in plant tissue due to biosolids and manure amendments, and wastewater irrigation. Environ Int 75:223–233

Sablayrolles C, Gabrielle B, Montrejaud-Vignoles M (2010) Life cycle assessment of biosolids land application and evaluation of the factors impacting human toxicity through plant uptake. J Ind Ecol 14:231–241. https://doi.org/10.1111/j.1530-9290.2010.00222.x

Scaria J, Anupama KV, Nidheesh PV (2021) Tetracyclines in the environment: an overview on the occurrence, fate, toxicity, detection, removal methods, and sludge management. Sci Total Environ 771

Sellier A, Khaska S, Le Gal La Salle C (2022) Assessment of the occurrence of 455 pharmaceutical compounds in sludge according to their physical and chemical properties: a review. J Hazard Mater 426

Sharma A, Sharma P, Sharma A et al (2017) Hazardous effects of petrochemical industries: a review. Adv Petrochem Sci

Singh RP, Singh P, Araujo ASF et al (2011) Management of urban solid waste: vermicomposting a sustainable option. Resour Conserv Recycl 55:719–729. https://doi.org/10.1016/j.resconrec.2011.02.005

Spinosa L, Ayol A, Baudez JC et al (2011) Sustainable and innovative solutions for sewage sludge management. Water (Switzerland) 3:702–717. https://doi.org/10.3390/w3020702

Snyder E, O'connor G, Mcavoy D (2010) Measured physicochemical characteristics and biosolids-borne concentrations of the antimicrobial triclocarban (TCC). Sci Total Environ 408:2667–2673

Tarpani RRZ, Alfonsín C, Hospido A, Azapagic A (2020) Life cycle environmental impacts of sewage sludge treatment methods for resource recovery considering ecotoxicity of heavy metals and pharmaceutical and personal care products. J Environ Manage 260. https://doi.org/10.1016/j.jenvman.2019.109643

Thong O, Ibrahim A, Abdul Rahim H et al (2021) Review of crude oil extraction methods from petroleum sludge. Int J Eng Technol Sci 8:40–50. https://doi.org/10.15282/ijets.8.1.2021.1006

Vishnuraj MR, Kandeepan G, Rao KH et al (2016) Occurrence, public health hazards and detection methods of antibiotic residues in foods of animal origin: a comprehensive review. Cogent Food Agric 2

Yoshida H, Christensen TH, Scheutz C (2013) Life cycle assessment of sewage sludge management: a review. Waste Manage Res 31:1083–1101

Yu CL, Wang SF, Pan PC et al (2006) Residential exposure to petrochemicals and the risk of leukemia: using geographic information system tools to estimate individual-level residential exposure. Am J Epidemiol 164:200–207. https://doi.org/10.1093/aje/kwj182

# Clay-Polymer Nanocomposites for Advanced Wastewater Sludge Management

Baiju John

**Abstract**

Industrial effluents are a major source of water pollution, necessitating effective wastewater treatment methods. Clay minerals and polymers, while individually limited, synergistically enhance water treatment as clay-polymer nanocomposites (CPNs). This paper highlights diverse CPN types, their synthesis, and factors influencing their efficiency, including pH, temperature, contact time, pollutant concentration, and adsorbent dose. Emphasizing maximum adsorption capacity, adsorption mechanisms, and kinetics, the study underscores CPNs' potential as a top-tier long-term solution for wastewater treatment. Wastewater sludge management constitutes a critical facet of wastewater treatment processes, with the potential for innovative solutions through the utilization of clay-polymer nanocomposites. These materials, which amalgamate the advantageous properties of polymers and nanoscale clays, hold promise for enhancing the treatment and disposal of wastewater sludge. This chapter offers a comprehensive overview of the multifaceted application of polymer nanocomposite clays in managing wastewater sludge, addressing various challenges through a systematic approach.

**Keyword**

Dyes (azo dyes, reactive dyes) · Bleaching (chlorine compounds) · Detergents, surfactants · Heavy metals (cadmium, lead, nickel, chromium, zinc) · Chemical intermediates (phenols, formaldehyde) · Polycyclic aromatic hydrocarbons (PAHs) · High organic content (fats, proteins, sugars) · Lignin and cellulose residues · Pesticides · Herbicides

B. John (✉)
Duann Innovations a Systems, Houston, TX, USA
e-mail: jonsbaiju@gmail.com

Polymer Nanocomposites Laboratory, Toyota Technological Institute, Nagoya, Japan

## 1 Introduction

A pivotal challenge in wastewater management lies in the artful pursuit of efficient dewatering, delicately orchestrated to diminish sludge volume by meticulously eliminating water content. In wastewater management, the breakthrough emerges through the utilization of polymer nanocomposite clays. These innovative materials showcase remarkable efficacy as flocculants and coagulants, orchestrating the aggregation of sludge particles and enhancing the separation of water from the solid phase during dewatering. The integration of nanoscale clay particles amplifies the flocculation process, leading to a substantial improvement in dewatering efficiency. Addressing the inherent challenges of wastewater sludge, laden with organic matter and potential hazards, these polymer nanocomposite clays play a transformative role in stabilizing organic components. Through adsorption and immobilization, they mitigate risks associated with odour generation, pathogen dissemination, and pollutant leaching. Moreover, their adsorptive properties efficiently capture heavy metals, restraining their mobility and averting environmental repercussions during disposal. The profound impact of polymer nanocomposite clays extends to economic viability, as effective dewatering significantly reduces sludge volume, facilitating easier and more economical transportation and disposal. Employing these materials in solidification processes transforms dewatered sludge into a stable, manageable solid material, crucial for safe disposal in landfills or containment facilities. Beyond economic considerations, the adoption of polymer nanocomposite clays contributes substantially to environmental sustainability by minimizing the demand for expansive landfill spaces and reducing pollutant discharge. Ultimately, the exploration and optimization of diverse nanoclays and polymers, harnessing their synergistic effects

© The Author(s), under exclusive license to Springer Nature Switzerland AG 2025

A. Mandpe and M. P. Shah (eds.), *Wastewater Sludge Nexus*, Sustainable Landscape Planning and Natural Resources Management, https://doi.org/10.1007/978-3-031-45758-6_9

in dewatering, stabilization, and solidification, underscore the potential for tailored and efficient solutions in the realm of wastewater sludge management, charting a course towards a more sustainable future.

## 2 Basic Principle of Pollution Removal: Adsorption

The demand for clean water increases year by year so does the need for wastewater resource management, the removal of pollutants from wastewater emerges as a pivotal pursuit, not only to meet escalating water demands but also to safeguard the delicate balance of our environment from the deleterious effluents it carries. Singapore, a beacon of innovative solutions, has notably adopted and perfected methods to use wastewater treatment for domestic purposes. Industrial wastewater, notorious for harbouring inorganic pollutants such as lead, cadmium, chromium, and arsenic, poses a persistent threat to ecosystems. In response, a myriad of techniques spanning the realms of physical, chemical, and biological methods have been employed. Amidst these, adsorption emerges as a beacon of efficiency and affordability. This method not only distinguishes itself by its cost-effectiveness but also by its ability to stand resolute against the drawbacks associated with other techniques, such as the production of secondary pollutants. In the symphony of wastewater treatment, adsorption orchestrates a harmonious melody of efficacy, sustainability, and economic feasibility, offering a promising cadence for the continued pursuit of environmental stewardship (Hnamte and Pulikkal 2022).

## 3 Pollution Removal

Pollutant removal is a critical aspect of ensuring the safety and sustainability of water resources. Various methods can be employed for this purpose, broadly categorized into physical, chemical, and biological methods, each with its distinct advantages and limitations, rendering their suitability context-dependent.

(i) Physical Methods:

Physical methods leverage natural forces such as gravity, particle size, and temperature to eliminate pollutants from water. Sedimentation, filtration, and degasification are examples of these processes. They play a crucial role in removing particulate matter and certain contaminants from water bodies.

(ii) Chemical Methods:

Chemical approaches involve the use of substances to induce reactions that facilitate pollutant removal. Flocculation, coagulation, chemical precipitation, ozonation, adsorption, and ion exchange are extensively employed in treating water contaminated with industrial wastes. However, it's essential to note that while these methods can be effective, their application should consider specific environmental conditions.

(iii) Biological Methods:

Biological methods harness living organisms or their processes to break down pollutants and purify water. Bioremediation, stabilization ponds, activated sludge, and aerobic/anaerobic treatments exemplify this category. They are particularly useful in degrading organic pollutants and fostering a more sustainable approach to wastewater treatment.

The significance of pollutant removal cannot be overstated, especially considering that water pollution stems from industrial wastes containing inorganic pollutants such as lead, cadmium, chromium, and arsenic. The persistence of these pollutants in the environment necessitates comprehensive strategies to safeguard water quality. While various physical, chemical, and biological methods exist, it's crucial to acknowledge the limitations of some conventional approaches. Methods like reverse osmosis, electrolysis, and ion exchange, though effective, are often economically impractical for widespread use. Additionally, chemical precipitation and coagulation, while promising for wastewater treatment, generate secondary pollutants in the form of sludge, adding to their drawbacks.

In this context, adsorption emerges as a compelling method due to its high efficiency and cost-effectiveness. Its ability to selectively remove pollutants makes it a viable solution for water treatment, offering a balance between effectiveness and economic feasibility. Efforts in ongoing research continue to explore and enhance existing methods while seeking innovative approaches to address the complex challenges posed by water pollution. In doing so, we aim not only to meet basic water demands but also to safeguard our environment from potential harm (Hnamte and Pulikkal 2022).

## 4 Adsorption

Adsorption is a fascinating surface phenomenon in which molecules, atoms, or ions from a fluid bulk selectively adhere to the surface of a solid material. This intriguing transfer can be attributed to various forces, encompassing both physical and chemical interactions. Remarkably, this process is inherently reversible, and its counterpart, where the adsorbate is released from the surface, is termed desorption.

The effectiveness of an adsorbent is a comprehensive measure that considers its porosity, surface area, desorption capacity, and the number of regeneration cycles it can undergo. Diverse adsorbents are available for wastewater treatment, with conventional options like activated carbon, zeolites, ion-exchange resins, silica gel, molecular sieves, and activated alumina being commercially prevalent. Beyond these, unconventional adsorbents derived from clay minerals (natural or modified), agricultural wastes, and polymeric resins, including cellulose, lignin, and hemicellulose extracted from agricultural biomass, are gaining increased attention.

There exist two distinctive types of adsorptions—physisorption and chemisorption—depending on the nature of the interactions between the adsorbate and the adsorbent as shown in Fig. 1. In physisorption, the adsorbates form bonds with the adsorbent surface through weak van der Waals forces. Conversely, chemisorption involves stronger interactions, such as chemical bonding or electrostatic forces, between the adsorbates and the adsorbent (Hnamte and Pulikkal 2022).

1. Physisorption: Physisorption is a type of adsorption where the adsorbate molecules are held onto the surface of the adsorbent by weak van der Waals forces, such as London dispersion forces, dipole-dipole interactions, and hydrogen bonding. This type of adsorption is usually reversible and does not involve any significant chemical bonding between the adsorbate and the adsorbent.
2. Chemisorption: Chemisorption is a more significant form of adsorption, involving a chemical reaction between the adsorbate and the adsorbent surface. It results in the formation of strong chemical bonds or interactions, such as covalent bonds, ionic bonds, or coordination bonds. Chemisorption is generally not reversible without the application of significant energy, as it involves the formation of chemical compounds between the adsorbate and the adsorbent.

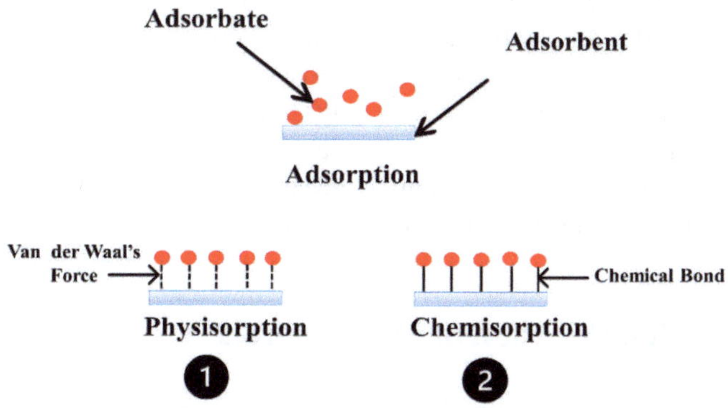


**Fig. 1** Illustrates the mechanism of adsorption with adsorbent and adsorbate; (1) shows physisorption where Van der Waal's force is the major factor, while (2) demonstrates chemisorption where the chemical bond is the primary player

Understanding the distinction between these two mechanisms is crucial in determining the appropriate adsorbent for a specific application, as well as in designing effective and efficient adsorption processes for wastewater treatment and other environmental remediation applications (Hnamte and Pulikkal 2022).

Several crucial variables influence adsorption processes, including the dosage of adsorbent, initial concentration of the pollutant, contact time, pH, and temperature. Details of the factors affecting the efficiency of adsorption process as shown in Fig. 2 when interacted with the CPNs are as follows:

- Adsorbent Dose: The amount of adsorbent added to the water affects the efficiency of the adsorption process. An optimal dose is required for effective removal of pollutants without wasting the adsorbent.
- Initial Concentration of the Pollutant: The concentration of the pollutant in the water directly impacts the adsorption capacity of the adsorbent. Higher initial pollutant

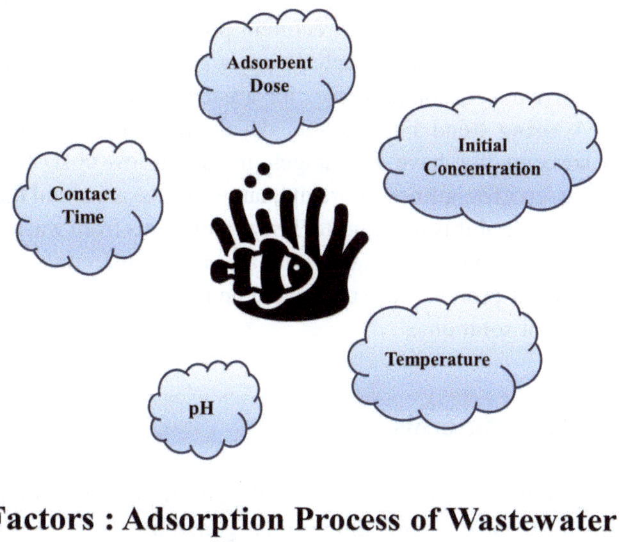


**Fig. 2** Illustrates the factors influencing the adsorption of wastewater when it interacts with the CPNs

concentrations may require a higher adsorbent dose or more efficient adsorbents.

- Contact Time: The duration of contact between the adsorbent and the wastewater affects the extent of adsorption. Longer contact times can lead to more effective removal of pollutants.
- pH: The pH of the solution can influence the surface charge of the adsorbent and the ionization of the pollutant, thereby affecting the adsorption process. Optimal pH conditions are often required for maximum efficiency.
- Temperature: Temperature can impact the rate of adsorption, with some processes being exothermic or endothermic. Understanding the temperature dependence can provide insights into the energetics of the adsorption process.

The efficacy and nature of adsorption can be elucidated through detailed adsorption and kinetic studies. Additionally, thermodynamic analyses can reveal whether the process is spontaneous or non-spontaneous, as well as whether it is exothermic or endothermic. Thermodynamic evaluation provides the details of the adsorption process like whether it is a spontaneous or non-spontaneous one or and whether it is exothermic or endothermic. Adsorption studies provide valuable information about the feasibility and energy changes associated with the adsorption process. The efficiency of an adsorbent is characterized by its porosity, surface area, desorption capacity, and the number of regenerations cycles it can undergo without significant loss of efficacy. Various types of adsorbents, such as activated carbon, zeolites, and various types of modified materials, can be used for wastewater treatment based on the specific contaminants present. The choice of adsorbent depends on factors like the type of pollutants, cost, and the scale of the treatment process. Understanding these variables and characteristics is crucial for designing effective adsorption processes for wastewater treatment.

A rising trend involves the development of adsorbents or sorbents that have advantages, including low cost, abundance, eco-friendliness, porosity, large surface area, and their remarkable ability to sequester heavy metal ions from wastewater. The exploration clay-based adsorption materials hold significant promise for sustainable and effective wastewater treatment solutions.

## 5 Clay Minerals

Clay minerals, a diverse assemblage of hydrous aluminosilicate layers with dimensions generally below 4 μm, weave a tapestry of geological significance across Earth's surface. Comprising tetrahedral $(SiO_4)^{4-}$ and octahedral $[Al_2O_3(OH)_3]^{3-}$ sheets, these minerals emerge as the elegant

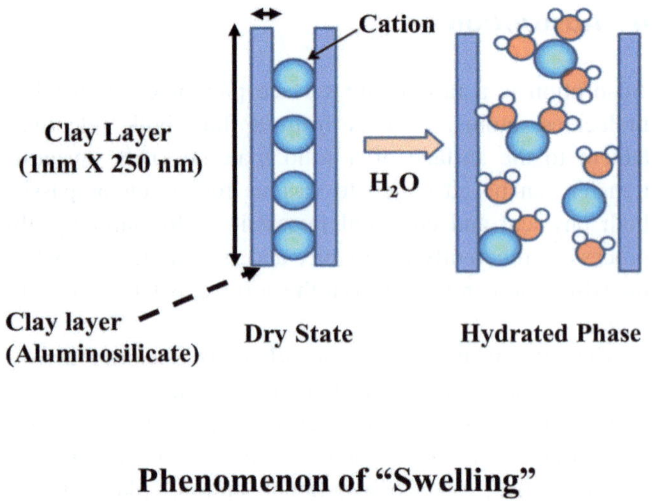


**Fig. 3** Shows the property of swelling; here, water molecules interact with the cations and hydrate the clay layers

result of chemical weathering, gracefully transforming from parent silicate minerals. Their diminutive size and expansive surface area imbue clay minerals with distinctive attributes, notably a high cation exchange capacity (CEC), catalytic prowess, and a malleable plasticity when moist. Within soil and sediment realms, clay minerals assert their importance, orchestrating pivotal roles in geological, environmental, and industrial arenas. Their exceptional CEC, ion exchange, and adsorption capabilities render them indispensable in fostering soil fertility and undertaking pollutant remediation. Notably, the enchanting property of "swelling," a graceful dance dictated by hydration, takes centre stage as shown in Fig. 3. Here water molecules interact with the cations and hydrate the clay layers. This unique attribute finds its place in the construction industry, sculpting ceramics, and waltzes into the oil and gas domain as a key player in drilling muds. Beyond their role as geological artisans, clay minerals gracefully choreograph the ballet of water and nutrient adsorption, curating the very foundation for soil fertility and flourishing plant life. As catalysts, they perform a symphony of chemical reactions, gracing industrial processes with their catalytic finesse. In the realm of water sludge treatment, understanding the nuances of clay mineral behaviour becomes the key to unlocking their full potential, revealing the hidden beauty within their layered structures (Okamoto and John 2013).

## 6 Classification of Clay Minerals

Clay minerals, fascinating in their diverse structural compositions, can be broadly categorized into two types based on their intricate arrangements. The first, known as the 1:1 type, boasts a singular tetrahedral sheet harmoniously paired with an octahedral counterpart. Contrastingly, the 2:1

type unfolds a captivating narrative where an octahedral sheet finds itself delicately ensconced between two tetrahedral sheets, creating a layered tableau of geological marvel. All the more, clay minerals unveil their secrets through seven distinctive groups, each defined by its expanding capacity. Kaolin, halloysite, and serpentine, members of the 1:1 category, gracefully abstain from expansion, their interlayer units bound in a dance of unyielding hydrogen bonds. Meanwhile, mica and illite, steadfast in their non-expanding 2:1 allure, remain untouched by the influences of solvation or heat, owing to the absence of exchangeable interlayer cations. Vermiculite, with its expanding 2:1 allure, beckons with exchangeable cations, while the smectites, exemplified by montmorillonite, stand as highly expanding 2:1 wonder, showcasing an impressive swelling index. Talc and pyrophyllite, resilient in their uncharged 2:1 identity, reveal subtle layer charge nuances amid minor substitutions. Chlorites, an embodiment of 2:1:1 elegance, harmoniously blend a negative charge within their 2:1 layer, balanced by an additional octahedral sheet. Lastly, the fibrous clays, embodied by palygorskite and sepiolite, weave a narrative of intermediate expanding capacities, gracefully refraining from swelling in the presence of aqueous or organic solvents. In this tapestry of geological intricacies, clay minerals unfurl their geological poetry, inviting exploration into the boundless realms of Earth's ancient formations (Okamoto and John 2013).

Smectite-based clay possess moderate charge densities and interlayer swelling ability, are suitable for creating organoclays (OCs) that easily intercalate organic compounds. These OCs serve as precursors for clay-polymer composites/nanocomposites, adjusting the surface properties of hydrophilic clay minerals to accommodate both hydrophilic and hydrophobic polymers. Common organic modifiers include quaternary ammonium cations/surfactants, alkyl amines, alkyl carbazol, dimethylsiloxane, glycerol carbonate, and oligomers. These modifiers replace hydrated cations in clay minerals, imparting a hydrophobic nature suitable for intercalating non-polar molecules (Mizuno et al. 2013).

## 7 Clay-Polymer Nanocomposites (CPNs)

The development and utilization of clay-polymer nanocomposites (CPNs) for wastewater treatment have garnered substantial interest, offering a promising avenue to leverage the synergies between natural clay minerals and synthetic polymers. CPNs hold the potential to overcome individual limitations, resulting in enhanced efficiency and cost-effectiveness in wastewater treatment. By intimately integrating clay minerals and polymers at the nanoscale, these composites exhibit synergistic effects, manifesting in improved adsorption capacity, mechanical strength, and stability compared to their isolated components. The increased surface area and optimized pore structure of CPNs contribute to heightened affinity for organic pollutants, thereby improving the overall effectiveness of wastewater treatment. Additionally, the regeneration capacity of CPNs allows for their repeated use across multiple cycles, bolstering their economic viability and sustainability (Mizuno et al. 2013).

To propel advancements in CPNs for wastewater treatment, it is imperative to focus on innovative synthesis approaches. These approaches should optimize the interaction between clay minerals and polymers, tailoring CPN properties for specific applications in wastewater treatment. Addressing potential environmental implications associated with CPN fabrication and use is crucial to ensuring sustainable practices align with commercial applications. Research on CPN-based wastewater treatment should emphasize comprehensive analyses, including unique synthesis processes, comparative studies, economic viability, and reaction mechanisms involving CPNs for pollutant removal. The performance of polymers is notably enhanced with the addition of small amounts of inorganic clay nanoparticles. Materials in the nano or micro scale, known as clay-polymer nanocomposites or composites, can be categorized into intercalated nanocomposites, exfoliated nanocomposites, and phase-separated micro-composites. Various polymers, such as polyvinyl chloride, epoxies, polyurethanes, polystyrenes, chitosan, polyesters, and polypropylene, have been utilized in CPN synthesis. The diverse applications of CPNs extend beyond wastewater treatment, as the addition of clay nanoparticles can enhance properties like mechanical strength, thermal stability, gas barrier properties, and flame retardancy in a wide range of materials.

The performance of polymers significantly improves with the addition of small amounts of inorganic clay nanoparticles. When a small amount of clay mineral (<5%) is dispersed in a polymer, the significant properties of the polymer are dramatically enhanced. Such materials at the nano or micro scale are known as clay-polymer nanocomposites or composites. Polymers such as polyvinyl chloride, epoxies, polyurethanes, polystyrenes, chitosan, polyesters, and polypropylene have been utilized for CPN synthesis (Okamoto and John 2013).

Clay-polymer nanocomposites (CPNs) materials have gained significant attention due to their improved properties compared to conventional polymers. CPNs are prepared by two-step processes. First the clay nanoplatelets are modified with organomodifier for better cationic interaction with the incoming polymer chains. The modified clay platelets

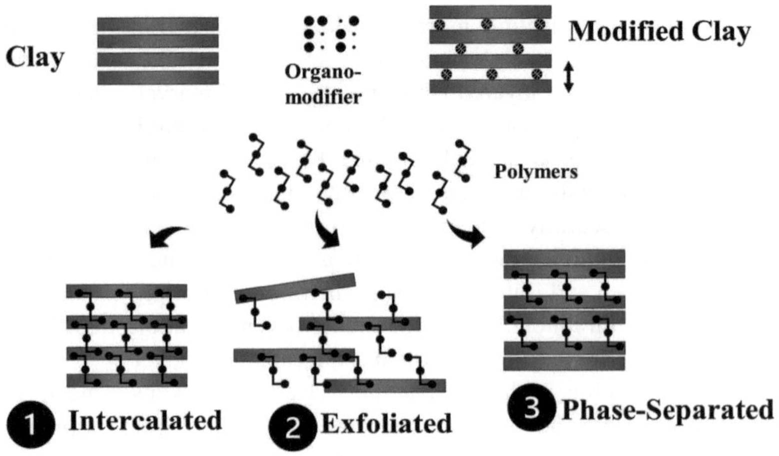


**Fig. 4** shows the preparation of CPNs in two steps. Firstly, the clay nanoplatelets are modified with an organomodifier, and subsequently, the modified clay platelets are processed with a small addition of polymers (<5% wt), resulting in CPNs. There are three broad types of CPNs based on the way polymers are attached to clay platelets: (1) intercalated CPNs, (2) exfoliated CPNs, and (3) phase-separated micro-CPNs

are processed with small addition of polymers (<5%wt) resulting in CPNs. The addition of small amounts of polymers to clay nanoparticles can lead to enhancements in various properties, including mechanical strength, thermal stability, gas barrier properties, and flame retardancy, among others. There are three broad types of CPNs based on the way polymers are attached to clay platelets: (1) intercalated nanocomposites, (2) exfoliated nanocomposites, and (3) phase-separated micro-composites, as shown in Fig. 4.

1. Intercalated nanocomposites: In these composites, the polymer chains are inserted into the interlayer space of the clay mineral, leading to an increase in the spacing between the clay layers. This arrangement often results in improved mechanical properties.
2. Exfoliated nanocomposites: These composites have individual clay layers dispersed within the polymer matrix at the nanometer scale. The increased surface area of the clay layers contributes to improved barrier properties and mechanical strength.
3. Phase-separated micro-composites: These composites involve the formation of distinct phases of polymer and clay, resulting in a microscale dispersion. This can lead to a balance between improved properties and processability (Mizuno et al. 2013).

## 7.1 Chitosan CPNs

Chitosan-based CPNs are becoming popular in wastewater treatment and adsorption processes due to their bio-origin, derived from biowaste of shrimp hatcheries/processing factories. These composites are being extensively studied for their potential in environmental remediation, thanks to their unique properties and high adsorption capacities. Chitosan-clay nanocomposites, particularly those using low molecular weight chitosan and sodium montmorillonite clay, have demonstrated remarkable efficiency in wastewater treatment. The process involves a series of steps, including dispersion, stirring, centrifugation, washing, and drying, to obtain the final nanocomposite material. The nitrate adsorption efficiency of different chitosan-based clay nanocomposites, such as Nano-CTS/Clino, Nano-CTS/Clino@H, and Nano-CTS/Clino@PEHA, when compared, shows the influence of functionalization on the adsorption capabilities of the nanocomposites (Yazdi et al. 2019). The chitosan-montmorillonite/polyaniline nanocomposite, prepared by intercalating the CTS biopolymer into MMT clay layers and subsequent impregnation of aniline, followed by polymerization to form polyaniline, has shown good performance in the removal of methylene blue from aqueous solutions. The combination of chitosan nanoparticles with methyl cellulose and kaolin clay showed efficient removal of metal ions such as Pb (Kanchana et al. 2012). Thus, chitosan CPN-based wastewater treatment shows good promise in the removal of different contaminants and is headed in the right direction for the development of efficient and eco-friendly adsorbents.

## 7.2 Organo CPNs

Organoclay-based CPN membranes represent a promising class of adsorbents employed in wastewater remediation due to their exceptional adsorptive properties for heavy metals such as arsenic and chromium. The organoclay-embedded polysulphone (Psf) nanocomposite membranes were found to be highly efficient in removing arsenic (As) from contaminated water. The Psf membranes were prepared using the non-solvent-induced separation (NIPS) method in which the Cloisite 30B nanoclay was dispersed in *N*-methyl-2-pyrrolidone via sonication, and Psf was added to form a homogeneous solution. The resulting mixture was then applied to a flat glass plate to ensure a uniform film thickness. After immersion in water, the film was rinsed with distilled water and dried at 70 °C. Along the similar lines a poly(p-phenylenediamine) or (poly-pPD) organoclay-based CPNs too demonstrated high efficiency in removing Cr (VI) complexes from contaminated water (Shokri et al. 2016). Similarly, Montmorillonite (MMT) was used as the organoclay with a modifier called dimethyl dehydrogenated tallow ammonium salt. To enhance adsorption capacity, MMT organoclay was functionalized with poly-pPD using an in-situ chemical oxidative polymerization technique. The resulting CPN exhibited superior adsorption capabilities for Cr (VI) complexes (Setshedi et al. 2013). These organoclay-based composites show great promise in advancing the field of wastewater treatment, providing effective solutions for the removal of heavy metal contaminants.

## 7.3 Magnetic CPNs

Various CPNs with magnetic properties have shown remarkable effectiveness in removing toxic heavy metals and dyes from contaminated water. For instance, a magnetic montmorillonite-based CPNs functionalized with polyethylenimine (MMT-$Fe_2O_3$-PEI) was synthesized through a one-pot hydrothermal process, demonstrating its efficacy in removing Cr (VI), particularly at low pH (Fayazi and Ghanbarian 2020). Similarly, chitosan-based CPNs, incorporating $Fe_3O_4$ through a chemical deposition method, have proven to be effective adsorbents for Cr (VI) removal in water. Additionally, magnetic polystyrene-palygorskite-based CPNs were developed for oil separation from water, utilizing a magnetized palygorskite surface and polymerization with styrene. Carboxymethyl cellulose-acrylic acid/magnetite cloisite 30B (CMC-cl-pAA/$Fe_3O_4$–C30B) emerged as a magnetic clay-polymer nanocomposite with excellent adsorption capabilities for methylene blue dye removal (Malatji et al. 2020). Furthermore, polyvinyl alcohol/laponite RD-based CPNs demonstrated efficacy in removing methylene blue from aqueous solutions by incorporating magnetite nanoparticles into the clay-polymer matrix. These diverse magnetic clay-polymer nanocomposites hold promise for addressing water contamination challenges through efficient adsorption processes (Mahdavinia et al. 2017).

## 7.4 Polystyrene CPNs

Polystyrene-based CPNs are becoming popular these days due to their ease of preparation. One such enchanting tale is woven through the meticulous dance of molecules in acetonitrile, where its role as a solvent orchestrates a mesmerizing 24.5 Å expansion. The ballet continues as polystyrene and organo-montmorillonite (MMT) CPNs take centre stage, their creation choreographed by the artful technique of melt intercalation. Here, a symphony of ingredients unfolds, with MMT organoclay prepared through the alchemy of a protonated 2-[(5-methyl-3-isoxazolyl)methyl)-1H-hexadecyl-benzimidazole, merging with malleable polystyrene at a crescendo of 300 °C. In another poetic manifestation, a polystyrene/organo-MMT CPN emerges through the ethereal realm of emulsion polymerization (Yu et al. 2018). Sodium lauryl sulphate, a virtuoso anionic surfactant, guides the transformation. The clay-modifier dances in harmony with styrene and divinylbenzene at 70 °C, creating a sublime mixture. The initiation, akin to the striking of a conductor's baton, is executed with potassium persulphate, and the finale involves the ethereal touch of ethanol for demulsification. A delicate encore unfolds as the product is meticulously washed, unveiling a masterpiece upon vacuum drying at low pressure. Meanwhile, maghnite clay, a luminary in the MMT family, undergoes a metamorphosis through the embrace of cetyltrimethyl ammonium bromide. The balletic integration of modified-maghnite clay into styrene polymer reaches its climax in the oil bath at 70 °C over a span of 72 h, revealing a composition that echoes the artistry of nature itself (Mrah and Meghabar 2021). The resultant polystyrene-based CPN is ready for wastewater treatment.

## 7.5 Epoxy CPNs

Epoxy CPNs, heralding a remarkable advancement in material science, were meticulously crafted through the infusion of highly expanded onium-modified clay minerals into epoxy matrices. This innovative synthesis technique imparted enhanced thermal stability and solvent resistance to the resulting materials. Among the triad of methods employed for their fabrication—mechanical mixing, shear mixing, and ultrasonic mixing—the shear mixing technique emerged as the preeminent choice. Employing a three-roll mill, the process involved the strategic placement of epoxy resin between the feed and centre rolls, with Cloisite 30B (C30B), an organically modified montmorillonite (MMT) clay, meticulously dispersed onto the epoxy surface. The ensuing compounding at a controlled temperature and RPM facilitated the exfoliation of clay nanoparticles until a discernible clear solution materialized, signalling completion. Subsequent steps involved the hardening, degassing, and drying of the resultant product, rendering it ready for versatile applications (Yasmin et al. 2003). Another noteworthy variant of epoxy-clay nanocomposite, featuring layered double hydroxides (LDHs)/epoxy, showcased a sophisticated preparation method. The amalgamation of epoxy (EPON 828) and LDH intercalated with amino laurate, employing the curing agent Jeffamine

D400, resulted in X-ray diffraction and TEM-confirmed well-dispersed and exfoliated LDHs/epoxy nanocomposites (Hsueh and Chen 2003). This ground-breaking research underscores the transformative potential of these advanced materials in diverse industrial and technological arenas.

## 7.6 Polyurethane CPNs

Polyurethanes (PUs) represent a fascinating class of polymers with distinctive chemical and physical attributes, owing to their dual-segmented structure. Comprising soft segments derived from polyols and hard segments from isocyanates and chain extenders, PUs exhibit a unique combination of strength and elasticity. The integration of organoclays into PU nanocomposites has proven to be a transformative approach, imparting enhanced mechanical properties and remarkable optical transparency. In a noteworthy study, montmorillonite (MMT) clay was incorporated into distilled water, followed by meticulous processing involving diethanolamine and dilute hydrochloric acid. The resulting product, subjected to filtration, washing, and vacuum drying, showcased the potential of PU nanocomposites. A recent breakthrough further explored novel polyurethane network/organoclay nanocomposites, strategically manipulating the soft segments within the PU matrix. By varying the soft segment content between 30 and 60 wt%, the study utilized a poly(dimethylsiloxane)-based macrodiol as the soft segment and 4,4-methylenediphenyldiisocyanate and hyper-branched polyester as the hard segment. Cloisite 30B served as the nanofiller, intricately dispersed into the PU network (Pergal et al. 2022). Comprehensive analyses, including FT-IR, X-ray scattering, thermogravimetric analysis, differential scanning calorimetry, atomic force microscopy, and scanning electron microscopy, revealed intriguing insights. The investigation highlighted that a reduction in soft segment content led to heightened strength, increased plasticity, and a more pronounced degree of phase separation in the microstructure. Simultaneously, the cross-linking density and hydrophobicity of the PU-based CPNs exhibited a discernible decrease. This research not only expands our understanding of PU-based CPNs but also paves the way for tailoring their properties through strategic manipulation of soft segments, offering exciting prospects for advanced materials with superior performance characteristics.

## 7.7 Polyimide CPNs

Polyimide-based CPNs are widely known for their exceptional thermal properties. One notable synthesis involved the fusion of 7,7-bis(4-aminophenoxy)-4,4,4,4′tetramethyl-2,2-spirobichroman and 4,4-hexafluoroisopropylidenediphthalic anhydride in m-cresol, enriched with isoquinoline. The resulting polymer was seamlessly blended with organo-modified montmorillonite (MMT), achieved by introducing the clay to a solution of ammonium salt of dodecylamine. This unique polyimide-clay nanocomposite exhibited heightened thermal stability, owing to the well-dispersed silicate layers of MMT clay, a feat achieved through dual intercalating agents like hexadecyl trimethyl ammonium bromide (CTAB)-4,4-oxydianiline (Mansoori et al. 2013). Interestingly, the thermal properties of polyimide CPNs were optimized when dual intercalating agents were employed, showcasing the finesse of such tailored polymer architectures. In a parallel exploration, the synthesis of poly(amic acid)-based CPN made with 3,3,4,4-benzophenone tetracarboxylic dianhydride (BTDA) and the organic diamine 2-(5-(3,5-diaminophenyl)-1,3,4-oxadiazole-2-yl) pyridine (POBD) in dimethylacetamide (DMAc) showed exemplified elevated tensile strength and Young's modulus (Mansoori et al. 2013). Polyimide CPNs formed by solvating 4,4-oxydianiline (ODA) and pyromellitic dianhydride (PMDA) in DMAc, incorporated vermiculite (VMT) exhibited superior oxidative, thermal, and insulation properties, showcasing the potential of tailored nanocomposites in diverse applications. Polyimide CPNs are useful hot industrial wastewater effluent treatment purposes (Sheikh et al. 2021).

## 7.8 Polyacrylate CPNs

Polyacrylate CPNs are versatile due to their ease of preparation and flexibility like an elastomer. The in-situ polymerization method is used to integrate ethyl acrylate (EA) and butyl acrylate (BA) in the synthesis of polyacrylate-CPNS, with a prepared MMT slurry and aqueous 2,2-Azobis(2-amidinopropane) dihydrochloride. This method allowed for the synthesis of nanocomposites with distinct compositions, including EA, BA, and a tailored EA/BA mixture. The resulting products underwent thorough washing and drying to achieve a constant weight, highlighting the precision of the process (Xu and Brittain 2006). Similarly, in the hyper-coagulation method, use cationic latex and clay slurry to create nanocomposites through careful evaporation, ensuring a constant weight. The innovative use of double modified MMT clay, intercalated with amino acids and surface-modified with methyl triethoxy silane, further exemplified the commitment to optimizing thermal stability in polyacrylate CPNs (Effenberger et al. 2009). The resulting materials exhibited superior thermal properties with increasing clay content and is very useful in the removal of heavy metals from tannery wastewater which are an environmental issue in third world developing economies.

## 8 Removal of Pollutants with CPNs from Wastewater

The removal of pollutants by CPNs primarily relies on adsorption as the key mechanism. The deposition of pollutants onto adsorbents involves van der Waals forces, covalent bonding, and electrostatic attraction. The specific nature of interaction depends on various factors like pH, temperature, and functional groups. The pollutants for removal from wastewater by CPNs can be classified into inorganic and organic pollutants. The inorganic part involves heavy metals, metals, and metalloids while organic ones are composed of dyes, pesticides, and biological components.

## 9 Treatment of Inorganic Pollutants

In the ongoing quest to safeguard our precious water resources and protect ecosystems from the pervasive threat of inorganic pollution, innovative solutions have emerged through the development of CPNs. These inorganic pollutants are composed of heavy metals, metals, and metalloids. Among the myriad pollutants, the insidious presence of copper (Cu), nickel (Ni), cadmium (Cd), lead (Pb), arsenic (As), antimony (Sb), etc., in wastewater continues to pose a formidable challenge to aquatic life and human health. These heavy metals can be from the industrial by-products or domestic sewage. Notably, the focus on lead (Pb (II)), a notorious water pollutant with severe health implications, has yielded remarkable results in the form of a clay-based poly(methoxyethyl)acrylamide (PMEA) polymer composite. This composite exhibited an impressive maximum adsorption capacity of 81.02 mg $g^{-1}$, operating optimally within a pH range of 4–6. Its robust stability, with only a 6% decrease in adsorption capacity after five regeneration cycles, underscores its potential as a sustainable solution (Solener et al. 2008). In parallel endeavours, binary nanocomposites, such as nanochitosan combined with methyl cellulose (NC + MC) and nanochitosan paired with kaolin clay (NC + KC), have demonstrated efficacy in Pb (II) removal, with (NC + KC) standing out with nearly 80% removal efficiency. A polyvinyl alcohol (PVA)-modified kaolinite clay nanocomposite has showcased simultaneous removal capabilities for both Pb (II) and Cd (II), achieving impressive desorption efficiency of 99% using hydrochloric acid. Furthermore, ion-imprinted clay/polymer nanocomposites, chitosan-palygorskite combinations, and lignin-based poly(acrylic acid) nanocomposites have each contributed distinctive approaches, highlighting their potential in the intricate landscape of heavy metal removal (Kanchana et al. 2012). Turning attention to chromium (Cr), another formidable environmental challenge, various adsorbents such as polypyrrole/organo-modified montmorillonite (PPy/OMMT) and cross-linked chitosan/aluminium-pilarred montmorillonite nanocomposites have shown promise, emphasizing the diversity of effective strategies (Mdlalose et al. 2017). These studies collectively underscore the significant strides made in developing efficient adsorbents, offering hope for a cleaner, healthier future through the intersection of science and environmental stewardship.

## 10 Treatment of Organic Pollutants

In addressing the critical issue of organic contaminants in wastewater, including dyes, pesticides, acids, and phenols, it becomes imperative to mitigate their presence due to their potential to disrupt biological systems even in trace amounts. Preserving the health of water bodies and sustaining a robust hydrological system necessitates the effective removal of these chemical species. Recent advancements in the field highlight the promising role of CPNs in wastewater treatment. The integration of CPNs not only presents a compelling solution for the removal of organic pollutants but also showcases their versatility and efficacy in water purification processes. This innovative approach underscores the ongoing efforts to develop sustainable technologies for safeguarding the quality of water resources and ensuring the well-being of aquatic ecosystems.

### 10.1 Treatment of Dyes

In the realm of environmental stewardship, a tapestry of innovative solutions unfolds to address the pressing issue of dye contamination in water especially from textile industry effluents. Extensive research has heralded the advent of chitosan-based CPNs, where the marriage of chitosan (CTS) with organo-modified montmorillonite (MMT) unveils a formidable ally in the battle against organic pollutants. Similarly, a novel CPN was reported, showcasing a remarkable maximum adsorption capacity of 290.80 mg $g^{-1}$ for the removal of Congo red (CR), an anionic dye. The orchestration of pH and temperature variations wielded insightful nuances into the dynamic performance of the CTS/MMT nanocomposite (Wang and Wang 2008). Tailoring CPNs through judicious molar ratios of CTS to MMT further unveiled a realm of possibilities, shaping the adsorption capacities with precision. Simultaneously, a symphony of studies converged on diverse CPNs, including N,O-carboxymethyl-chitosan/montmorillonite (N,O-CMC-MMT), chitosan-g-poly(acrylic acid)/montmorillonite (CTS-g-PAA/MMT), and magnetic polyvinyl alcohol/laponite (mPVALap), each showcasing their prowess in selectively extracting dyes like methylene blue (MB) (Wang et al. 2008). The narratives of polyamide/vermiculite and chitosan-montmorillonite/polyaniline (CTS-MMT/PANI) nanocomposites added layers to this rich tapestry, underscoring the versatility of tailored materials in shaping advanced adsorption technologies (Minisy et al. 2021). A harmonious amalgamation of chitosan (CTS) and carboxymethyl chitosan (CMC), embellished with nanoclay, illustrated superior dye adsorption capacities, with CTS-grafted nanocomposites taking the spotlight. Conversely, cationic starch (CST)-montmorillonite (MMT) nanocomposites danced into the scene, achieving optimal adsorption capacities through meticulous adjustments of MMT to CST ratios (Minisy et al. 2021). A crescendo of creativity unfolded with a chitosan-modified kaolinite clay and *Azadirachta indica* leaves powder nanocomposite, showcasing remarkable efficacy in removing Congo Red (CR) and methylene blue (MB) (Minisy et al. 2021). As the narrative unfolded further, the stage expanded to accommodate superadsorbent hydrogel nanocomposites, exemplified by the lignocellulose-g-poly/montmorillonite (acrylic acid) (LNC-g-PAA/MMT) nanocomposite, boasting an extraordinary adsorption capacity for methylene blue (MB) (Shi et al. 2013). The tale continued with magnetic core-shell composites, weaving together layered double hydroxide, L-phenylalanine, pyromellitic dianhydride, and polyvinyl alcohol for the efficient removal of methyl orange (MO). Clay-polymer nanocomposites, including polyaniline/montmorillonite (PANI/MMT) and cellulose-clay hybrids, exhibited prowess in mitigating the environmental risks posed by acid dyes and Drimarine Yellow HF-3GL dye, respectively (Shi et al. 2013). In a captivating denouement, the synthesis of cross-linked chitosan-coated bentonite (CCB) composite beads emerged as a beacon of hope for mitigating the environmental impact of tartrazine. The orchestration of optimal conditions and the unveiling of promising adsorption capacities herald a new chapter in the quest for sustainable solutions in water purification. These diverse narratives, each a testament to human ingenuity, converge to form a symphony of possibilities, promising a brighter, cleaner future for our precious water resources.

## 10.2 Treatment of Pesticides, Phenolic Compounds, and Acids

In the realm of environmental remediation, the synergistic marriage of cationic polymers with clay minerals has proven to be a promising avenue for the removal of diverse organic pollutants such as pesticides, phenolic compounds and acids from aqueous solutions. These pollutants can be from industrial sludge or agriculture industry. One such noteworthy endeavour involved the modification of calcium-rich Arizona montmorillonite (SAz-1) with the cationic polymer hexadimethrine (HEXAD), resulting in the creation of SA-HEXAD nanocomposites. These nanocomposites exhibited a distinct affinity for anionic pesticides, including mecoprop (MCPP), clopyralid, and 2-methyl-4-chlorophenoxyacetic acid, showcasing adsorption percentages ranging between 50 and 75%. The interaction dynamics were further elucidated using the pesticide MCPP as a reference, revealing that ionic forces predominantly dictated the adsorption phenomenon. Contrastingly, hexadecyltrimethylammonium-intercalated SAz-1 montmorillonite (SA-HDTMA) showcased robust adsorption percentages (>75%) for most pesticides, underscoring its efficacy (Celis et al. 2007). In a parallel exploration, the removal efficiency of the herbicide atrazine from aqueous solutions was investigated using 4-vinylpyridine in conjunction with organo-modified sepiolite clays. The findings highlighted the superior performance of phenyltrimethylammonium chloride-modified clays in achieving a remarkable 95% removal efficiency, surpassing their hexadecyltrimethylammonium bromide (HDTMA-Br) counterparts (Ramírez-Gomez et al. 2019). The adsorption kinetics, well-described by the pseudo-first-order model, revealed a reduction in efficiency with an increasing ratio of acrylamide monomer. In addressing priority water pollutants, the design of CPNs utilizing polydiallyldimethylammonium chloride and poly-4-vinylpyridine-co-styrene (PVPcoS) for trinitrophenol (picric acid) and trichlorophenol removal displayed intriguing insights. PDADMAC-modified montmorillonite exhibited heightened affinity for the more anionic picric acid, while PVPcoS-modified montmorillonite displayed an enhanced adsorption capacity for the non-ionic trichlorophenol. The differential binding characteristics were attributed to the varying charge distributions and structural compatibilities. Shifting focus to thiamine (vitamin B1)-based organoclay nanocomposites, their efficacy was demonstrated across different clay types, with B1-Swy-2 organoclay proving exceptionally effective in batch and column experiments (Hernandez-Hernandez et al. 2018). The dual-mode mechanism of adsorption suggested potential applications in food-grade phenol adsorption. Additionally, alginate-clay nanocomposites emerged as proficient adsorbents for phenolic compounds, showcasing varying adsorption capacities depending on the type of clay incorporated. Lastly, addressing the persistence of polychlorinated biphenyls (PCBs), alginate-MMT composite beads demonstrated enhanced sorption capacities with increasing PCB concentration, aligning with a pseudo-second-order kinetic model. This multifaceted exploration underscores the versatility and promise of clay-polymer nanocomposites in mitigating the impact of diverse organic pollutants in aqueous environments.

## 11 Treatment of Biological Pollutants

In the realm of water purification, the challenge of eliminating biological pollutants has spurred innovative research into alternative techniques. Contamination from domestic sewage and animal husbandry are major sources of biological pollutants of our times. Traditionally, chlorination has proven effective against pathogens, yet its widespread use is hindered by the formation of harmful by-products. Ozonization, another promising method, faces limitations due to ozone's instability and the creation of undesirable by-products. The use of CPNs at present is becoming very popular in the removal of biological contaminants in water bodies. Cationic starch/Na-bentonite composites and poly(acrylamide)/saponite nanocomposites with silver nanoparticles demonstrated potent antibacterial properties against *Escherichia coli* and other bacterial strains. Microwave-assisted preparation of chitosan-based nanocomposites incorporating silver-bentonite, zinc oxide-bentonite, and their combination showcased remarkable antibacterial efficacy against *E. coli* and *Enterococcus faecalis*. Nylon-6 and cationic surfactant-modified montmorillonite organoclays exhibited biocidal activity against *Staphylococcus aureus* and *E. coli*, with Cloisite 30B standing out as the most effective. This study shows the inherent antimicrobial potential of clay-polymer composites, often overlooked but crucial in water purification. Furthermore, the incorporation of silver into organo-montmorillonite proved successful in reducing *E. coli* in nanocomposite films (Nigmatullin et al. 2008). Similarly, sodium alginate polymer/bentonite nanocomposites encapsulating antibacterial zinc oxide demonstrated promising results against *S. aureus*. A recent investigation involving xanthan gum grafted with poly(vinylimidazole) and montmorillonite clay exhibited excellent antibacterial activity against *E. coli* and *S. aureus* (Abu Elella et al. 2021). While the focus on the regenerability of these composites after use remains limited, it is evident that clay-polymer nanocomposites hold immense potential in combatting waterborne pathogens, surpassing conventional polymers as antimicrobial agents. Future research should prioritize regeneration studies and adsorption-desorption dynamics to fully realize the practicality of these composites in water purification.

## 12 Adsorption Mechanism of CPNs with Pollutants

The adsorption mechanism employed by CPNs for the removal of pollutants involves the deposition of contaminants onto adsorbents through a combination of van der Waals forces, covalent bonding, and electrostatic attraction. The nature of interaction is influenced by various factors, including the pH of the solution, temperature, and the functional groups present in the interacting species. The adsorption of organic pollutants, including dyes and pesticides, onto CPNs is intricately linked to both the pH of the system and the functional groups present on the adsorbents with electrostatic and non-electrostatic interactions, including van der Waals forces, hydrophobic interactions, and π–π stacking. The nature of interaction of pollutants onto CPNs depends on various factors, emphasizing the need for specific CPN fabrication protocols targeting adsorbate species. Computational simulations are recommended for optimizing synthetic and reaction processes, facilitating efficient pollutant removal.

The evaluation of adsorption through adsorption isotherms helps in better understanding the mechanism involved in adsorbent and adsorbate. The most used adsorption isotherms used for this purpose are Langmuir, Freundlich, Temkin, and Redlich-Peterson models as shown in Table 1. These models describe the relationship between the equilibrium adsorbate concentration in the liquid phase and the equilibrium adsorption amount on the surface of the solid phase adsorbent at a specific temperature (Wang and Guo 2020).

1. Langmuir Isotherm Model: The Langmuir model describes how molecules adsorb (stick) to a surface.

**Table 1** Shows the equations for calculating the amount of adsorption using (1) Langmuir isotherm, (2) Freundlich isotherm, (3) Temkim isotherm, and (4) Redlich-Peterson isotherm

| | | |
|---|---|---|
| ❶ | Langmuir isotherm<br>$q_e = \frac{b \cdot q_{\max} \cdot C_e}{1 + b \cdot C_e}$ | $q_e$—amount of adsorbed molecules (mg $g^{-1}$)<br>$b$—Langmuir equilibrium constant (L $mg^{-1}$)<br>$q_{\max}$—the maximum adsorption capacity (mg $g^{-1}$)<br>$C_e$—the adsorbate concentration at equilibrium (mg $L^{-1}$) |
| ❷ | Freundlich isotherm<br>$q_e = K_F \cdot C_e^{1/n}$ | $K_F$—Freundlich partition coefficient (L $g^{-1}$)<br>$n$—empirical constant ranging between 0 and 1<br>$R$—universal gas constant |
| ❸ | Temkin isotherm<br>$q_e = RT/b_T \ln A_T \cdot C_e$ | $T$—temperature (K)<br>$b_T$—Temkin isotherm constant (J $mol^{-1}$)<br>$A_T$—equilibrium binding constant (L $mg^{-1}$) |
| ❹ | Redlich-Peterson isotherm<br>$q_e = \frac{K_R \cdot C_e}{1 + a_R \cdot C_e^g}$ | $K_R$ (L $g^{-1}$) and $a_R$ ($L^g$ $mg^{-g}$)—constants<br>$g$—an exponent ($0 \le g \le 1$) |

Imagine you have a surface, and some molecules want to stick to it. Langmuir helps us understand how many molecules can stick and how they do it.

2. Freundlich Isotherm Model: The Freundlich model is another way to describe adsorption, especially for surfaces that don't have uniform sticking abilities. Not all parts of the surface are equally good for sticking molecules, hence assessed in a more flexible way.
3. Temkin Isotherm Model: The Temkin model looks at the interaction between adsorbate (molecules sticking) and adsorbent (the surface). Temkin is interested in how friendly or unfriendly the surface is to the molecules trying to stick.
4. Redlich-Peterson Isotherm Model: The Redlich-Peterson model is a mix of Langmuir and Freundlich, combining their features. Imagine taking the best parts of Langmuir and Freundlich and putting them together. Redlich-Peterson is a compromise that works well for different types of surfaces and sticking molecules. All these models help scientists and researchers understand and predict how molecules interact with surfaces, especially when it comes to adsorption. Each model has its own way of looking at these interactions, offering insights into different aspects of the process.

Langmuir isotherm assumes monolayer adsorption on a homogeneous surface, while Freundlich isotherm is suitable for heterogeneous surfaces. Temkin and Redlich-Peterson models consider factors like indirect adsorbate-adsorbate interactions and non-uniform adsorption energies. In all these adsorption models following factors are enumerated with their unit values (Wang and Guo 2020). $q_e$ is the amount of adsorbed molecules (mg $g^{-1}$), $b$ is the Langmuir equilibrium constant (L $mg^{-1}$), $q_{\max}$. is the maximum adsorption capacity (mg $g^{-1}$), $C_e$ is the adsorbate concentration at equilibrium (mg $L^{-1}$), $K_F$ is the Freundlich partition coefficient (L $g^{-1}$), $n$ is empirical constant ranging between 0 and 1, $R$ is the universal gas constant, $T$ is the temperature (K), $b_T$ is the Temkin isotherm constant (J $mol^{-1}$), $A_T$ is the equilibrium binding constant (L $mg^{-1}$), $K_R$ (L $g^{-1}$) and $a_R$ ($L^g$ $mg^{-g}$) are constants, and $g$ is an exponent ($0 \leq g \leq 1$).

## 13 Future Prospect of CPNs in Wastewater Treatment

In conclusion, the exploration of CPNs for water treatment holds tremendous promise and presents a compelling avenue for addressing the ever-growing challenges of wastewater management. The amalgamation of clay and polymers not only capitalizes on their individual strengths but also creates synergies that outperform conventional materials in terms of cost-effectiveness, abundance of raw materials, and selectivity towards diverse pollutants. Despite the encouraging properties demonstrated by CPNs, it is essential to acknowledge the current limitations, such as the predominance of laboratory-scale studies and the need for further optimization in real-world scenarios. To propel CPNs into mainstream application, emphasis should be placed on addressing challenges like regeneration capacity and fouling, potentially through innovative approaches like photocatalytic degradation. Additionally, future research endeavours should prioritize the development of standard protocols ensuring sustainability, environmental friendliness, and the utilization of greener polymer sources. By optimizing existing products, exploring novel catalyst incorporations, and evaluating performance in practical settings, CPNs stand poised to evolve into the next generation of adsorbents for wastewater treatment. As we navigate the intricate intersection of economic viability and environmental consciousness, CPNs emerge as a beacon of promise, offering a transformative solution to water pollution challenges and paving the way for a cleaner and more sustainable future.

## References

Abu Elella MH, Goda ES, Abdallah HM, Shalan AE, Gamal H, Yoon KR (2021) Innovative bactericidal adsorbents containing modified xanthan gum/montmorillonite nanocomposites for wastewater treatment. Int J Biol Macromol 167:1113–1125. https://doi.org/10.1016/j.ijbiomac.2020.11.065

Celis R, Trigo C, Facenda G, Hermosín MDC, Cornejo J (2007) Selective modification of clay minerals for the adsorption of herbicides widely used in olive groves. J Agric Food Chem 55:6650–6658. https://doi.org/10.1021/jf070709q

Effenberger F, Schweizer M, Mohamed WS (2009) Synthesis and characterization of some polyacrylate/montmorillonite nanocomposites by in situ emulsion polymerization using redox initiation system. J Appl Polym Sci 112:1572–1578. https://doi.org/10.1002/app.29605

Fayazi M, Ghanbarian M (2020) One-pot hydrothermal synthesis of polyethylenimine functionalized magnetic clay for efficient removal of noxious Cr(VI) from aqueous solutions. SILICON 12:125–134. https://doi.org/10.1007/s12633-019-00105-9

Hernandez-Hernandez KA, Illescas J, Díaz-Nava M, del C, Martínez-Gallegos S, Muro-Urista C, Ortega-Aguilar RE, Rodríguez-Alba E, Rivera E (2018) Preparation of nanocomposites for the removal of phenolic compounds from aqueous solutions. Appl Clay Sci 157:212–217. https://doi.org/10.1016/j.clay.2018.01.020

Hnamte M, Pulikkal AK (2022) Clay-polymer nanocomposites for water and wastewater treatment: a comprehensive review. Chemosphere 307:135869. https://doi.org/10.1016/j.chemosphere.2022.135869

Hsueh H-B, Chen C-Y (2003) Preparation and properties of LDHs/epoxy nanocomposites. Polymer 44:5275–5283. https://doi.org/10.1016/S0032-3861(03)00579-2

Kanchana V, Gomathi T, Geetha V, Sudha P (2012) Adsorption analysis of Pb (II) by nanocomposites of chitosan with methyl cellulose and clay. Der Pharm Lett 4:1071–1079. https://www.scholarsresearchlibrary.com/articles/adsorption-analysis-of-pbii-by-nanocomposites-of-chitosan-with-methyl-cellulose-and-clay.pdf

Mahdavinia GR, Soleymani M, Sabzi M, Azimi H, Atlasi Z (2017) Novel magnetic polyvinyl alcohol/laponite RD nanocomposite hydrogels for efficient removal of methylene blue. J Environ Chem Eng 5:2617–2630. https://doi.org/10.1016/j.jece.2017.05.017
Malatji N, Makhado E, Ramohlola KE, Modibane KD, Maponya TC, Monama GR, Hato MJ (2020) Synthesis and characterization of magnetic clay-based carboxymethyl cellulose-acrylic acid hydrogel nanocomposite for methylene blue dye removal from aqueous solution. Environ Sci Pollut Res 27:44089–44105. https://doi.org/10.1007/s11356-020-10166-8
Mansoori Y, Sanaei SS, Zamanloo MR, Imanzadeh G, Atghia SV (2013) Synthesis and properties of new polyimide/clay nanocomposite films. Bull Mater Sci 36:789–798. https://doi.org/10.1007/s12034-013-0542-4
Mdlalose L, Balogun M, Setshedi K, Tukulula M, Chimuka L, Chetty A (2017) Synthesis, characterization and optimization of poly(p-phenylenediamine)-based organoclay composite for Cr(VI) remediation. Appl Clay Sci 139:72–80. https://doi.org/10.1016/j.clay.2017.01.015
Minisy IM, Salahuddin NA, Ayad MM (2021) Adsorption of methylene blue onto chitosan–montmorillonite/polyaniline nanocomposite. Appl Clay Sci 203:105993. https://doi.org/10.1016/j.clay.2021.105993
Mizuno C, John B, Okamoto (2013) Percolated network structure formation and rheological properties in Nylon 6/clay nanocomposites. Macromol Mater Eng 298(4):400–411. https://doi.org/10.1002/mame.201200065
Mrah L, Meghabar R (2021) In situ polymerization of styrene–clay nanocomposites and their properties. Polym Bull 78:3509–3526. https://doi.org/10.1007/s00289-020-03274-5
Nigmatullin R, Gao F, Konovalova V (2008) Polymer-layered silicate nanocomposites in the design of antimicrobial materials. J Mater Sci 43:5728–5733. https://doi.org/10.1007/s10853-008-2879-4
Okamoto M, John B (2013) Synthetic biopolymer nanocomposites for tissue engineering scaffolds. Prog Polym Sci 38:1487–1503. https://doi.org/10.1016/j.progpolymsci.2013.06.001
Pergal MV, Gojgić-Cvijović G, Steinhart M, Manojlović D, Ostojić S, Pezo L, Spírkova M (2022) Novel polyurethane network/organoclay nanocomposites: microstructure and physicochemical properties. Prog Org Coating 163:106664. https://doi.org/10.1016/j.porgcoat.2021.106664
Ramírez-Gómez JA, Illescas J, Díaz-Nava M, del C, Muro-Urista C, Martínez Gallegos S, Rivera E (2019) Synthesis and characterization of clay polymer nanocomposites of P(4VP-co-AAm) and their application for the removal of Atrazine. Polymers 11. https://doi.org/10.3390/polym11040721
Setshedi KZ, Bhaumik M, Songwane S, Onyango MS, Maity A (2013) Exfoliated polypyrrole-organically modified montmorillonite clay nanocomposite as a potential adsorbent for Cr(VI) removal. Chem Eng J 222:186–197. https://doi.org/10.1016/j.cej.2013.02.061
Sheikh ZA, Ali N, Ali F, Ali A, Shabir G, Bilal M (2021) Fabrication, morphological, structural and electrochemical characterization of sulfonated polyimide/clay-based hybrid nanocomposite membranes for energy application. J Polym Res 28:1–12. https://doi.org/10.1007/s10965-020-02306-y
Shi Y, Xue Z, Wang X, Wang L, Wang A (2013) Removal of methylene blue from aqueous solution by sorption on lignocellulose-g-poly(acrylic acid)/montmorillonite three-dimensional cross-linked polymeric network hydrogels. Polym Bull 70:1163–1179. https://doi.org/10.1007/s00289-012-0898-4
Shokri E, Yegani R, Pourabbas B, Kazemian N (2016) Preparation and characterization of polysulfone/organoclay adsorptive nanocomposite membrane for arsenic removal from contaminated water. Appl Clay Sci 132:611–620. https://doi.org/10.1016/j.clay.2016.08.011
Sölener M, Tunali S, Özcan AS, Özcan A, Gedikbey T (2008) Adsorption characteristics of lead(II) ions onto the clay/poly(methoxyethyl)acrylamide (PMEA) composite from aqueous solutions. Desalination 223:308–322. https://doi.org/10.1016/j.desal.2007.01.221
Wang J, Guo X (2020) Adsorption isotherm models: classification, physical meaning, application and solving method. Chemosphere 258:127279. https://doi.org/10.1016/j.chemosphere.2020.127279
Wang L, Wang A (2008) Adsorption behaviors of Congo red on the N, O-carboxymethylchitosan/montmorillonite nanocomposite. Chem Eng J 143:43–50. https://doi.org/10.1016/j.cej.2007.12.007
Wang L, Zhang J, Wang A (2008) Removal of methylene blue from aqueous solution using chitosan-g-poly(acrylic acid)/montmorillonite superadsorbent nanocomposite. Colloids Surf A Physicochem Eng Asp 322:47–53. https://doi.org/10.1016/j.colsurfa.2008.02.019
Xu Y, Brittain WJ (2006) Preparation of polyacrylate-clay nanocomposites by in-situ polymerization and heterocoagulation. J Appl Polym Sci 101:3850–3855
Yasmin A, Abot JL, Daniel IM (2003) Processing of clay/epoxy nanocomposites by shear mixing. Scripta Mater 49:81–86. https://doi.org/10.1016/S1359-6462(03)00173-8
Yazdi F, Anbia M, Salehi S (2019) Characterization of functionalized chitosan clinoptilolite nanocomposites for nitrate removal from aqueous media. Int J Biol Macromol 130:545–555. https://doi.org/10.1016/j.ijbiomac.2019.02.127
Yu C, Ke Y, Deng Q, Lu S, Ji J, Hu X, Zhao Y (2018) Synthesis and characterization of polystyrene-montmorillonite nanocomposite particles using an anionic-surfactant-modified clay and their friction performance. Appl Sci 8. https://doi.org/10.3390/app8060964

# Toxicity and Ecotoxicology of Sludge

Prasann Kumar

**Abstract**

The proper handling and control of sludge, which is a byproduct generated from the treatment of wastewater and various industrial activities, presents notable environmental and human health concerns due to its potential toxicity and ecotoxicological impacts. This extensive study delves into the intricate aspects of sludge toxicity and ecotoxicology, offering a profound comprehension of the diverse pollutants and their effects on both land-based and water-based ecosystems. This study investigates the mechanisms contributing to sludge toxicity, its interactions with various environmental matrices, and its impacts on biodiversity and ecosystem functioning. The research employs a combination of laboratory experiments, field studies, and modeling approaches to gain a comprehensive understanding of these phenomena. Furthermore, this study examines various mitigation strategies and sustainable management practices aimed at reducing the detrimental effects of sludge on the environment. This study serves as an essential reference for policymakers, environmental experts, and researchers who seek to tackle the intricate challenges related to the toxicity of sludge and the field of ecotoxicology.

**Keywords**

Toxicity · Pollution · Sludge · Experiment · Environment · Field · No poverty · Zero hunger · Mitigation

P. Kumar (✉)
Department of Agronomy, School of Agriculture, Lovely Professional University, Phagwara, Punjab, India
e-mail: prasann0659@gmail.com

## 1 Introduction

The wastewater treatment process is of utmost importance as it is critical in protecting public health and maintaining environmental integrity (Ahmad Bhat et al. 2023; Ghosh and Biswas 2023; Grunst et al. 2023; Nawaz et al. 2023; Tran et al. 2023). The process entails the elimination of pollutants and contaminants from both domestic and industrial wastewater to guarantee that the discharged effluent into natural water bodies meets the required standards of quality. Nevertheless, an unavoidable consequence of this procedure is the formation of sludge. This substance exists in a semi-solid or solid state and comprises the remaining pollutants, organic substances, and microbial biomass extracted from the wastewater. Although there have been notable advancements in wastewater treatment, the ongoing concern of managing and disposing of sludge presents significant challenges. The topic of sludge management is a complex matter that involves various aspects, including the physical control and disposal of sludge, as well as its potential toxicity and ecotoxicological impacts. Sludge composition is characterized by a diverse array of pollutants, such as heavy metals, organic compounds, pathogens, and emerging contaminants, indicating that it is not a harmless substance. The comprehensive assessment of the environmental consequences of wastewater treatment processes necessitates a thorough understanding of the toxicity and ecotoxicology of sludge. This introductory statement establishes the foundation for a thorough investigation into the intricate and interrelated subjects of sludge toxicity and ecotoxicology (Osuoha et al. 2023; Li et al. 2023b; Cipriani-Avila et al. 2023). The treatment of wastewater is a crucial component of contemporary infrastructure, serving to safeguard both public health and the natural environment. The treatment of municipal and industrial sewage involves a sequence of physical, chemical, and biological procedures aimed at the elimination of contaminants and

© The Author(s), under exclusive license to Springer Nature Switzerland AG 2025 
A. Mandpe and M. P. Shah (eds.), *Wastewater Sludge Nexus*, Sustainable Landscape Planning and Natural Resources Management, https://doi.org/10.1007/978-3-031-45758-6_10

pathogens. The processes above give rise to sludge formation, which gradually accumulates as a concentrated amalgamation of pollutants, organic substances, and microorganisms. In numerous instances, the presence of sludge is inevitable, as it arises as a byproduct of the treatment procedure, thereby necessitating the prioritization of its management. Managing sludge entails making choices regarding the handling, disposal, and potential value extraction from this substance. The present scenario requires the consideration of issues about sludge toxicity and ecotoxicology. Gaining knowledge about the characteristics of contaminants present in sludge, their potential ramifications on the environment and human well-being, and strategies for mitigating these consequences holds significant significance. To gain a comprehensive understanding of the intricate nature of sludge toxicity and ecotoxicology, it is imperative to thoroughly examine the composition of sludge and its wide array of origins. The design of sludge exhibits variability, contingent upon its source, thereby rendering it a non-uniform substance. There exist two principal categories of sludge, namely wastewater sludge and industrial sludge. Wastewater sludge is obtained from treating domestic sewage, while diverse manufacturing and processing activities generate industrial sludge (Quang et al. 2023; Zhang et al. 2023; Khalaf et al. 2023; Moret et al. 2023; Zhou and Gao 2023; Dat et al. 2023; Yusuf et al. 2023). The composition of sludge is subject to the influence of various factors, including the inherent characteristics of the source wastewater and the specific treatment processes employed. Sludge has the potential to harbor a diverse array of pollutants, encompassing heavy metals such as lead, cadmium, and mercury, organic compounds including polycyclic aromatic hydrocarbons, pharmaceuticals, and pesticides, as well as pathogens like bacteria and viruses. Additionally, emerging contaminants like nanoparticles and microplastics may also be present in sludge. The origins of these contaminants can vary, and their occurrence in sludge underscores the importance of comprehending the toxicity and ecotoxicology of sludge. Methods for assessing toxicity testing is a crucial aspect of evaluating the potentially harmful effects of various substances. This process involves using different methods to evaluate the toxicity of sludge, which presents a formidable undertaking. Toxicity can manifest as acute or chronic, with the effects contingent upon the recipient organisms, which may exist in aquatic or terrestrial environments. To comprehend the harmful properties of sludge, it is imperative to employ specialized methodologies for testing, including bioassays, ecotoxicological assessment, and standardized protocols. Within the laboratory setting, researchers and environmental professionals use various tests to assess the potential detrimental effects of sludge on living organisms. The battery of tests may encompass acute toxicity assessments, which evaluate immediate lethal effects, and chronic toxicity evaluations, which gauge long-term consequences. Bioassays, a valuable methodology for assessing the toxicity of sludge, involve the utilization of living organisms to detect and measure toxic effects (Mohan et al. 2023; Dupouy 2023; Ansari et al. 2023; Singh et al. 2023a; Walch et al. 2023; Kumar et al. 2023; Kaur et al. 2023; Green et al. 2023). Furthermore, the ecotoxicological evaluation encompasses the examination of the impacts of sludge on complete ecosystems, thereby offering a more comprehensive comprehension of its ecological ramifications. Gaining a thorough understanding of the mechanisms that contribute to sludge toxicity is crucial for elucidating the potential environmental and human health consequences. The toxicity of sludge is influenced by various factors, including bioavailability, which refers to the capacity of contaminants to be assimilated by living organisms. Contaminants in sludge can exhibit two distinct characteristics that affect their availability for uptake by organisms. Firstly, these contaminants may be tightly bound to particles or organic matter, reducing accessibility. Alternatively, they may possess higher solubility, enabling organisms to absorb them readily. The mechanisms by which contaminants impact ecosystems exhibit variations between terrestrial and aquatic environments. The presence of sludge in soils has the potential to modify both the physical and chemical properties of the earth, thereby impacting the availability of nutrients and the growth of plants. Contaminants have the potential to infiltrate aquatic systems and undergo bioaccumulation as they traverse the trophic levels within food webs. The potential for interactions among various contaminants can give rise to synergistic or antagonistic effects, thereby introducing complexity to the comprehension of sludge toxicity. The ramifications of sludge toxicity transcend the confines of laboratory settings and directly affect the ecosystems in which sludge is utilized or discarded. Applying sludge can affect terrestrial ecosystems, encompassing agricultural lands and forested areas. These effects may include soil contamination, alterations in plant communities, and impacts on wildlife. The disposal of sludge in aquatic ecosystems has the potential to result in water contamination, thereby causing disturbances to marine life and exerting adverse effects on biodiversity. The presence of contaminants in sludge has the potential to undergo bioaccumulation in aquatic organisms, including fish, which may consequently pose risks to human consumers. The ecological ramifications of sludge are diverse, encompassing disturbances in the cycling of nutrients, alterations in the composition of species, and modifications in the provision of ecosystem services (Al-Jubouri et al. 2023; Patel et al. 2023; Menya et al. 2023; Moreira et al. 2023; Mohamed et al. 2023; Casabella-Font et al. 2023; Méndez-Loranca et al. 2023; GodvinSharmila et al. 2023). A rigorous approach to risk assessment and modeling is necessary to effectively manage the toxicity of sludge and address its ecotoxicological implications. The process of human health risk assessment entails the examination of potential exposures to contaminants present in sludge, followed by the quantification of the associated risks across different exposure scenarios. The primary objective of environmental risk assessment is to comprehensively evaluate the effects of sludge on ecosystems and the broader environment. Models and simulations are highly advantageous instruments for forecasting the destiny and conveyance of contaminants from sludge into the surrounding environment. These tools contribute to comprehending the enduring consequences associated with the disposal or application of sludge and can facilitate formulating strategies for managing associated risks. Implementing mitigation and remediation strategies is essential in addressing and resolving various challenges and issues. These strategies aim to minimize the negative impacts and consequences of certain events or circumstances, as well as to rectify any. The mitigation of the adverse effects stemming from the toxicity of sludge is a matter of utmost importance for individuals working in the field of environmental science and those involved in making policy decisions. The primary objective of mitigation and remediation strategies is to decrease the toxicity of sludge and minimize its ecological ramifications. The prioritization of beneficial reuse, mainly through land application, is becoming increasingly crucial in sustainable sludge management practices (Sunyer-Caldú et al. 2023; Mu et al. 2023; Peivasteh-roudsari et al. 2023; Cong et al. 2023; He et al. 2023; Beschorner and Randolph 2023; Long et al. 2023; Gulamhussein et al. 2023). Furthermore, there is a growing interest in the utilization of phytoremediation and bioremediation techniques to remediate soils and sediments that have been contaminated. These approaches involve the use of plants or microorganisms to either absorb or degrade the contaminants present in the affected areas. The establishment of regulatory frameworks and the formulation of policy recommendations are crucial to guide the responsible management of sludge while simultaneously ensuring the protection of the environment. The analysis of empirical cases offers significant scholarly contributions by shedding light on the pragmatic obstacles and achievements encountered in sludge toxicity and ecotoxicology. Case studies serve as valuable tools for examining the intricate and multifaceted aspects of sludge management, as well as its consequential effects on both the natural environment and local communities. Through analyzing various geographical areas, scholars and professionals can acquire a more profound comprehension of efficacious methodologies and valuable insights. The present study's conclusion regarding the toxicity and ecotoxicology of sludge delves into the existing challenges, emerging patterns, and prospective avenues for future research within this domain. The significance of ongoing research and innovation in sludge management becomes apparent in light of the growing challenges posed by urbanization and industrialization. The consideration of the intricate nature of sludge toxicity and ecotoxicology is of utmost importance to attain sustainable and environmentally conscientious practices in the treatment of wastewater. In essence, the management of sludge, an intrinsic byproduct of wastewater treatment, encompasses a complex array of considerations that surpass the mere physical manipulation and elimination of this substance. Examining sludge's toxicity and ecotoxicology is paramount due to its significant implications for the environment, human health, and sustainability. The primary objective of this study is to offer a thorough comprehension of the toxicity and ecotoxicology of sludge, thereby providing valuable insights into the difficulties, possible remedies, and prospects within the significant domain of environmental science and engineering (Soto-Donoso et al. 2023; Dupouy and Popping 2023; Kurade et al. 2023; Laad and Ghule 2023; Iheanacho et al. 2023; Ettahiri et al. 2023; Zahmatkesh et al. 2023; Bashir et al. 2023; Pradhan et al. 2023).

## 2 Sludge Composition and Sources

The process of wastewater treatment is an essential and pivotal procedure that serves as a crucial safeguard for both public health and the environment. By employing a combination of physical, chemical, and biological processes, this method efficiently eliminates contaminants, organic substances, and disease-causing microorganisms from residential and industrial wastewater. Nevertheless, an inescapable consequence of this procedure is the formation of sludge, a multifaceted substance characterized by its varied origins and compositions. This discourse examines the various classifications of sludge, the origins of its production, and its chemical and biological attributes. The composition of sludge exhibits considerable variability, as it is contingent upon both its source and the specific methodologies employed during wastewater treatment. There exist two main classifications of sludge, namely wastewater sludge and industrial sludge. Wastewater sludge, or biosolids, is produced as a byproduct of the treatment process for domestic sewage or municipal wastewater. When subjected to further treatment to comply with regulatory requirements for secure land application or disposal, it is commonly known as biosolids. Wastewater sludge generally comprises a combination of organic substances, pathogens, nutrients, and a limited amount of inorganic pollutants. The composition of a given population can exhibit variability based on factors such as the demographic being served, the contributions of the local industry, and the treatment processes that are implemented. Industrial sludge, alternatively referred to as industrial wastewater treatment sludge, is produced as a byproduct of diverse manufacturing and processing activities. Industries such as textiles, pulp and paper, food and beverage, and chemical manufacturing are responsible for the generation of industrial sludge (Singh et al. 2023b; Hjort et al. 2023; Li et al. 2023a, d; Yuan et al. 2023; Mali et al. 2023; Clance et al. 2023; Rose 2023). The composition of industrial sludge exhibits significant diversity and is contingent upon the unique processes employed within a given industry. The presence of heavy metals, organic chemicals, and various contaminants associated with industrial activities is a possibility. Sludge production is inherently interconnected with the characteristics of the processed wastewater and the diverse procedures employed in wastewater treatment. The following is the primary sources of sludge generation: During the primary treatment phase of wastewater treatment, the removal of larger solids, grit, and debris from the wastewater is achieved through physical processes such as sedimentation and flotation. The substance produced during this phase is commonly referred to as primary sludge, which comprises both organic and inorganic particulate matter. Secondary treatment involves using biological mechanisms, such as activated sludge or trickling filters, to decompose organic substances and eliminate dissolved and suspended impurities. The microorganisms in these processes generate biomass subsequently isolated from the treated water. The resultant substance is referred to as secondary sludge or biological sludge. Tertiary treatment processes may be employed in certain wastewater treatment plants to enhance the quality of the effluent further. Tertiary treatment encompasses various processes, such as chemical coagulation, filtration, and advanced oxidation. The sludge produced as a byproduct of tertiary treatment is commonly known as chemical sludge, as it consists of chemical precipitates and residual substances. Industrial processes often involve the discharge of wastewater, and as a result, many industries have established their wastewater treatment facilities. The sludge generated by industrial wastewater treatment plants exhibits characteristics unique to the particular industrial processes and contaminants in the wastewater. The composition of industrial sludge can exhibit significant variability, contingent upon the specific sector of origin. In areas where combined sewer systems are implemented, the occurrence of Combined Sewer Overflows (CSOs) can arise due to intense precipitation or the melting of snow. These CSOs result in the release of a blend of stormwater and untreated sewage into aquatic environments. The discharge process can lead to the formation of sludge and contribute to the contamination of water bodies. Gaining knowledge about the chemical and biological properties of sludge is imperative to grasp its implications on the environment and human health fully (Sewwandi et al. 2023; Moneta et al. 2023; François et al. 2023; Ghosh et al. 2023; Wang et al. 2023; Chen et al. 2023b; Timilsina et al. 2023; Boutet et al. 2023). The composition of sludge exhibits significant variability, rendering it a multifaceted and ever-changing substance. Organic Matter: Both wastewater sludge and industrial sludge comprise a range of organic compounds, including proteins, fats, carbohydrates, and cellulose. The organic constituents serve as a primary energy source for microorganisms in the sludge. In addition to organic matter, sludge may also harbor various types of inorganic contaminants. In industrial sludge, the constituents may encompass heavy metals such as lead, cadmium, and mercury alongside additional elements such as phosphorus and nitrogen. The presence of inorganic contaminants in wastewater sludge is contingent upon the origins of pollution. Pathogens, such as bacteria, viruses, and parasites, can be present in wastewater sludge, particularly in untreated or partially treated sludge. If the sludge is not adequately treated and managed, these microorganisms can present substantial health hazards. Nutrients: Sludge frequently contains many essential elements, including nitrogen and phosphorus. Although applying these nutrients as biosolids on agricultural lands can have positive effects, their excessive release into the environment can result in water pollution and the eutrophication of water bodies. The topic of emerging contaminants has garnered significant attention in recent years, particularly regarding their potential presence in sludge. These emerging contaminants encompass a range of substances, including pharmaceuticals, personal care products, and microplastics. These contaminants can infiltrate the wastewater system and subsequently integrate into the composition of the sludge matrix. The solids content of sludge exhibits significant variability. The composition of wastewater encompasses both suspended solids, which are particles that can be readily separated through sedimentation, and dissolved solids, which are substances that have undergone dissolution within the wastewater (Amiri et al. 2023; James et al. 2023; Abedpour et al. 2023; Chen et al. 2023a; Puri et al. 2023; Li et al. 2023c; Akhtar et al. 2021; Kumar et al. 2021c). A high concentration of solids in sludge can pose difficulties in its management and treatment. The pH levels and alkalinity of sludge can vary depending on the treatment processes employed and the characteristics of the influent wastewater. The pH level can potentially impact the accessibility of specific contaminants within the sludge. An unpleasant aesthetic quality frequently accompanies the presence of odors in sludge. The olfactory sensations experienced are attributed to the microbial processes occurring within the sludge, leading to the decomposition of organic substances. In terms of aesthetics, the visual characteristics of sludge can exhibit a range of variations, encompassing a dense, opaque, and mire-like composition, as well as a comparatively coarser, desiccated consistency upon transformation into biosolids. Sludge is a residual substance generated from the treatment of wastewater and industrial operations, exhibiting diverse characteristics contingent upon its origin and the specific treatment techniques employed. A comprehensive comprehension of the various categories of sludge, their origins, and their chemical and biological attributes is crucial in effectively tackling the multifaceted environmental, health, and disposal issues linked to this intricate substance. In the pursuit of sustainable wastewater management and the preservation of the environment, the management of sludge continues to be a crucial factor in ensuring the protection of water resources and the well-being of the general public (Table 1).

## 3 Sludge Toxicants

The substance known as sludge, which is produced as a byproduct of wastewater treatment and various industrial processes, possesses characteristics that deviate significantly from those of a benign nature. The substance encompasses a diverse array of toxic substances that can cause adverse impacts on both the natural environment and human well-being. This discourse delves into the various classifications of poisonous substances in sludge, encompassing heavy metals, organic contaminants, pathogens, and emerging contaminants. Heavy metals represent a significant category of harmful substances commonly present in sludge. The metals above are chemical elements possessing a substantial atomic mass, frequently distinguished by their density and capacity to pose a potential hazard to biological entities. Several heavy metals are commonly detected in sludge. Namely, Lead (Pb) is classified as a toxic heavy metal, known for its detrimental impact on human health and ecosystems. The substance can potentially accrue within terrestrial and aquatic environments, resulting in persistent pollution. Cadmium (Cd) is an additional heavy metal with toxic properties identified in sludge samples. The substance in question is recognized for its carcinogenic characteristics and its capacity to cause harm to aquatic organisms (Amiri et al. 2023; James et al., 2023; Abedpour et al. 2023; Chen et al. 2023a; Puri et al. 2023; Li et al. 2023c; Akhtar et al. 2021; Kumar et al. 2021a, c; Adelodun et al. 2022; Kumar and Mistri 2020; Das et al. 2022; Kumari et al. 2022; Upadhyay et al. 2023; Goud et al. 2022; Aley et al. 2022). Mercury (Hg): Mercury, in its diverse states, exhibits a high degree of toxicity. Methylmercury, a variant of mercury that can be generated within aquatic ecosystems, has the propensity to accumulate in fish and give rise to potential health hazards for individuals who consume fish that have been contaminated. Copper (Cu) is a vital micronutrient for both plants and animals. However, the presence of excessive copper in sludge can result in toxicity within delicate ecosystems. Excessive amounts of copper can harm aquatic organisms, especially in bodies of water where sludge is deposited. Zinc (Zn) is classified as an indispensable micronutrient, similar to copper, due to its vital role in various physiological processes. Nevertheless, the presence of excessive quantities of zinc in sludge has the potential to induce detrimental impacts on both soil organisms and aquatic ecosystems. Arsenic (As) is widely recognized as a potent carcinogen and toxic element. The presence of sludge can lead to the accumulation of this substance in soils and water bodies, posing potential risks to both human health and the environment. The occurrence of these high concentrations of heavy metals in sludge is commonly attributed to the discharge of industrial waste, the runoff of stormwater, and the introduction of diverse substances from the original wastewater. The necessity of effectively managing sludge arises from its capacity for bioaccumulation and its persistence in the environment, which pose a significant risk of long-term contamination (Chakraborty et al. 2021; Kotia et al. 2022; Kumar et al. 2021b; Intisar et al. 2023; Kumar et al. 2020). Organic contaminants encompass a heterogeneous array of compounds present in sludge, comprising a variety of chemicals, several of which are recognized for their inherent toxicological characteristics. Several primary classifications of organic contaminants found in sludge encompass polycyclic aromatic hydrocarbons (PAHs), which are a class of organic compounds that have the potential to be generated through incomplete combustion processes. These substances are widely recognized carcinogens and can accumulate in sludge from diverse sources, including industrial emissions and vehicular runoff. Pharmaceuticals and personal care products (PPCPs) encompass a wide range of substances, comprising both prescription and non-prescription drugs, alongside personal care items. These compounds can infiltrate the wastewater system and subsequently accumulate in sludge, raising concerns regarding their potential impact on the environment and the development of antibiotic resistance. The application of pesticides in agriculture and landscaping can result in their infiltration into sludge, posing a potential threat to non-target organisms within terrestrial and aquatic ecosystems. Polychlorinated biphenyls (PCBs) are a class of persistent organic pollutants that were extensively utilized in both electrical equipment and industrial contexts. These

**Table 1** Sludge components, its composition, and sources

| Sludge component | Composition details | Sources of sludge |
|---|---|---|
| Organic matter | Decomposed waste, microorganisms, and plant materials | Municipal wastewater treatment |
| Heavy metals | Elements like lead, cadmium, and mercury | Industrial processes |
| Pathogens | Disease-causing microorganisms | Wastewater treatment facilities |
| Organic contaminants | Synthetic chemicals and pharmaceuticals | Agricultural runoff |
| Sediments | Suspended particles and settled solids | Stormwater management |
| Nutrients | Nitrogen and phosphorus compounds | Agricultural runoff |
| Inorganic salts | Compounds like chlorides and sulfates | Deicing activities |
| Grease and oils | Lipids and fats from cooking and industrial sources | Food processing industries |
| Microplastics | Tiny plastic particles | Urban and industrial runoff |
| pH | Acidity or alkalinity | Industrial effluents |
| Inorganic salts | Compounds like chlorides and sulfates | Deicing activities |
| Grease and oils | Lipids and fats from cooking and industrial sources | Food processing industries |
| Microplastics | Tiny plastic particles | Urban and industrial runoff |
| pH | Acidity or alkalinity | Industrial effluents |
| Grit and sand | Coarse particles and abrasives | Stormwater and wastewater treatment |
| Ammonium ($NH_4^+$) | Nitrogen compound | Agriculture and wastewater treatment |
| Phosphates ($PO_4^-$) | Phosphorus compound | Agricultural runoff and wastewater |
| Sulfides ($S^-$) | Sulfur compounds | Industrial discharges |
| Algae and plants | Aquatic vegetation and algae | Natural water bodies |
| Volatile organics | Easily evaporated organic compounds | Industrial discharges |

*Source* Rose 2023; Boutet et al. 2023; Sewwandi et al. 2023; François et al. 2023

substances possess a high level of toxicity and have the potential to accumulate within the environment, including sludge. Volatile organic compounds (VOCs) refer to a class of organic chemicals that possess the propensity to undergo evaporation and disperse into the atmosphere readily. Certain compounds have been identified in sludge, which can contribute to concerns regarding air pollution in the context of sludge management. Endocrine-disrupting chemicals (EDCs) refer to a class of chemicals that possess the ability to disrupt the normal functioning of the endocrine system. This interference can cause detrimental health effects on both humans and wildlife. Certain endocrine-disrupting chemicals (EDCs) have been identified in sludge samples. The occurrence of these organic contaminants in sludge is linked to sources originating from both domestic and industrial activities (Chakraborty et al. 2021; Kotia et al. 2022; Kumar et al. 2021a; Intisar et al. 2023; Kumar et al. 2020; Mali et al. 2023; Moneta et al. 2023; James et al. 2023; Ghosh et al. 2023; François et al. 2023; Wang et al. 2023; Timilsina et al. 2023; Amiri et al. 2023; Chen et al. 2023b; Li et al. 2023a). The presence of these contaminants in sludge can endure over time and, if not appropriately controlled, could give rise to hazards for both ecosystems and human well-being. Pathogens and microbiological hazards constitute a distinct category of toxicants found in sludge. Sludge has the potential to harbor a diverse range of microorganisms, comprising bacteria, viruses, protozoa, and helminths, thereby presenting substantial health hazards. The introduction of pathogens into sludge can occur either through the original wastewater source or due to cross-contamination during the treatment process. Several essential factors concerning pathogens in sludge are as follows: Bacterial Pathogens: Sludge can harbor pathogenic bacteria, including *Escherichia coli* (*E. coli*), *Salmonella*, and *Clostridium perfringens.* If humans come into contact with the sludge or if it is improperly disposed of in the environment, these pathogens have the potential to cause gastrointestinal illnesses. Viral pathogens, such as enteric viruses and hepatitis A virus, may also be detected in sludge. Insufficient management and disposal practices can lead to the transmission of viral infections. Protozoa, such as Giardia and Cryptosporidium, and helminths, parasitic worms, have been present in sludge. These microorganisms can potentially induce waterborne illnesses and present a significant risk to human well-being. Implementing appropriate disinfection and treatment methods for sludge is of utmost importance to effectively reduce the potential for pathogen transmission. Strict adherence to guidelines and regulations is imperative to safeguard public health when biosolids are designated for land application. Emerging

**Table 2** Heavy metal, disadvantages, and its presence in sludge

| Heavy metal | Disadvantages | Presence in sludge |
|---|---|---|
| Arsenic (As) | Carcinogenic, neurotoxic, water pollution | Commonly found in sludge from mining and smelting |
| Cadmium (Cd) | Kidney damage, bone weakening | Present in sludge from industrial and urban sources |
| Chromium (Cr) | Carcinogenic water contamination | Found in sludge from tanneries and electroplating |
| Lead (Pb) | Neurological damage, developmental issues | Widespread presence in urban and industrial sludge |
| Mercury (Hg) | Neurotoxic bioaccumulation in the food chain | It is found in sludge from industrial and coal-fired plants |
| Copper (Cu) | Environmental toxicity, aquatic impacts | Familiar with wastewater sludge from various sources |
| Zinc (Zn) | Environmental toxicity, soil contamination | Present in sludge from industrial discharges |
| Nickel (Ni) | Carcinogenic, skin sensitization | Present in sludge from electroplating and alloys |
| Cobalt (Co) | Allergenic, potential carcinogen | Found in sludge from various industrial processes |
| Manganese (Mn) | Neurotoxic water contamination | It is found in sludge from mining and industrial activities |

*Source* Nawaz et al. 2023; Cipriani-Avila et al. 2023; Quang et al. 2023; Yusuf et al. 2023; Mohan et al. 2023; Zhang et al. 2023; Moret et al. 2023; Zhou and Gao 2023; Dat et al. 2023; Khalaf et al. 2023

contaminants refer to a class of toxic substances found in sludge that have garnered growing interest in recent times owing to their potential implications for the environment and human health. The substances above comprise a range of contaminants. The presence of pharmaceuticals, encompassing both prescription and over-the-counter drugs, within sludge, has elicited apprehension regarding their potential ecological ramifications and the emergence of antibiotic resistance (Chakraborty et al. 2021; Kotia et al. 2022; Kumar et al. 2021a; Intisar et al. 2023; Kumar et al. 2020; Mali et al. 2023; Moneta et al. 2023; James et al. 2023; Ghosh et al. 2023; François et al. 2023; Wang et al. 2023; Timilsina et al. 2023; Amiri et al. 2023; Chen et al. 2023b; Li et al. 2023a; Dupouy 2023; Singh et al. 2023a; Green et al. 2023; Moreira et al. 2023; Casabella-Font et al. 2023; Al-Jubouri et al. 2023; Kaur et al. 2023; Kumar et al. 2023; Walch et al. 2023). Personal Care Products (PCPs): Substances present in personal care products, including cosmetics and sunscreens, have the potential to infiltrate the wastewater system and ultimately accumulate in sludge. The concern arises due to their persistent nature and potential to disrupt endocrine function. Nanoparticles, deliberately designed at the nanoscale, have found applications across a wide range of industries and consumer goods. The investigation of their occurrence in sludge has emerged as a recent area of scholarly inquiry aimed at evaluating potential environmental and human health hazards. The issue of microplastics in sludge is becoming increasingly worrisome. Microplastics refer to tiny plastic particles smaller than 5 mm in diameter. These particles are present in personal care products and can infiltrate the wastewater system, posing a risk of contaminating both sludge and water bodies. Per- and Polyfluoroalkyl Substances (PFAS) encompass a collection of artificial compounds that find application in a diverse range of consumer goods, such as firefighting foam, nonstick cookware, and textiles. The compounds above exhibit a high degree of persistence and have been detected in sludge, giving rise to apprehensions regarding their prolonged presence in the environment and potential impacts on human health. The investigation of the occurrence and conduct of emerging contaminants in sludge is a subject of continuous scholarly inquiry as the comprehension of the potential hazards linked to these substances progresses. It can be inferred that sludge comprises a diverse array of harmful substances that can potentially present hazards to the natural ecosystem and the well-being of individuals. Heavy metals, organic contaminants, pathogens, and emerging contaminants are discrete classifications of toxic substances in sludge, each presenting unique difficulties and potential ramifications. Effectively managing sludge by regulatory guidelines and advancing research to mitigate the impact of these toxicants are essential measures in safeguarding both ecosystems and public health amidst the ongoing challenges posed by wastewater treatment and industrial processes (Table 2).

## 4 Toxicity Testing Methods

The assessment of toxicity is an essential element in evaluating the risks posed to both the environment and human health. The process aids in the evaluation of the detrimental impacts of diverse substances, including chemicals, pollutants, and environmental samples, on both living organisms and ecosystems. This discourse aims to explore the fundamental elements of toxicity testing methodologies, encompassing acute and chronic toxicity assessments, bioassays, and ecotoxicological evaluations, as well as standardized test protocols. Additionally, we will examine the challenges and limitations associated with these approaches. The assessment of acute toxicity is a pivotal approach employed to examine the immediate deleterious impacts of a substance on living organisms. The scope of this testing is centered on the examination of short-term exposure, typically spanning a duration of hours or days. The primary objective of this study is to ascertain the concentration or dosage of a substance that induces prompt and severe detriment to a particular organism or species. Acute toxicity testing encompasses several pivotal characteristics: The measurement of lethal concentration (LC50) is commonly employed in sensitive toxicity tests to determine the concentration at which 50% of the test organisms experience mortality due to toxic effects during a designated exposure period. Brief Exposure Duration: In acute experiments, the duration of exposure is typically brief, typically spanning from a few hours to a few days. This methodology is appropriate for evaluating immediate and profound impacts. Limited Information: Although acute tests offer valuable insights into the toxicological properties of a substance, they may not adequately capture the long-term, sublethal consequences or the wider ecological ramifications. Chronic toxicity testing is a scientific procedure that investigates the enduring and sublethal effects of a substance on living organisms, as opposed to acute testing, which focuses on immediate effects. This methodology offers valuable insights into the possible long-term and cumulative effects of exposure. The primary characteristics of chronic toxicity testing encompass the following. The No-Observed-Adverse-Effect Concentration (NOAEC), which pertains to regular testing, seeks to determine the maximum concentration of a substance that does not elicit any detectable adverse effects during a prolonged duration (Ghosh and Biswas 2023; Nawaz et al. 2023; Li et al. 2023b; Khalaf et al. 2023; Dat et al. 2023; Mohan et al. 2023; Ansari et al. 2023; Moreira et al. 2023; Casabella-Font et al. 2023; Sunyer-Caldú et al. 2023; Peivasteh-roudsari et al. 2023; Cong et al. 2023; He et al. 2023d). The term used to refer to this particular concentration is the No-Observed-Adverse-Effect Concentration (NOAEC). The exposure time is extended in chronic toxicity tests, typically from weeks to several months. This approach aims to replicate more realistic conditions to evaluate the ecological and health effects. A thorough comprehension: The implementation of chronic testing is of utmost importance in assessing the wider ecological and health ramifications of substances, thereby facilitating a more comprehensive grasp of their potential impacts on ecosystems. Bioassays refer to experimental methodologies that employ living organisms or biological systems to evaluate the toxicological effects of a given substance. Bioassays are a valuable tool for assessing the toxicity of specific substances toward particular species and are employed in both acute and chronic testing scenarios. Bioassays possess several key characteristics. Species-specific bioassays are commonly used in scientific research, as they utilize test organisms that accurately represent the target species or ecosystems under investigation. This methodology enables the implementation of customized evaluations. The sensitivity of bioassays is notable, as they can detect even subtle effects when exposed to low concentrations of a toxicant. The sensitivity above proves to be particularly advantageous in the timely detection of environmental issues. Bioassays are a reliable method for assessing the effects of toxicants on organisms, as they accurately capture the responses observed in real-life scenarios by considering the intricate interactions between the toxicant and the biological system. Ecotoxicological Evaluation: The evaluation of ecotoxicology surpasses the scope of individual-species experimentation and instead emphasizes the effects of toxic substances on entire ecosystems. This approach considers the interrelationships among various species and the ecological mechanisms operating within a specific ecosystem. Several important factors are involved in the ecotoxicological assessment: Ecotoxicological assessments are conducted to comprehend the impact of toxicants on populations, communities, and the overall functioning of ecosystems. Both direct and indirect effects are taken into consideration. Complex Interactions: These assessments consider the intricate interrelationships among species, trophic levels, and abiotic factors within ecosystems. Field studies are frequently conducted in natural environments as part of ecotoxicological assessments to obtain a comprehensive understanding of the effects of substances on ecosystems in realistic settings. To guarantee the uniformity and dependability of toxicity testing, standardized test protocols have been formulated by numerous national and international regulatory agencies and organizations. The protocols above offer comprehensive instructions, methodologies, and guidelines for executing toxicity tests. The utilization of standardized protocols presents numerous benefits: The implementation of standardized protocols facilitates the ability of researchers to perform experiments with a notable level of reproducibility (Mali et al. 2023; Yuan et al. 2023; Li et al. 2023a, d; Singh et al. 2023b; Hjort et al. 2023; Bashir et al. 2023; Pradhan et al. 2023; Ettahiri et al. 2023; Zahmatkesh et al. 2023; Iheanacho et al. 2023; Laad and Ghule 2023). This ensures that the results remain consistent and can be compared reliably across various laboratories and studies. Regulatory compliance necessitates the utilization of standardized protocols to evaluate the toxicity of substances, specifically in the context of environmental risk assessments and product registrations, as mandated by various regulatory authorities. Quality Control: Standardized protocols commonly incorporate stringent quality control measures to guarantee the dependability of outcomes and reduce experimental variability. Interlaboratory Comparisons: The utilization of standardized protocols enables the facilitation of interlaboratory comparisons and the exchange of data, thereby playing a crucial role in validating the robustness and reliability of the tests. Nevertheless, it is imperative to recognize that standardized test protocols may only sometimes align flawlessly with every circumstance. Certain organisms or ecosystems may possess distinct characteristics that necessitate tailored test methodologies. Although toxicity testing methods are important, they are not exempt from various challenges and limitations. Species-Specificity: Bioassays are frequently customized to particular species, potentially limiting their ability to reflect the entirety of an ecosystem or human health comprehensively. It is important to note that a substance that exhibits non-toxic properties in one species may possess a significantly elevated level of toxicity when exposed to another species. The extrapolation process poses challenges when applying findings derived from laboratory tests to real-world scenarios. Laboratory experiments frequently employ concentrations that deviate from realistic levels, while toxicity can be substantially influenced by variables such as temperature, salinity, and nutrient levels. Resource Intensive: Chronic toxicity tests are characterized by high resource requirements and time-consuming nature, rendering them less feasible for numerous regulatory applications (Mali et al. 2023; Yuan et al. 2023; Li et al. 2023a, d; Singh et al. 2023b; Hjort et al. 2023; Bashir et al. 2023; Pradhan et al. 2023; Ettahiri et al. 2023; Zahmatkesh et al. 2023; Iheanacho et al. 2023; Laad and Ghule 2023; Menya et al. 2023; Casabella-Font et al. 2023; Mu et al. 2023; He et al. 2023b; Beschorner and Randolph 2023; Cong et al. 2023; Peivasteh-roudsari et al. 2023). Toxicity testing primarily centers around examining individual substances; however, in natural environments, organisms are frequently subjected to mixtures of contaminants. Comprehending the interactive effects of multiple substances poses a multifaceted challenge. Ecological Significance: The comprehensive understanding of environmental interactions and the impacts on ecosystems in natural environments may need to be adequately captured by laboratory-based toxicity tests. Ethical considerations arise when employing live organisms in toxicity testing, raising concerns about animal welfare and environmental ethics. There is a growing trend among researchers to investigate alternative approaches, such as in vitro testing and computer modeling, to decrease dependence on live animals. The presence of emerging contaminants, such as nanoparticles and microplastics, presents significant obstacles in terms of devising suitable testing methodologies and comprehending their enduring impacts on organisms and ecosystems. Using toxicity testing methods is imperative to evaluate the detrimental impacts of substances on both organisms and ecosystems. Acute and chronic testing, bioassays, ecotoxicological assessment, and standardized test protocols are crucial components in comprehending the potential hazards linked to diverse contaminants. Nevertheless, it is of utmost importance to acknowledge the difficulties and constraints associated with these methodologies, tackle ethical considerations, and persist in devising groundbreaking strategies to improve the dependability and ecological applicability of toxicity testing within the framework of contemporary assessments of environmental and human health risks.

## 5 Mechanisms of Sludge Toxicity

The issue of sludge toxicity, which arises from a wide range of pollutants, is a multifaceted problem that can have significant consequences for both land-based and water-based ecosystems. To obtain a comprehensive understanding of sludge toxicity, it is imperative to investigate the fundamental mechanisms that govern the uptake of contaminants by organisms, the pathways by which these contaminants exert their toxic effects, and the potential for synergistic or antagonistic interactions among them. The present discourse explores the intricacies of sludge toxicity, emphasizing the significance of bioavailability, toxicological pathways, and the interplay between contaminants. The bioavailability and uptake of pollutants refer to the extent to which these substances can be absorbed and assimilated by living organisms (Mali et al. 2023; Yuan et al. 2023; Li et al. 2023a, b, d; Singh et al. 2023b; Hjort et al. 2023; Bashir et al. 2023; Pradhan et al. 2023; Ettahiri et al. 2023; Zahmatkesh et al. 2023; Iheanacho et al. 2023; Laad and Ghule 2023; Menya et al. 2023; Casabella-Font et al. 2023; Mu et al. 2023; He et al. 2023c; Beschorner and Randolph 2023; Cong et al. 2023; Peivasteh-roudsari et al. 2023; Osuoha et al. 2023; Zhang et al. 2023; Khalaf et al. 2023; Dat et al. 2023; Moret et al. 2023; Quang et al. 2023; Zhou and Gao 2023; Yusuf et al. 2023; Dupouy 2023; Mohan et al. 2023; Singh et al. 2023a; Ansari et al. 2023; Cipriani-Avila et al. 2023; Tran et al. 2023; Nawaz et al. 2023; Grunst et al. 2023). The term "bioavailability" pertains to the fraction of a contaminant present in sludge that is readily accessible and capable of being absorbed by organisms in their living state. The comprehension of sludge toxicity relies on a fundamental concept wherein the extent of bioavailability plays a crucial role in determining the capacity of contaminants to induce toxic effects. Various critical factors influence the variability of contaminant bioavailability in sludge. Chemical Composition: Contaminants within sludge can manifest in diverse chemical compositions, impacting their solubility and mobility characteristics. As an illustration, certain heavy metals can potentially exist as insoluble precipitates, reducing their bioavailability. In contrast, others may be present in a more soluble state, rendering them more readily accessible for biological uptake. The physical properties of sludge, including its texture and structure, can potentially influence the process of contaminant diffusion into the surrounding environment. Sludge possessing a greater concentration of organic matter has the potential to enhance its capacity for retaining contaminants, thereby diminishing their bioavailability. The solubility of pollutants can be influenced by the pH of the sludge or the receiving environment as it pertains to pH and redox conditions. For example, variations in pH levels have the potential to modify the speciation of heavy metals, thereby impacting their bioavailability. The presence of microorganisms can influence the bioavailability of sludge (Mali et al. 2023; Yuan et al. 2023; Li et al. 2023a, b, d; Singh et al. 2023a, b; Hjort et al. 2023; Bashir et al. 2023; Pradhan et al. 2023; Ettahiri et al. 2023; Zahmatkesh et al. 2023; Iheanacho et al. 2023; Laad and Ghule 2023; Menya et al. 2023; Casabella-Font et al. 2023; Mu et al. 2023; He et al. 2023; Beschorner and Randolph 2023; Cong et al. 2023; Peivasteh-roudsari et al. 2023; Osuoha et al. 2023; Zhang et al. 2023; Khalaf et al. 2023; Dat et al. 2023; Moret et al. 2023; Quang et al. 2023; Zhou and Gao 2023; Yusuf et al. 2023; Dupouy 2023; Mohan et al. 2023; Ansari et al. 2023; Cipriani-Avila et al. 2023; Tran et al. 2023; Nawaz et al. 2023; Grunst et al. 2023). Microorganisms can alter the chemical composition and solubility of contaminants, thereby affecting their speciation. Consequently, these microbes can have a significant impact on the bioavailability of pollutants, either by enhancing or reducing them. Over time, the bioavailability of pollutants has the potential to undergo alterations. Some compounds have the potential to exhibit changes in their bioavailability as a result of undergoing chemical and biological transformations. The uptake of contaminants by organisms is contingent upon the bioavailability of the toxicants present in the sludge, as well as the inherent characteristics of the microorganisms. Several factors can influence the uptake of contaminants, including the physiological traits of organisms, their feeding habits, and the specific types of contaminants present. Several important uptake mechanisms include Root Uptake: Within terrestrial ecosystems, plants can assimilate pollutants from sludge using their root systems. The process is significantly influenced by the bioavailability of contaminants in soil. Filter feeding is a prevalent strategy observed in various aquatic organisms, such as mussels and clams. These organisms possess the ability to accumulate contaminants present in sludge found in sediments. This accumulation occurs due to their filtration process, wherein they extract particles and plankton from the surrounding water. Direct ingestion of sludge particles containing contaminants is a phenomenon observed in both terrestrial and aquatic ecosystems, wherein organisms consume these particles. Respiratory Uptake: Aquatic microorganisms can uptake certain pollutants present in water through their gills as part of the respiration process. The mechanisms by which pollutants from sludge affect ecosystems can be intricate and encompass various stages. Comprehending these pathways is of utmost importance to evaluate the potential hazards and formulate efficient measures for risk reduction. Toxicological pathways exhibit variability contingent upon the specific nature of the contaminants and the contextual factors within the environment. Nonetheless, they typically encompass the subsequent components: The release of contaminants from sludge into the environment leads to exposure of organisms to these harmful substances. Exposure can manifest through various pathways, namely ingestion, dermal contact, or respiration, contingent upon the specific organism and the nature of the environment, be it terrestrial or aquatic. Bioaccumulation refers to the process by which contaminants accumulate within the tissues of organisms due to their uptake from the surrounding environment. The method of bioaccumulation holds particular importance for persistent contaminants, such as certain heavy metals and organic compounds. Biotransformation refers to the capacity exhibited by certain organisms to undergo the process of converting specific contaminants into alternative chemical forms. The method of biotransformation has the potential to either mitigate or enhance the toxicity of pollutants using detoxification or activation, respectively. Contaminants elicit toxic effects on organisms across multiple levels, encompassing molecular, cellular, and organismal dimensions. These effects above may lead to compromised growth, reproduction, and survival. Ecological Consequences: The adverse impacts on individual organisms due to toxicity can result in wider ecological ramifications. These may encompass modifications in population dynamics, transitions in community structure, and modifications in ecosystem functioning. One notable characteristic of sludge toxicity is the capacity for interactions between various contaminants, resulting in synergistic or antagonistic effects. Synergistic effects manifest when the collective influence of multiple impurities surpasses the cumulative impact of their products. Conversely, negative effects manifest when the existence of one contaminant diminishes the toxic properties of another. These interactions possess a high degree of complexity. They are subject to influence from various factors, such as Chemical Interactions. Certain contaminants have the potential to engage in chemical interactions, thereby influencing their speciation and bioavailability. One metal's presence can interfere with another metal's ability to bind to organisms' binding sites. Biological Interactions: Introducing a single contaminant can potentially modify the physiological reactions of organisms toward another pollutant. As an illustration, a particular contaminant has the potential to either facilitate or impede the absorption of another contaminant. Physiological Responses: The concurrent influence of contaminants can result in synergistic or antagonistic physiological reactions within organisms. These responses may be observed as alterations in metabolism, mechanisms of detoxification, and various other physiological processes. Comprehending these interactions is paramount in evaluating the genuine ecological and health hazards linked to sludge toxicity. Furthermore, this emphasizes the significance of adopting a comprehensive methodology for assessing environmental risks, considering the amalgamation of pollutants present in sludge and their potential synergistic impacts. The intricate and diverse nature of sludge toxicity mechanisms encompasses various aspects, such as the accessibility and absorption of pollutants, toxicological processes in soil and water, and the possibility of combined or opposing impacts among contaminants. A thorough comprehension of these mechanisms is imperative for the successful evaluation of risks, implementation of measures to reduce risks, and the establishment of sustainable practices for managing sludge. The issue of sludge toxicity poses a considerable obstacle in achieving a harmonious equilibrium between the imperative for efficient wastewater treatment and preserving both ecosystems and human well-being (Table 3).

## 6 Ecological Impacts of Sludge

The ecological consequences of sludge, a byproduct generated from wastewater treatment and industrial activities, can be substantial for terrestrial and aquatic ecosystems. Sludge discharge into the environment introduces a variety of pollutants, such as heavy metals, organic pollutants, and pathogens, which can cause disturbances in natural ecosystems and yield significant ramifications. This discourse delves into the ecological ramifications of sludge on both terrestrial and aquatic ecosystems, as well as the intricacies of food chains and the consequential loss of biodiversity and alterations in ecosystem services. Terrestrial ecosystems refer to the various ecological systems that exist on land. These ecosystems encompass a wide range of habitats, including forests; soil contamination can occur as a result of the application of sludge to agricultural lands, mainly due to the presence of heavy metals and organic contaminants (Li et al. 2023b; Beschorner and Randolph 2023; Sunyer-Caldú et al. 2023; Casabella-Font et al. 2023; Moreira et al. 2023; Mu et al. 2023; Peivasteh-roudsari et al. 2023). Over time, the gradual accumulation of these contaminants can have a detrimental impact on soil quality, thereby diminishing its capacity to sustain the growth of plants and other organisms residing within the earth. Plant Uptake: Certain plant species possess the ability to accumulate heavy metals from sludge, thereby facilitating their potential transfer into the food chain. Phytoremediation facilitates the movement of pollutants from the soil into plants, progressing through the various trophic levels. The impact on soil microbes is significant as these microorganisms play a crucial role in the cycling of nutrients and the decomposition of organic matter within the soil. The presence of sludge contamination has the potential to adversely affect microorganisms, leading to disturbances in soil processes and a decline in overall soil health. The fall of insects and invertebrates: Insects and invertebrates play crucial roles within terrestrial ecosystems. Sludge contamination can result in detrimental consequences for these organisms, resulting in a reduction in pollinators, decomposers, and other species that play a crucial role in the functioning of ecosystems. Exposure of Avian and Mammalian Species: Avifauna and mammalian fauna that engage in foraging activities within agricultural fields or areas contaminated with sludge may encounter potential exposure to various contaminants. The disclosure above may potentially result in detrimental consequences for these species, impacting their overall well-being and reproductive efficacy. The discharge of sludge into aquatic ecosystems has the potential to introduce various contaminants, such as heavy metals, organic compounds, and pathogens, into water bodies, resulting in water contamination. The presence of these contaminants can result in direct toxicity to aquatic organisms. The degradation of water quality can occur due to the fact of elevated nutrient levels in sludge, which has the potential to induce eutrophication in marine ecosystems. These phenomena can lead to detrimental algal blooms, decreased levels of oxygen, and alterations in the composition of aquatic communities. Bioaccumulation refers to the process by which marine organisms can potentially accumulate contaminants, hefty metals, in their tissues, primarily from sludge. The presence of these contaminants has the potential to undergo biomagnification within the food web, resulting in elevated concentrations at higher trophic levels. The health of fish, which play a crucial role in aquatic ecosystems, can be significantly compromised by sludge contamination (Goud et al. 2022; Das et al. 2022; Adelodun et al. 2022; Li et al. 2023c; Chen et al. 2023a; James et al. 2023; Boutet et al. 2023; Timilsina et al. 2023). The presence of heavy metals and organic contaminants has the potential to exert adverse effects on the health of fish, encompassing various aspects such as reproductive success and the broader dynamics of fish populations. The impacts of invertebrates: Aquatic invertebrates, including insects and crustaceans, exhibit sensitivity to alterations in water quality caused by the presence of sludge contamination. The decline of these organisms has the potential to cause disturbances within food webs, consequently impacting the accessibility of food resources for various aquatic species, including fish. Bioaccumulation refers to the process by which contaminants present in sludge accumulate in both marine and terrestrial ecosystems, resulting in elevated concentrations of toxic substances in organisms occupying higher trophic levels. Predatory species occupying the apex of the food chain may exhibit elevated levels of contaminants. Top-down effects refer to the ecological

**Table 3** Mechanism of sludge toxicity

| Mechanism of sludge toxicity | Description and key points |
|---|---|
| Bioavailability and uptake | Contaminants in sludge become accessible to organisms, leading to uptake. This can occur through ingestion, inhalation, or dermal contact |
| Chemical reactions | Contaminants may react with other environmental substances, altering their toxicity or form |
| Bioaccumulation and biomagnification | Organisms can accumulate contaminants in their tissues, increasing in concentration as they move up the food chain |
| Soil contamination | Contaminants in sludge can infiltrate soil, leading to decreased soil quality and potential crop uptake |
| Water contamination | Sludge can contaminate water bodies, affecting aquatic life and water quality |
| Airborne transport | Contaminants in sludge can become airborne through dust and particulate matter, posing risks to respiratory health |
| Microbial transformation | Microorganisms in sludge can transform contaminants, sometimes increasing their toxicity |
| Leaching and runoff | Contaminants can leach from sludge and be carried by runoff, potentially contaminating groundwater and surface water |
| Toxicological pathways | Specific routes through which contaminants interact with biological systems lead to adverse effects on health and ecosystems |
| Synergistic and antagonistic effects | Interaction between multiple contaminants can either amplify or reduce their combined toxicity |

*Source* Rose 2023; Boutet et al. 2023; Sewwandi et al. 2023; François et al. 2023; Kumar et al. 2023; Green et al. 2023; Mohamed et al. 2023

consequences that arise from the accumulation of contaminants in top predators within ecosystems. Predatory species may encounter diminished reproductive efficacy, population decrements, and modifications in community composition. Indirect Effects: The adverse impacts of sludge contaminants on a particular species can result in secondary consequences for other species coexisting within the same ecological system. As an illustration, the decrease in the population of a dominant predator can result in an upsurge in the abundance of prey species, consequently impacting plant populations and inducing modifications in the physical composition of habitats. The contamination of sludge can result in a reduction in biodiversity through its detrimental impact on the well-being and viability of diverse species (Kumar et al. 2020, 2021a; Intisar et al. 2023; Chakraborty et al. 2021; Kotia et al. 2022; Aley et al. 2022; Goud et al. 2022; Upadhyay et al. 2023). The diminishment of species diversity can be attributed to various factors, including the disturbance of food chains, degradation of habitats, and the direct impact of toxins. The provision of ecosystem services, which refers to the benefits that ecosystems offer to human populations, can be adversely affected by sludge contamination. The services above encompass pollination, water purification, and nutrient cycling. The diminishment of ecosystem services can occur due to the decline of crucial species and the disruption of ecosystem processes. Habitat alteration can occur as a result of the ecological impacts caused by the presence of sludge. For instance, alterations in plant populations or the depletion of crucial species can lead to modifications in the composition and arrangement of vegetation, thereby affecting the physical and ecological attributes of habitats. The stability of ecological communities can be diminished due to the environmental disturbance resulting from sludge contamination. Various factors, such as heightened susceptibility to invasive species, pest outbreaks, and fluctuations in population dynamics, can undermine the resilience of ecosystems. Sludge can exert substantial ecological consequences on terrestrial and aquatic ecosystems. The presence of contaminants originating from sludge has the potential to disrupt natural processes, inflict harm upon individual species, and modify the dynamics of ecosystems. The impacts above can result in shifts in biodiversity, declines in ecosystem services, and modifications in community stability. The proficient administration of sludge, encompassing its appropriate handling, elimination, and governance, is of utmost importance to mitigate its ecological impact and safeguard the well-being and functionality of ecosystems, as well as the societal benefits they offer (Kumar et al. 2020, 2021a, b; Intisar et al. 2023; Chakraborty et al. 2021; Kotia et al. 2022; Aley et al. 2022; Goud et al. 2022; Upadhyay et al. 2023; Kumari et al. 2022; Das et al. 2022; Kumar and Mistri 2020; Adelodun et al. 2022).

## 7 Risk Assessment and Modeling

Risk assessment and modeling are crucial in evaluating potential hazards linked to diverse processes, products, or activities, encompassing environmental and human health. These tools facilitate decision-making by enabling decision-makers to make well-informed choices, allocate resources effectively, and formulate strategies to mitigate potential risks. This discourse delves into the principles and techniques of human health risk assessment, environmental risk assessment, diverse modeling methodologies, and the significance of uncertainty and sensitivity analysis in risk assessment. The process of human health risk assessment is a methodical approach employed to determine the probability of detrimental outcomes arising from human contact with chemical, physical, or biological substances. The process generally encompasses several essential stages: The initial phase of hazard identification involves the identification and characterization of potential hazards that may be linked to a specific substance or exposure scenario. The process entails examining the existing scientific data to ascertain the toxicological characteristics of the essence. Exposure assessment is a process that involves the quantification of the scope, duration, and frequency of human exposure to a hazardous agent. This step consists of examining various pathways through which exposure can occur, such as ingestion, inhalation, and dermal contact, as well as identifying populations that have been exposed. The process of dose–response assessment aims to establish the correlation between the level of exposure and the probability of experiencing detrimental health effects. The process entails the identification of the threshold levels at which harm is observed, as well as the assessment of the magnitude of adverse effects across different levels of exposure. Risk characterization combines the outcomes obtained from the preceding stages to assess and quantify the level of risk. This frequently entails the comparison of estimated levels of exposure with established reference values, such as acceptable daily intake (ADI) or reference doses (RfD), to ascertain the potential risk to human health (Adelodun et al. 2022; Rose 2023; Moneta et al. 2023; Ghosh et al. 2023; Timilsina et al. 2023; James et al. 2023; Amiri et al. 2023; Boutet et al. 2023; Wang et al. 2023; Chen et al. 2023b). The primary objective of environmental risk assessment is to assess the potential negative impacts of different agents on ecosystems, with a specific emphasis on ecological well-being rather than human health. The components encompassed within this framework are as follows: The process of hazard assessment is akin to that of human health risk assessment, as it serves to identify and analyze the potential hazards that specific agents may pose to ecological components, including wildlife, plants, and aquatic ecosystems. Exposure assessment is a crucial component of environmental risk assessment, as it examines the various pathways through which agents can potentially enter the environment and subsequently affect ecological receptors. This assessment evaluates the magnitude, duration, and frequency of exposure across multiple species and ecosystems. The process of effects assessment involves the quantification of the probability and magnitude of negative impacts on ecological receptors. This stage may encompass both laboratory and field investigations aimed at assessing the levels of toxicity and ecological ramifications. The process of risk characterization involves the integration of findings from the preceding stages to evaluate the comprehensive risk posed to ecosystems. The assessment aims to determine a substantial risk of adverse ecological impacts and may result in suggestions for implementing risk management and mitigation strategies (Singh et al. 2023a; Kaur et al. 2023; Mohamed et al. 2023; GodvinSharmila et al. 2023; Menya et al. 2023; Kumar et al. 2023; Walch et al. 2023; Ansari et al. 2023; Mohan et al. 2023). Risk assessment frequently incorporates modeling methodologies to gauge and forecast potential risks. The complexity of these models may vary, and they are designed to suit specific risk assessment tasks. Several prevalent approaches include exposure modeling, which involves estimating the concentration or dose of a hazardous agent that individuals or ecosystems are expected to encounter. These models consider various factors, including environmental conditions, the release of pollutants, and the pathways through which individuals are exposed. Toxicological models are utilized to assess the potential detrimental impacts of a hazardous substance on both human and ecological receptors. These models consider the dose–response relationships and toxicological parameters derived from experimental or observational studies. Ecological models are computational tools that aim to replicate and analyze the intricate interactions and dynamic processes occurring within ecosystems. These models can predict the consequences of pollution exposure on populations, communities, and ecosystem dynamics. Human health risk models integrate orientation and toxicological information to assess the potential for detrimental health outcomes among particular people. Probabilistic models are frequently employed to accommodate and address the inherent variability and uncertainty present in a given system or phenomenon. Environmental fate models are computational tools that are used to forecast the behavior of contaminants in various environmental media, such as air, water, and soil. Contaminants can be better understood through the insights they provide regarding their transport, transformation, and persistence. Uncertainty constitutes an intrinsic component of the process of risk assessment and modeling. This phenomenon demonstrates imperfect knowledge and the inherent variability in data. Sensitivity analysis and uncertainty analysis are integral components of risk assessment employed to address uncertainties and bolster the dependability of outcomes. The uncertainty analysis process involves evaluating the extent and magnitude of uncertainty linked to different elements of risk assessment, such as data inputs, models, and assumptions. Frequently, probabilistic techniques, such as Monte Carlo simulations, are employed to quantify the level of uncertainty associated with risk assessments. Sensitivity analysis is a method that investigates the impact of variations in model parameters or inputs on the resulting outcomes (Cipriani-Avila et al. 2023; Osuoha et al. 2023; Zhang et al. 2023; Zhou and Gao 2023; Dupouy 2023; Yusuf et al. 2023; Moret et al. 2023; Khalaf et al. 2023; Quang et al. 2023). The purpose of this study is to determine the factors that exert the most significant influence on risk estimates. Sensitivity analysis is crucial in prioritizing data collection and research endeavors to mitigate uncertainty. By integrating uncertainty and sensitivity analysis techniques into risk assessment and modeling, decision-makers can enhance their comprehension of the dependability of risk estimates. This, in turn, enables them to make well-informed decisions about risk management and the implementation of measures aimed at mitigating potential risks. Risk assessment and modeling play a crucial role in the evaluation and management of risks to both human health and the environment. The processes above encompass a series of sequential stages, which contain the identification of potential hazards, assessment of exposure levels, evaluation of the relationship between dose and response, and the final determination of risk. Modeling methodologies are employed to assess and forecast potential hazards, while sensitivity and uncertainty analysis mitigate the inherent uncertainties associated with these procedures. The utilization of efficient risk assessment and modeling techniques plays a crucial role in facilitating well-informed decision-making processes and formulating strategies aimed at mitigating risks and safeguarding the well-being of individuals and the environment.

## 8 Mitigation and Remediation Strategies

Implementing sustainable practices in sludge management plays a pivotal role in ensuring the protection of both the environment and human well-being. Sludge, which is the residual substance resulting from the treatment of wastewater and various industrial processes, frequently encompasses a diverse array of impurities, such as heavy metals, organic pollutants, and pathogens. In response to the difficulties presented by sludge, a range of mitigation and remediation approaches have been devised. The strategies above prioritize the reduction of environmental and health hazards linked to sludge while concurrently investigating methods to utilize its potential as a valuable resource. This discourse aims to examine the fundamental approaches to achieving sustainable sludge management, encompassing the processes of sludge treatment and beneficial reuse, phytoremediation and bioremediation techniques, alongside the pertinent regulatory frameworks and policy recommendations. The objective of sustainable sludge management is to achieve a harmonious equilibrium between the imperative for efficient waste treatment and the imperative for safeguarding ecosystems and advancing resource recovery. The primary objective is to mitigate potential environmental and health hazards while simultaneously optimizing the advantageous utilization of sludge (Cipriani-Avila et al. 2023; Osuoha et al. 2023; Zhang et al. 2023; Zhou and Gao 2023; Dupouy 2023; Yusuf et al. 2023; Moret et al. 2023; Khalaf et al. 2023; Quang et al. 2023; Casabella-Font et al. 2023; Mu et al. 2023; Beschorner and Randolph 2023; Gulamhussein et al. 2023; Cong et al. 2023; Peivasteh-roudsari et al. 2023; Menya et al. 2023; Al-Jubouri et al. 2023; Kaur et al. 2023). The essential elements of sustainable sludge management encompass the following: The practice of sustainable sludge management commences with implementing measures to minimize sludge generation at its origin. This objective can be attained by implementing enhanced industrial procedures, adopting cleaner production methods, and minimizing the release of pollutants into wastewater systems. The appropriate collection and treatment of sewage and industrial effluents are imperative to effectively eliminate the most significant solids and contaminants before their introduction into the wastewater treatment system. Implementing efficient primary treatment methods results in a decrease in the overall burden placed on secondary treatment facilities. Implementing sustainable management practices facilitates the extraction of valuable resources from sludge, thereby promoting resource recovery. The extraction of energy, nutrients, and water from sludge can be achieved through various processes, including anaerobic digestion, thermal treatment, and nutrient recovery technologies. Beneficial reuse is a core principle in the realm of sustainable sludge management. The treatment and processing of sludge can result in the production of safe and valuable products that find application in agriculture, horticulture, and soil enhancement. This methodology effectively minimizes waste and preserves valuable resources. Implementing efficient sludge treatment methods is an essential aspect of ensuring sustainable sludge management. The principal objective of sludge treatment is to decrease its volume and mitigate its detrimental characteristics, thereby improving its potential for advantageous repurposing. Multiple treatment modalities are utilized for this objective: Anaerobic digestion is a biologically mediated process that facilitates the decomposition of organic matter present in sludge, resulting in the generation of biogas, a sustainable energy source, while concurrently achieving the stabilization of the sludge. The processed sludge can serve as a soil amendment or a source of plant nutrients. Aerobic digestion is a biological process that utilizes oxygen to facilitate the decomposition of organic substances present in sludge. Although less prevalent compared to anaerobic digestion, this process can also generate a stabilized sludge that is appropriate for advantageous reutilization. Thermal treatment methods, such as incineration and pyrolysis, employ elevated temperatures to eliminate organic contaminants and pathogens (Cipriani-Avila et al. 2023; Osuoha et al. 2023; Zhang et al. 2023; Zhou and Gao 2023; Dupouy 2023; Yusuf et al. 2023; Moret et al. 2023; Khalaf et al. 2023; Quang et al. 2023; Casabella-Font et al. 2023; Mu et al. 2023; Beschorner and Randolph 2023; Gulamhussein et al. 2023; Cong et al.

2023; Peivasteh-roudsari et al. 2023; Menya et al. 2023; Al-Jubouri et al. 2023; Kaur et al. 2023). These methods are highly effective in the reduction of sludge volume and the generation of energy as a secondary outcome. Composting is an inherent biological process that facilitates the conversion of organic matter in sludge into a final product that is stable and abundant in essential nutrients. The utilization of compost derived from sludge can serve as a means to enhance soil quality by acting as a soil conditioner, thereby ameliorating both the structure and fertility of the soil. Alkaline stabilization is a process that entails the introduction of alkaline substances, such as lime, into the sludge to elevate the pH level, diminish pathogen concentrations, and achieve the stabilization of heavy metals. The processed sludge has the potential to be utilized in agricultural practices. Phytoremediation and bioremediation represent novel methodologies for the management of sludge contamination. These techniques leverage the inherent capabilities of plants and microorganisms to address and restore polluted sites. In the realm of sludge management, these applications can be implemented as follows: Phytoremediation is a remediation technique that utilizes plants to effectively absorb, accumulate, or detoxify contaminants present in sludge or soils that have been contaminated. Certain plant species, commonly referred to as hyperaccumulators, exhibit a notable propensity for specific heavy metals and can be effectively employed for the remediation of metal-contaminated sludge. Bioremediation is a remediation technique that harnesses the metabolic capabilities of microorganisms, specifically bacteria and fungi, to degrade or convert pollutants in sludge. This can encompass the degradation of organic pollutants or the immobilization of heavy metals via microbial processes. Nature-based remediation techniques provide viable and economically efficient strategies for addressing sludge contamination, mainly when there is a significant concern for potential environmental and human health hazards. The implementation of robust regulatory frameworks and the formulation of policy recommendations are essential for the efficient management of sludge, with the primary objective of ensuring its environmentally sustainable and safe handling. Typically, these regulations encompass the following components: Regulations are in place to establish precise quality standards for treated sludge and to guarantee its safety for beneficial reuse in agricultural or other relevant applications. These criteria may encompass restrictions on heavy metals, pathogens, and organic contaminants. Regulatory bodies mandate the acquisition of permits for the operation of sludge management facilities, with operators obligated to provide reports on activities related to sludge treatment, disposal, and monitoring. Zoning and land use planning regulations play a crucial role in the allocation of appropriate areas for the establishment of sludge management facilities, thereby mitigating potential conflicts with residential or environmentally sensitive regions. Waste-to-Energy Regulations: If sludge is utilized for energy generation via methods such as anaerobic digestion or incineration, the regulations about facilities engaged in waste-to-energy activities are deemed applicable. Public engagement and transparency play a crucial role in ensuring the efficacy of sludge management practices. Public consultation may be mandated by regulations during the permitting process, and restrictions may require public access to information about sludge management practices. Research and Development Support: Policymakers have the potential to offer assistance for research and development endeavors focused on enhancing sludge treatment and beneficial reuse technologies, as well as exploring novel approaches to sludge management (Clance et al. 2023; Pradhan et al. 2023; Zahmatkesh et al. 2023; Li et al. 2023a; Yuan et al. 2023; Intisar et al. 2023; Kumar et al. 2021a, b, c; Kumar et al. 2021; Chakraborty et al. 2021; Kotia et al. 2022; Aley et al. 2022; Goud et al. 2022; Das et al. 2022; Kumar and Mistri 2020; Kumari et al. 2022; Upadhyay et al. 2023; Adelodun et al. 2022). The implementation of sustainable sludge management practices plays a pivotal role in safeguarding the environment, preserving valuable resources, and upholding human health and safety. Mitigation and remediation strategies encompass a range of treatment methods, including resource recovery and innovative approaches such as phytoremediation and bioremediation techniques. Regulatory frameworks and policy recommendations ensure that sludge management adheres to environmental standards and effectively mitigates risks to ecosystems and public health. Through the adoption of sustainable sludge management practices societies have the opportunity to effectively tackle the challenges associated with sludge while simultaneously harnessing its potential benefits as a valuable resource.

## 9 Future Perspectives and Concluding Remarks

The domain of sludge toxicity and ecotoxicology has made notable advancements in comprehending the environmental and human health hazards linked to this intricate waste substance. In concluding our discourse, we must contemplate prospective outlooks and the persistent obstacles encountered in mitigating the deleterious effects of sludge toxicity on ecosystems. The subsequent points elucidate several crucial factors and potential areas for further investigation and implementation.

1. The imperative for further investigation lies in enhancing the efficacy and sophistication of monitoring and assessment methodologies about the toxicity of sludge. This encompasses the utilization of emerging technologies, such as sensor networks, remote sensing, and high-throughput toxicological assays, to enhance the detection

and quantification of contaminants in sludge, as well as assessing their ecological implications.

2. Importance of Integrated Approaches: Implementing integrated approaches in sludge management is paramount. This necessitates the comprehensive examination of the complete life cycle of sludge, encompassing its inception, subsequent treatment, ultimate disposal, and potential advantageous utilization. Developing comprehensive solutions will require the essential participation of scientists, engineers, policymakers, and stakeholders through interdisciplinary collaboration.
3. Strategies for Mitigating Risk: It is imperative to prioritize the development of efficient risk mitigation strategies for sludge toxicity. This encompasses enhanced techniques for treating and disposing of sludge, encouragement of resource recovery and the advantageous utilization of sludge. Implementing sustainable sludge management practices is imperative to mitigate the adverse environmental and health consequences of sludge disposal.
4. Implementing circular economy principles, encompassing waste reduction, reuse, and recycling, can potentially revolutionize our approach to sludge management. By utilizing the energy and valuable resources inherent in sludge, it is possible to mitigate its environmental impact and make a significant contribution toward the goal of sustainability.
5. The investigation of ecosystem resilience should encompass an examination of the capacity of ecosystems to withstand and recover from the adverse effects of sludge contamination. A comprehensive comprehension of the inherent ability of natural systems to adapt and recover is imperative for attaining sustainable outcomes over an extended period.
6. The proliferation of emerging contaminants, including microplastics, nanoparticles, and pharmaceuticals, presents novel complexities in sludge toxicity and ecotoxicology. Subsequent investigations ought to prioritize the development of efficacious techniques aimed at identifying and alleviating these nascent perils.
7. Implementing comprehensive and current regulatory frameworks is imperative to guarantee the secure and effective management of sludge. Policymakers need to engage in ongoing evaluation and adjustment of regulations in order to effectively incorporate novel scientific insights and address the emergence of new contaminants.
8. Promoting public awareness and education is crucial in highlighting the potential hazards of sludge contamination and emphasizing the significance of adopting sustainable approaches to sludge management. The provision of public education has the potential to foster the endorsement of environmentally conscientious practices and policies.

In summary, sludge toxicity and ecotoxicology encompass various dimensions, necessitating a thorough and cohesive strategy. Through integrating scientific, technological, and policy advancements, a prospective trajectory can be established wherein sludge is effectively and ethically managed sustainably. The forthcoming agenda should emphasize the conservation of ecosystems, the safeguarding of human health, and the effective utilization of valuable resources found in sludge. Using continuous research, technological advancements, and collaborative efforts, it is possible to tackle the obstacles presented by sludge effectively and lay the foundation for a future characterized by sustainability and environmental friendliness.

## References

Abedpour H, Moghaddas JS, Borhani MN, Borhani TN (2023) Separation of toxic contaminants from water by silica aerogel-based adsorbents: a comprehensive review. J Water Process Eng 53:103676. https://doi.org/10.1016/j.jwpe.2023.103676

Adelodun B, Adewumi JR, Ajala OA, Ajibade FO, Ajibade TF, Akmal MH, Aley P et al (2022) List of contributors. In: Kumar A, Singh J, Ferreira LFR (eds) Microbiome under changing climate. Woodhead Publishing, pp 19–24. https://doi.org/10.1016/B978-0-323-90571-8.00028-6

Bhat A, Sartaj GC, Li W, Ameen F, Yaseera N, Wei Y, Li F (2023) Chapter eleven—fate of bio-contaminants in soil systems and available remediation methods. In: Huang K, Bhat SA, Cui G (eds) Fate of biological contaminants during recycling of organic wastes. Elsevier, pp 213–227. https://doi.org/10.1016/B978-0-323-95998-8.00014-5

Akhtar N, Amin-ul Mannan M, Banik RM, Baysal Ö, Bera S, Biswas P, Christina E et al (2021) List of contributors. In: Kumar A, Singh J, Samuel J (eds) Volatiles and metabolites of microbes. Academic Press, pp 19–21. https://doi.org/10.1016/B978-0-12-824523-1.00028-6

Al-Jubouri SM, Sabbar HA, Khudhair EM, Ammar SH, Al Batty S, Khudhair SY, Mahdi AS (2023) Silver oxide-zeolite for removal of an emerging contaminant by simultaneous adsorption-photocatalytic degradation under simulated sunlight irradiation. J Photochem Photobiol A: Chem 442:114763. https://doi.org/10.1016/j.jphotochem.2023.114763

Aley P, Singh J, Kumar P (2022) Chapter 23—adapting the changing environment: microbial way of life. In: Kumar A, Singh J, Ferreira LFR (eds) Microbiome under changing climate. Woodhead Publishing, pp 507–525. https://doi.org/10.1016/B978-0-323-90571-8.00023-7

Amiri MK, Zahmatkesh S, Emami MRS, Bokhari A (2023) Curve fitting model of polycarbonate $Al_2O_3$-nanoparticle membranes for removing emerging contaminants from wastewater: effect of temperature and nanoparticles. Chemosphere 322:138184. https://doi.org/10.1016/j.chemosphere.2023.138184

Ansari AA, Shamim MA, Khan AM, Anwar K, Wani AA (2023) Nanomaterials as a cutting edge in the removal of toxic contaminants from water. Mater Chem Phys 295:127092. https://doi.org/10.1016/j.matchemphys.2022.127092

Bashir M, Batool M, Arif N, Tayyab M, Zeng Y-J, Zafar MN (2023) Strontium-based nanomaterials for the removal of organic/inorganic contaminants from water: a review. Coord Chem Rev 492:215286. https://doi.org/10.1016/j.ccr.2023.215286

Beschorner KE, Randolph AB (2023) Friction performance of resilient flooring under contaminant conditions relevant to healthcare settings. Appl Ergon 108:103960. https://doi.org/10.1016/j.apergo.2022.103960

Boutet V, Dominique M, Eccles KM, Branigan M, Dyck M, van Coeverden de Groot P, Lougheed SC, Rutter A, Langlois VS (2023) An exploratory spatial contaminant assessment for polar bear (*Ursus maritimus*) liver, fat, and muscle from Northern Canada. Environ Pollut 316:120663. https://doi.org/10.1016/j.envpol.2022.120663

Casabella-Font O, Zahedi S, Gros M, Balcazar JL, Radjenovic J, Pijuan M (2023) Graphene oxide addition to anaerobic digestion of waste activated sludge: impact on methane production and removal of emerging contaminants. Environ Pollut 324:121343. https://doi.org/10.1016/j.envpol.2023.121343

Chakraborty S, Kumar P, Sanyal R, Mane AB, Prasanth DA, Patil M, Dey A (2021) Unravelling the regulatory role of Mirnas in secondary metabolite production in medicinal crops. Plant Gene 27: 100303. https://doi.org/10.1016/j.plgene.2021.100303

Chen J-Q, Sharifzadeh Z, Bigdeli F, Gholizadeh S, Li Z, Hu M-L, Morsali A (2023a) MOF composites as high potential materials for hazardous organic contaminants removal in aqueous environments. J Environ Chem Eng 11(2):109469. https://doi.org/10.1016/j.jece.2023.109469

Chen Q, Lü F, Zhang H, He P (2023b) Where should fenton go for the degradation of refractory organic contaminants in wastewater? Water Res 229:119479. https://doi.org/10.1016/j.watres.2022.119479

Cipriani-Avila I, Molinero J, Cabrera M, Medina-Villamizar EJ, Capparelli MV, Jara-Negrete E, Pinos-Velez V et al (2023) Occurrence of emerging contaminants in surface water bodies of a coastal province in Ecuador and possible influence of tourism decline caused by COVID-19 lockdown. Sci Total Environ 866:161340. https://doi.org/10.1016/j.scitotenv.2022.161340

Clance LR, Ziegler SL, Fodrie FJ (2023) Contaminants disrupt aquatic food webs via decreased consumer efficiency. Sci Total Environ 859:160245. https://doi.org/10.1016/j.scitotenv.2022.160245

Cong Y, Ye L, Zhang S, Zheng Q, Zhang Y, Lv S-W (2023) Efficient degradation of emerging contaminant by newly-constructed Ni-CoO yolk-shell hollow sphere in the presence of peroxymonosulfate: performance and mechanism. Process Saf Environ Protect 170:685–693. https://doi.org/10.1016/j.psep.2022.12.052

Das T, Saha SC, Sunita K, Majumder M, Ghorai M, Mane AB, Prasanth DA et al (2022) Promising botanical-derived monoamine oxidase (MAO) inhibitors: pharmacological aspects and structure-activity studies. South Afr J Bot 146:127–145. https://doi.org/10.1016/j.sajb.2021.09.019

Dat ND, Huynh QS, Tran KAT, Nguyen ML (2023) Performance of heterogeneous Fenton catalyst from solid wastes for removal of emerging contaminant in water: a potential approach to circular economy. Results Eng 18:101086. https://doi.org/10.1016/j.rineng.2023.101086

Dupouy E (2023) Chapter 24—trends in risk assessment of chemical contaminants in food. In: Knowles ME, Anelich LE, Boobis AR, Popping B (eds) Present knowledge in food safety. Academic Press, pp 320–328. https://doi.org/10.1016/B978-0-12-819470-6.00046-9

Dupouy E, Popping B (2023) Chapter 17—emerging contaminants. In: Knowles ME, Anelich LE, Boobis AR, Popping B (eds) Present knowledge in food safety. Academic Press, pp 267–269. https://doi.org/10.1016/B978-0-12-819470-6.00047-0

Ettahiri Y, Bouargane B, Fritah K, Akhsassi B, Pérez-Villarejo L, Aziz A, Bouna L, Benlhachemi A, Novais RM (2023) A state-of-the-art review of recent advances in porous geopolymer: applications in adsorption of inorganic and organic contaminants in water. Constr Build Mater 395:132269. https://doi.org/10.1016/j.conbuildmat.2023.132269

François M, Lin K-S, Rachmadona N, Khoo KS (2023) Advancement of nanotechnologies in biogas production and contaminant removal: a review. Fuel 340:127470. https://doi.org/10.1016/j.fuel.2023.127470

Ghosh S, Biswas A (2023) Chapter 9—emerging contaminants in municipal wastewater: occurrence, characteristics, and bioremediation. In: Haq I, Kalamdhad A, Pandey A (eds) Current developments in biotechnology and bioengineering. Elsevier, pp 153–178. https://doi.org/10.1016/B978-0-323-91902-9.00001-8

Ghosh S, Falyouna O, Onyeaka H, Malloum A, Bornman C, AlKafaas SS, Al-Sharify ZT et al (2023) Recent progress on the remediation of metronidazole antibiotic as emerging contaminant from water environments using sustainable adsorbents: a review. J Water Process Eng 51:103405. https://doi.org/10.1016/j.jwpe.2022.103405

GodvinSharmila V, Shanmugavel SP, Tyagi VK, Banu JR (2023) Microplastics as emergent contaminants in landfill leachate: source, potential impact and remediation technologies. J Environ Manage 343:118240. https://doi.org/10.1016/j.jenvman.2023.118240

Goud EL, Singh J, Kumar P (2022) Chapter 19—climate change and their impact on global food production. In: Kumar A, Singh J, Ferreira LFR (eds) Microbiome under changing climate. Woodhead Publishing, pp 415–436. https://doi.org/10.1016/B978-0-323-90571-8.00019-5

Green H, Kidd J, Jackson L (2023) Reference module in food science. In: Novel and emerging cleaning and sanitization technologies. Elsevier. https://doi.org/10.1016/B978-0-12-822521-9.00231-8

Grunst AS, Grunst ML, Fort J (2023) Contaminant-by-environment interactive effects on animal behavior in the context of global change: evidence from avian behavioral ecotoxicology. Sci Total Environ 879:163169. https://doi.org/10.1016/j.scitotenv.2023.163169

Gulamhussein MA, Saini B, Dey A (2023) Removal of pharmaceutical contaminants through membrane bioreactor. Mater Today: Proc 77:260–268. https://doi.org/10.1016/j.matpr.2022.11.299

He Q, Habib F, Tong T, Yu C (2023) Reference module in food science. In: Raman spectroscopy for detection of foodborne pathogens, chemical contaminants and nanoparticles. Elsevier. https://doi.org/10.1016/B978-0-12-822521-9.00188-X

Hjort RG, Pola CC, Soares RRA, Oliveira DA, Stromberg L, Claussen JC, Gomes C (2023) Reference module in food science. In: Advances in biosensors for detection of foodborne microorganisms, toxins, and chemical contaminants. Elsevier. https://doi.org/10.1016/B978-0-12-822521-9.00187-8

Iheanacho S, Ogbu M, Bhuyan MS, Ogunji J (2023) Microplastic pollution: an emerging contaminant in aquaculture. Aquac Fish 8(6):603–616. https://doi.org/10.1016/j.aaf.2023.01.007

Intisar A, Ramzan A, Hafeez S, Hussain N, Irfan M, Shakeel N, Gill KA, Iqbal A, Janczarek M, Jesionowski T (2023) Adsorptive and photocatalytic degradation potential of porous polymeric materials for removal of pesticides, pharmaceuticals, and dyes-based emerging contaminants from water. Chemosphere 336:139203. https://doi.org/10.1016/j.chemosphere.2023.139203

James CA, Sofield R, Faber M, Wark D, Simmons A, Harding L, O'Neill S (2023) The screening and prioritization of contaminants of emerging concern in the marine environment based on multiple biological response measures. Sci Total Environ 886:163712. https://doi.org/10.1016/j.scitotenv.2023.163712

Kaur M, Ghosh D, Guleria S, Arya SK, Puri S, Khatri M (2023) Microplastics/nanoplastics released from facemasks as contaminants of emerging concern. Marine Pollut Bull 191:114954. https://doi.org/10.1016/j.marpolbul.2023.114954

Khalaf EM, Jabbar HS, Romero-Parra RM, Al-Awsi GRL, Budi HS, Altamimi AS, Abdulfadhil Gatea M, Falih KT, Singh K, Alkhuzai KA (2023) Smartphone-assisted microfluidic sensor as an intelligent device for on-site determination of food contaminants: developments and applications. Microchem J 190:108692. https://doi.org/10.1016/j.microc.2023.108692

Kotia A, Rutu P, Singh V, Kumar A, Dhoke S, Kumar P, Singh DK (2022) Rheological analysis of rice husk-starch suspended in water for sustainable agriculture application. Mater Today: Proc 50:1962–1966. https://doi.org/10.1016/j.matpr.2021.09.325

Kumar M, Sridharan S, Sawarkar AD, Shakeel A, Anerao P, Mannina G, Sharma P, Pandey A (2023) Current research trends on emerging contaminants pharmaceutical and personal care products (PPCPs): a comprehensive review. Sci Total Environ 859:160031. https://doi.org/10.1016/j.scitotenv.2022.160031

Kumar P, Devi P, Dey SR (2021a) Chapter 6—fungal volatile compounds: a source of novel in plant protection agents. In: Kumar A, Singh J, Samuel J (eds) Volatiles and metabolites of microbes. Academic Press, pp 83–104. https://doi.org/10.1016/B978-0-12-824523-1.00001-8

Kumar P, Kumar T, Singh S, Tuteja N, Prasad R, Singh J (2020) Potassium: a key modulator for cell homeostasis. J Biotechnol 324:198–210

Kumar P, Mistri TK (2020) Transcription factors in SOX family: potent regulators for cancer initiation and development in the human body. Semin Cancer Biol 67:105–113. https://doi.org/10.1016/j.semcancer.2019.06.016

Kumar P, Sharma K, Saini L, Dey SR (2021b) Chapter 8—role and behavior of microbial volatile organic compounds in mitigating stress. In: Kumar A, Singh J, Samuel J (eds) Volatiles and metabolites of microbes. Academic Press, pp 143–161. https://doi.org/10.1016/B978-0-12-824523-1.00010-9

Kumar V, Dwivedi P, Kumar P, Singh BN, Pandey DK, Kumar V, Bose B (2021c) Mitigation of heat stress responses in crops using nitrate primed seeds. South Afr J Bot 140: 25–36. https://doi.org/10.1016/j.sajb.2021.03.024

Kumari P, Singh J, Kumar P (2022) Chapter 21—impact of bioenergy for the diminution of an ascending global variability and change in the climate. In: Kumar A, Singh J, Ferreira LFR (eds) Microbiome under changing climate. Woodhead Publishing, pp 469–487. https://doi.org/10.1016/B978-0-323-90571-8.00021-3

Kurade MB, Mustafa G, Zahid MT, Awasthi MK, Chakankar M, Pollmann K, Khan MA et al (2023) Integrated phycoremediation and ultrasonic-irradiation treatment (IPUT) for the enhanced removal of pharmaceutical contaminants in wastewater. Chem Eng J 455:140884. https://doi.org/10.1016/j.cej.2022.140884

Laad M, Ghule B (2023) Removal of toxic contaminants from drinking water using biosensors: a systematic review. Groundwater Sustain Develop 20:100888. https://doi.org/10.1016/j.gsd.2022.100888

Li H, Cheng B, Zhang J, Zhou X, Shi C, Zeng L, Wang C (2023a) Recent advances in the application of bismuth-based catalysts for degrading environmental emerging organic contaminants through photocatalysis: a review. J Environ Chem Eng 11(5):110371. https://doi.org/10.1016/j.jece.2023.110371

Li N, He X, Ye J, Dai H, Peng W, Cheng Z, Yan B, Chen G, Wang S (2023b) $H_2O_2$ activation and contaminants removal in heterogeneous Fenton-like systems. J Hazard Mater 458:131926. https://doi.org/10.1016/j.jhazmat.2023.131926

Li X, Yang H, Ma H, Zhi HG, Li D, Zheng X (2023c) Carbonaceous materials applied for cathode electro-Fenton technology on the emerging contaminants degradation. Process Saf Environ Protect 169:186–198. https://doi.org/10.1016/j.psep.2022.11.037

Li Y, Liu Y, Feng L, Zhang L (2023d) A review: manganese-driven bioprocess for simultaneous removal of nitrogen and organic contaminants from polluted waters. Chemosphere 314:137655. https://doi.org/10.1016/j.chemosphere.2022.137655

Long Y, Song L, Shu Y, Li B, Peijnenburg W, Zheng C (2023) Evaluating the spatial and temporal distribution of emerging contaminants in the pearl river basin for regulating purposes. Ecotoxicol Environ Saf 257:114918. https://doi.org/10.1016/j.ecoenv.2023.114918

Mali H, Shah C, Raghunandan BH, Prajapati AS, Patel DH, Trivedi U, Subramanian RB (2023) Organophosphate pesticides an emerging environmental contaminant: pollution, toxicity, bioremediation progress, and remaining challenges. J Environ Sci 127:234–250. https://doi.org/10.1016/j.jes.2022.04.023

Méndez-Loranca E, Vidal-Ruiz AM, Martínez-González O, Huerta-Aguilar CA, Gutierrez-Uribe JA (2023) Beyond cellulose extraction: recovery of phytochemicals and contaminants to revalorize agricultural waste. Bioresour Technol Rep 21:101339. https://doi.org/10.1016/j.biteb.2023.101339

Menya E, Jjagwe J, Kalibbala HM, Storz H, Olupot PW (2023) Progress in deployment of biomass-based activated carbon in point-of-use filters for removal of emerging contaminants from water: a review. Chem Eng Res Design 192:412–440. https://doi.org/10.1016/j.cherd.2023.02.045

Mohamed BA, Hamid H, Montoya-Bautista CV, Li LY (2023) Circular economy in wastewater treatment plants: treatment of contaminants of emerging concerns (CECs) in effluent using sludge-based activated carbon. J Cleaner Prod 389:136095. https://doi.org/10.1016/j.jclepro.2023.136095

Mohan B, Priyanka GS, Chauhan A, Pombeiro AJL, Ren P (2023) Metal-organic frameworks (MOFs) based luminescent and electrochemical sensors for food contaminant detection. J Hazard Mater 453:131324. https://doi.org/10.1016/j.jhazmat.2023.131324

Moneta G, Benedetta SEA, Barbaro E, Capriotti AL, Cerrato A, Laganà A, Montone CM et al (2023) Untargeted analysis of environmental contaminants in surface snow samples of Svalbard islands by liquid chromatography-high resolution mass spectrometry. Sci Total Environ 858:159709. https://doi.org/10.1016/j.scitotenv.2022.159709

Moreira RG, Branco GS, Lo Nostro FL (2023) Chapter 3.2.3—effects of aquatic contaminants in female fish reproduction. In: Carnevali O, Hardiman G (eds) Environmental contaminants and endocrine health. Academic Press, pp 257–268. https://doi.org/10.1016/B978-0-12-824464-7.00004-0

Moret S, Conchione C, Barp L (2023) Reference module in food science. In: Food packaging contaminants with a special focus on hydrocarbon contaminants and nanoparticles. Elsevier. https://doi.org/10.1016/B978-0-12-822521-9.00180-5

Mu J, Feng J, Wang X, Liu B (2023) Oxygen vacancy boosting peroxymonosulfate activation over nanosheets assembled flower-like $CoMoO_4$ for contaminant removal: performance and activity enhancement mechanisms. Chem Eng J 459:141537. https://doi.org/10.1016/j.cej.2023.141537

Nawaz S, Tabassum A, Muslim S, Nasreen T, Baradoke A, Kim TH, Boczkaj G, Jesionowski T, Bilal M (2023) Effective assessment of biopolymer-based multifunctional sorbents for the remediation of environmentally hazardous contaminants from aqueous solutions. Chemosphere 329:138552. https://doi.org/10.1016/j.chemosphere.2023.138552

Osuoha JO, Anyanwu BO, Ejileugha C (2023) Pharmaceuticals and personal care products as emerging contaminants: need for combined treatment strategy. J Hazard Mater Adv 9:100206. https://doi.org/10.1016/j.hazadv.2022.100206

Patel P, Nandi A, Verma SK, Kaushik N, Suar M, Choi EH, Kaushik NK (2023) Zebrafish-based platform for emerging bio-contaminants and virus inactivation research. Sci Total Environ 872:162197. https://doi.org/10.1016/j.scitotenv.2023.162197

Peivasteh-roudsari L, Barzegar-bafrouei R, Sharifi KA, Azimisalim S, Karami M, Abedinzadeh S, Asadinezhad S et al (2023) Origin, dietary exposure, and toxicity of endocrine-disrupting food chemical contaminants: a comprehensive review. Heliyon e18140. https://doi.org/10.1016/j.heliyon.2023.e18140

Pradhan B, Chand S, Chand S, Rout PR, Naik SK (2023) Emerging groundwater contaminants: a comprehensive review on their health hazards and remediation technologies. Groundwater Sustain Develop 20:100868. https://doi.org/10.1016/j.gsd.2022.100868

Puri M, Gandhi K, Kumar MS (2023) Emerging environmental contaminants: a global perspective on policies and regulations.

J Environ Manage 332:117344. https://doi.org/10.1016/j.jenvman.2023.117344
Quang HHP, Dinh DA, Dutta V, Chauhan A, Lahiri SK, Gopalakrishnan C, Radhakrishnan A, Batoo KM, Phan Thi L-A (2023) Current approaches, and challenges on identification, remediation and potential risks of emerging plastic contaminants: a review. Environ Toxicol Pharmacol 101:104193. https://doi.org/10.1016/j.etap.2023.104193
Rose M (2023) Reference module in food science. In: Dioxins and other environmental contaminants in meat products, including fish and sea food. Elsevier. https://doi.org/10.1016/B978-0-323-85125-1.00199-X
Sewwandi M, Wijesekara H, Rajapaksha AU, Soysa S, Vithanage M (2023) Microplastics and plastics-associated contaminants in food and beverages; global trends, concentrations, and human exposure. Environ Pollut 317:120747. https://doi.org/10.1016/j.envpol.2022.120747
Singh AW, Soni R, Pal AK, Tripathi P, Lal JA, Tripathi V (2023a) Chapter 12—microalgal-based bioremediation of emerging contaminants in wastewater: a sustainable approach. In: Shukla P (eds) Progress in biochemistry and biotechnology. Microbial bioprocesses. Academic Press, pp 275–297. https://doi.org/10.1016/B978-0-323-95332-0.00015-6
Singh A, Chaurasia D, Khan N, Singh E, Bhargava PC (2023b) Efficient mitigation of emerging antibiotics residues from water matrix: integrated approaches and sustainable technologies. Environ Pollut 328:121552. https://doi.org/10.1016/j.envpol.2023.121552
Soto-Donoso N, Favier L, Villalobos SF, Paredes-García V, Bataille T, Marco JF, Hlihor RM, Le Fur E, Venegas-Yazigi D (2023) Lanthanum hydroxichloride/anatase composite and its application for effective UV-light driven oxidation of the emergent water contaminant cetirizine. Chem Eng Res Design. https://doi.org/10.1016/j.cherd.2023.06.032
Sunyer-Caldú A, Golovko O, Kaczmarek M, Asp H, Bergstrand K-J, Gil-Solsona R, Gago-Ferrero P, Diaz-Cruz MS, Ahrens L, Hultberg M (2023) Occurrence and fate of contaminants of emerging concern and their transformation products after uptake by Pak Choi (*Brassica rapa* Subsp. Chinensis). Environ Pollut 319:120958. https://doi.org/10.1016/j.envpol.2022.120958
Timilsina A, Adhikari K, Yadav AK, Joshi P, Ramena G, Bohara K (2023) Effects of microplastics and nanoplastics in shrimp: mechanisms of plastic particle and contaminant distribution and subsequent effects after uptake. Sci Total Environ 894:164999. https://doi.org/10.1016/j.scitotenv.2023.164999
Tran H-T, Bolan NS, Lin C, Binh QA, Nguyen M-K, Luu TA, Le V-G, Pham CQ, Hoang H-G, Vo DVN (2023) Succession of biochar addition for soil amendment and contaminants remediation during co-composting: a state of art review. J Environ Manage 342:118191. https://doi.org/10.1016/j.jenvman.2023.118191
Upadhyay SK, Devi P, Kumar V, Pathak HK, Kumar P, Rajput VD, Dwivedi P (2023) Efficient removal of total arsenic (As3+/5+) from contaminated water by novel strategies mediated iron and plant extract activated waste flowers of marigold. Chemosphere 313:137551. https://doi.org/10.1016/j.chemosphere.2022.137551
Walch H, Praetorius A, von der Kammer F, Hofmann T (2023) Generation of reproducible model freshwater particulate matter analogues to study the interaction with particulate contaminants. Water Res 229:119385. https://doi.org/10.1016/j.watres.2022.119385
Wang R, Luo J, Li C, Chen J, Zhu N (2023) Antiviral drugs in wastewater are on the rise as emerging contaminants: a comprehensive review of spatiotemporal characteristics, removal technologies and environmental risks. J Hazard Mater 457:131694. https://doi.org/10.1016/j.jhazmat.2023.131694
Yuan Y, Jia H, Xu DY, Wang J (2023) Novel method in emerging environmental contaminants detection: fiber optic sensors based on microfluidic chips. Sci Total Environ 857:159563. https://doi.org/10.1016/j.scitotenv.2022.159563
Yusuf A, Amusa HK, Eniola JO, Giwa A, Pikuda O, Dindi A, Bilad MR (2023) Hazardous and emerging contaminants removal from water by plasma-based treatment: a review of recent advances. Chem Eng J Adv 14:100443. https://doi.org/10.1016/j.ceja.2023.100443
Zahmatkesh S, Karimian M, Pourhanasa R, Ghodrati I, Hajiaghaei-Keshteli M, Ismail MA (2023) Wastewater treatment with algal based membrane bioreactor for the future: removing emerging containments. Chemosphere 335:139134. https://doi.org/10.1016/j.chemosphere.2023.139134
Zhang R, Chen B, Zhang H, Tu L, Luan T (2023) Stable isotope-based metabolic flux analysis: a robust tool for revealing toxicity pathways of emerging contaminants. TrAC Trends Anal Chem 159:116909. https://doi.org/10.1016/j.trac.2022.116909
Zhou J-L, Gao F (2023) Phytohormones: novel strategy for removing emerging contaminants and recovering resources. Trends Biotechnol. https://doi.org/10.1016/j.tibtech.2023.02.008

# Assessing Ecological and Human Health Risks Associated with Sludge

Vipin Nautiyal, Tanushree Prasad, and Sweta Kumari

**Abstract**

The pervasive presence of hazardous material raises environmental and public health concerns across the world. Heavy metals associated with wastewater treatment plant sludge are considered to be one of the most principal causes of harmful environmental and health consequences. However, it has been reported that the determination of total components does not provide a precise assessment of the upcoming environmental impact. In addition to this, the environmental as well as health problems associated with wastewater treatment plants are not only restricted to solid waste but also linked to pollutant emissions (both liquid and gaseous) that might negatively affect urban air quality. Universally, the primary approaches of sludge management include landfill disposal and incineration. The landfilling technique is one of the major sources of groundwater contamination and demands plenty of landfill space. While the leftover residues after incineration and gaseous air pollutants released into the atmosphere is responsible for environmental degradation. A large number of microbes are concentrated in sludge and the conventional method of wastewater treatment do not completely destroy such microbes. Hence, composting (a thermophilic aerobic decomposition) of the sludge should be subjected before the land application. The temperature at which the heat is generated for composting is the ideal proliferation criteria for thermophilic microorganism including actinomycetes and *Aspergillus fumigatus*. Thus, the compost workers not only exposed to the enteric pathogens present in sludge but also to the thermophilic fungi, spores, and actinomycetes. Additionally, fungus like *A. flavus* produces aflatoxin, which is among the potent human carcinogens. The hydrocarbon compounds released from the volatilization of the oily sludge are also primary risk factors to the workers or receptors. A wide range of microbial and chemical pollutants, including thermo-tolerant coliforms, *E. coli*, Salmonella spp., Helminth eggs, and heavy metals like cadmium, copper, iron, lead, chromium, and zinc are also associated with several infections.

**Keywords**

Sludge · Wastewater treatment · Health concern · Environmental risk

## 1 Introduction

Life without water would be impossible, as it is one of the most indispensable natural resources of Earth. Due to anthropogenic activities, like sewage, the water resources are getting contaminated, which results in pollution. Sewage sludge is the insoluble solid residue remaining after the sewage treatment. Therefore, landfilling and incineration of the sludge are considered safe disposal alternatives, as it is one of the major environmental concerns throughout the world (Sanchez et al. 2004).

Sludge is a form of solid or semi-solid residues produced during the wastewater treatment process and is frequently regarded as a toxic product. The sludge is classified into two types based on its origin: (i) primary sludge (ii) secondary sludge. Primary sludge is chemically inorganic in nature and is generated during primary sewage treatment procedures such as, sedimentation, coagulation, filtration, and flotation. Secondary sludge is organic in nature and is produced during the secondary wastewater treatment procedures like activated

V. Nautiyal
Pt. Ram Sumer Shukla Government Medical College, Rudrapur 263153, Uttarakhand, India

T. Prasad · S. Kumari (✉)
Department of Microbiology, Uttaranchal (P.G.) College of Biomedical Sciences and Hospital, Dehradun 248001, Uttarakhand, India
e-mail: sweta828113@gmail.com

© The Author(s), under exclusive license to Springer Nature Switzerland AG 2025

A. Mandpe and M. P. Shah (eds.), *Wastewater Sludge Nexus*, Sustainable Landscape Planning and Natural Resources Management, https://doi.org/10.1007/978-3-031-45758-6_11

sludge treatment. During secondary wastewater treatment, bacteria reduced the biological oxygen demand of the sewage by decomposing the organic matter which results in the formation of microbial biomass that accumulates as secondary sludge. Due to the organic nature of secondary sludge, it is difficult to dewater and therefore it is combined with primary sludge before disposal to enhance its dewatering ability (Devi and Saroha 2017).

The sewage sludge is a complex heterogeneous mixture of detergents, pesticides, various salts, dyes, undigested inorganic and organic components (Sommers et al. 1976). The undigested organic matter is originated from the lipids, peptides, and carbohydrate complex from the sewage while, the inorganic matter is originated from the man-made polymers and soil. Since the high protein content of the microorganisms utilized in the treatment of wastewater, sewage sludge is high in nitrogenous substances (Samolada and Zabaniotou 2014).

Sewage sludge is an undesirable and inevitable by-product of wastewater treatment procedures. It is produced by sedimentation, before and after wastewater treatment. On the account of developing industries, and agriculture, and improving lifestyle, progressively wastewater is being produced therefore, relative legislation and regulation needs to be improved and implemented for environmental protection (Wang 1997). It is a major source of pollution that include odor, water pollution, contamination by heavy metals and veterinary hazards (Yuan et al. 2011). The most common sludge disposal areas include landfill incineration, but these processes are impermissible due to land limitation and restrictive legislation (Cao et al. 2010). Industrial emissions containing dioxins and heavy metals are the consequences of air pollution, while on the other hand, landfill disposal causes soil pollution that effects the natural microflora of the soil and decrease the soil fertility (Mondala et al. 2009).

Sewage sludge is the by-product produced due to the treatment of municipal wastewater required to remove pollutants, which is mainly obtained by sedimentation of the organic matter of wastewater in sewage treatment plants and is allowed to settle in specifically designed basins. It is chiefly composed of organic matter such as human excreta, which is further modified by biological and non-biological stabilization treatments. It is easily fermentable and undergoes stabilization processes, i.e., chemical treatments, aerobic stabilization, anaerobic stabilization (with bio-gas production), drying (air or heat), and composting (solid-state), before any kind of utilization. Composting reduces the volume of processed sewage sludge and produces a storable solid that is hygienically safe. In contrast, aerobic and anaerobic stabilization procedures produce the fluid product, without any volume reduction substantially, and are characterized as poorly stabilized organic matter, and inadequate destruction of pathogens (Havelaar et al. 1983).

Sludge has potential risks of causing heavy-metal pollution since it usually contains heavy metals, such as chromium (Cr), copper (Cu), nickel (Ni), cadmium (Cd), mercury (Hg), zinc (Zn), arsenic (As), and lead (Pb). To reduce or stabilize the heavy metals present in sewage sludge, some pre-treatment procedures should be espoused before disposal. Moreover, comprehending the properties of heavy metals, during the treatment is also important for the control of heavy metal pollution. Even though the overall level of heavy metals is indicated by the total metal concentration that are toxic to plants and animals and their bioavailability mostly depend on their ways of binding and specific chemical forms (Chen et al. 2008).

The composition of sludge is determined by the pollution load, nature of the underlying sewage treated and the type of treatment used. Industrial sludge is generated by the treatment of wastewater in industries like paper, textile, tannery, fertilizer, distillery, food, pharmaceutical, steel, electroplating, and petrochemical. The paper industry produces enormous amount of heavy metals, celluloses, fibers, lignin, chlorinated substances, and chromatophores. The sludge from the acid mines is produced from the acid mining treatment plants comprise of a high concentration of dissolved metals. In addition, the sludge generated from the electroplating, acid, and steel industries consists of high metal oxides or hydroxides and less solid content (Wei et al. 2008).

## 2 Requirement for an Ecotoxicological Assessment

The European Union Hazardous Waste Directive expresses their concern and gives emphasis on the need for assessing the environmental risk caused by various waste management practices. It has advised that the chemical composition, general qualities of a waste and leachability of potential hazardous substances must be understood before the development of any disposal strategy. Basic assessment, on-site verification, and compliance testing are frequently included in common characterization and monitoring of waste. The ecotoxicity of the waste and its sludge appears to be an important factor for waste and filling (Mantis et al. 2005).

The ecotoxicological examination of waste can be used to control the detrimental effects of sludge associated with agriculture. For this, a tiered system could be utilized, including avoidance test as the first tier which is the screening level test accompanied by plant growth and reproduction as the second tier (Natal-da-Luz et al. 2009). For waste characterization, several countries employed a variety of single leaching tests. Over the past few years, Europe has been working on the harmonizing leaching tests (Mantis et al. 2005).

## 3 Proper Sludge Treatment

The treatment of sludge has challenged the environmental protection, it majorly threatens water and soil resources and causes air pollution. The directive 2018/851/EC, i.e., environmental legislation provides strict action for waste prevention and management for instance prevention, preparing for reuse, recycling, other recovery, and disposal (Collivignarelli et al. 2019), energy and material recovery are desirable methods of cleaning sludge under the premise method of minimization. One of the traditional methods of sludge disposal, i.e., landfill, is the liable cause of secondary pollution of soil and water resource (Yang et al. 2017). Sewage sludge used as an organic fertilizer or soil improver in agriculture menace human health and the natural environment since harmful pollutants and microbial pathogens continue to accumulate in food (Krzyzanowski et al. 2016). Incineration adopts high temperatures to oxidize organic matter and uses an incinerator to heat and dry it to stabilize and reduce its harmful effects on sludge. It is one of the effective and universal methods to reduce the quantity and toxicity of sludge and to recover its energy, due to the limitation of land resources and the increment of energy demand (Zhang et al. 2020). However, the emission of NOx (mainly NO and $NO_2$) and other gaseous pollutants is an important problem in sludge incineration, the nitrogen content of sludge is higher than other fuels such as coal and biomass, which results in high emission of nitrogen oxides (Sänger et al. 2001), which results in high content of nitrogen and metal oxides (iron oxide and magnesium oxide) in feedstock. However, control of pollution emissions ($SO_2$ and NOx) is required along with the reduction of disposal costs due to the high energy consumption of drying processes.

Most conventional wastewater treatment does not destroy the microorganisms completely, where the pathogenic microorganisms are concentrated in the sludge. Therefore, a stabilized method such as composting which is a thermophilic aerobic decomposition process before land application is accompanied. The number of viable microorganisms is efficaciously reduced during the process of heat generation (Epstein et al. 1976). Preceding to composting sludge and wood chips are assorted in a ratio of 1:3 by volume. The stationary compost pile constructed above a porous pipe network attached to a blower constitutes the forced aeration pile system, which draws air through the pile. One of the foremost objectives of composting is to obtain high temperatures of 55–65 °C with proper mixing and aeration throughout the system for a prolonged period to heat the entire mass. Due to this, most of the microorganisms including viruses will become inactivated, however, the lipopolysaccharide layer in Gram-negative bacteria and various other microbial products could not be detoxified due to the heat generated during composting. Moreover, it provides an ideal environment for the proliferation of numerous thermophilic microorganisms such as Mucorales, actinomycetes and particularly *Aspergillus fumigates*, owing to this most of the enteric pathogenic microorganisms along with thermophilic actinomycetes and fungi that proliferate during composting are exposed to compost workers, which could initiate an infection apparently in a healthy individual (Clark et al. 1983).

Aspergillus sp. is ubiquitous in most countries of the world, the fungus flourishes on a variety of substrates, including stored grain or hay, dung, decaying vegetation, and soil. *A. fumigatus* is the most common microorganism at compost sites as the optimum temperature for its growth is 45 °C. Aspergillus sp. has been exhibited to cause disease in both normal and compromised individuals (Millner et al. 1977). Therefore, there is a potential health risk because of the exposure of *A. fumigatus* spores exposed in higher concentration to individuals involved in the composting process. Furthermore, the aflatoxin produced by the fungus *Aspergillus flavus* is one of the most potent human carcinogens, known, and could increase during the process of composting.

Aspergillosis disease is attributed to the antigenic stimulation, colonization, or tissue invasion by *Aspergillus* sp., which is acquired by the inhalation of spores. Since the minute size of the airborne spores (conidia), could migrate toward alveoli, or procure entrance to paranasal sinuses. The disease varies in extremity, i.e., from an accidental, saprophytic relationship with the host to a fulminating, fatal infection. Moreover, the major aspects in the course of infections caused by Aspergillus species have been contemplated in the dose, i.e., number of colony-forming unit (CFU), mode of entry, and immune status of the host. *A. fumigatus* causes serious and fatal infections mostly in immunosuppressed patients receiving chemotherapy for lymphoma or leukemia and who have had kidney transplants, together with individuals taking antibiotics or adrenal cortical hormones also have been delineated to have a higher incidence of mycotic infections. Furthermore, predisposing factors include malignant or other debilitating diseases, like granulocytopenia, pneumonitis, and other underlying pulmonary diseases (Rippon 1982). Besides, patients with a history of asthma, exposed to *A. fumigatus* spores could develop allergic bronchopulmonary aspergillosis, as colonization of aspergillus in the bronchi of these patients causes episodic bronchial plugging, which leads to secular bronchiectasis. During the microscopical examination, these patients could also produce sputum plugs, which reveal fungal mycelia. Significant eosinophilia of blood and sputum signifies elevation of total serum immunoglobulin, i.e., IgE and serum precipitating antibody to aspergillus antigen as abnormalities in the laboratory test. Therefore, persistent bronchopulmonary aspergillosis causes chronic bronchial and peribronchial inflammation and possibly causes irreversible complications such as bronchiectasis and pulmonary fibrosis.

The dust at the composting site also contains significant quantities of lipopolysaccharide layer derived from gram-negative microorganisms (Clark et al. 1983). Clinical symptoms ascribed to endotoxins include fever, diarrhea, nausea, nasal and eye irritation, headache, and chest tightness (Mattsby and Rylander 1978).

## 4 Pathogens Present in Sludge

The type of pathogens present in the sludge mostly depends on the state of health of the population, as well as on the presence of tanneries, abattoirs, meat-processing factories, and hospitals in that particular area (Bruce and Davis 1983). The degree of parasitological and microbiological contamination is generally noteworthy despite the soaring hygienic standards of developed countries.

The concentration of virus in the sludge is estimated to be 103 $Kg^{-1}$ and almost 100 % of sludge encompass enteric viruses (Schwartzbrod and Mignotte 1996). Pathogenic bacteria are present in high concentration in sewage sludge for instance 107 Shigella spp., 106 Salmonella spp., 106 Yersinia spp., 105 Staphylococcus coagulase positive, 103 Brucella spp., is present in $g^{-1}$ of dry matter of raw and digested liquid sludge (Fig. 1). Pathogenic yeast and fungi are also regarded as human pathogens, and *Aspergillus fumigatus* is a medically important fungal opportunist and respiratory allergen that deliberately contaminates the atmosphere of composting plants and is always present in the sludge (Millner et al. 1977). Moreover, the sludge is also contaminated by the eggs of parasites in particular lumbricoides eggs and other parasitic helminths, which is probably underestimated notwithstanding the increasing interest in parasitology is meeting nowadays (Havelaar et al. 1983).

## 5 Reduction of Pathogens from Sludge

The reduction of pathogens from the sludge can be divided into biological and non-biological procedures (Fig. 2). The non-biological processes include

Lime sanitation—sewage sludge pathogens are inactivated by the addition of sufficient lime to reach a pH of 12 after 2 h of contact, the inactivation and destruction of pathogens are contributed by the generation of heat in the lime-sludge mixture (Strauch 1991).

Air and heat drying—it provides a product with high organic matter and is effective in the storage of dried sewage sludge for a longer duration. Imperceptible variation in moisture during the storage could have a significant effect on the persistence of pathogens that were partially inactivated through drying (Yeager and Ward 1981).

Furthermore, heat and irradiation are also used as the primary treatment for the reduction of pathogens and precede the sludge digestion nevertheless it is more extortionate.

The biological process includes

Sludge lagooning—it is a low-tech treatment in which the sludge resides in a basin or a lagoon for a period between 1 month and 3 years to ensure the reduction of pathogens in acceptable amounts. Sludge lagooning is both space and time-consuming and is less frequently used in industrialized countries as the number of available sites is declining.

Anaerobic digestion—The process of sludge sanitation is based on the temperatures during the treatment, and the competition between mineralizing microflora and the pathogens, it also depends on the inhospitable nature of the compost mass (Stentiford 1986). Precisely, the key feature of the aerobic and anaerobic sanitation procedure is the time/temperature ratio.

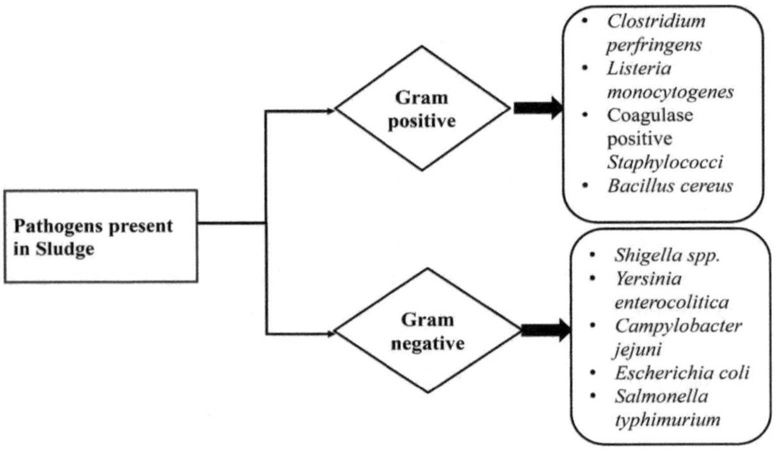


**Fig. 1** Bacterial pathogens found in sludge

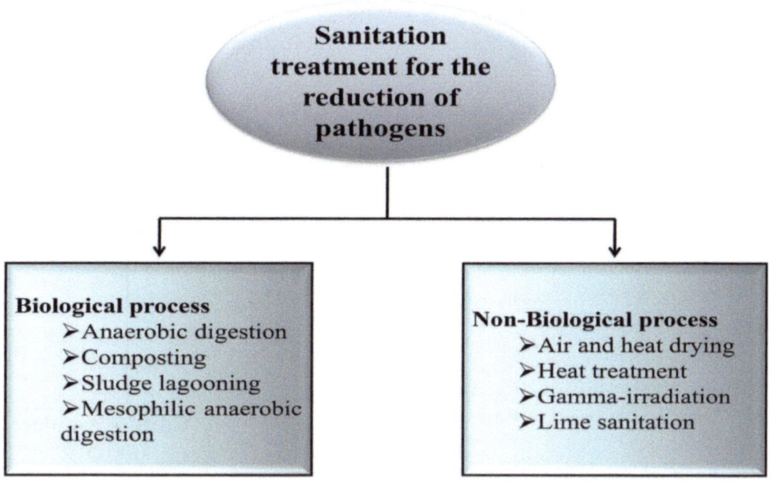


**Fig. 2** Biological and non-biological process of pathogen reduction

Composting—the composting process involves both the antagonistic organisms and the temperature to cooperate in the pathogen inactivation. The inactivation of the virus was higher in raw than in digested sewage sludge at a temperature under 45 °C, while the inactivation times were very close for both digested and raw sewage sludge at a temperature above 45 °C. The viruses were effectively inactivated at the temperature of 60 °C after 20 min of exposure, while the embryonation of the *A. tumbricoides* eggs was prevented with a treatment of 2 h at 50 °C.

Anaerobic digestion—the technology was designed for biogas production, where an anaerobic digester is the main characteristic of the treatment, which is fed at intervals of 1–2 days with raw sewage to supply fresh substrate continuously to methane-producing microflora, which results in the contamination of the processed sludge continuously by the addition of raw sludge. However, anaerobic digestion is considered an effective method in comparison with conventional anaerobic storage of sludge to reduce or eliminate pathogenic bacteria (Olsen and Larsen 1987). The process of anaerobic digestion could be operated at mesophilic (30–35 °C) or thermophilic (50–55 °C) temperatures. The operational parameters such as pH (Henry et al. 1983), volatile fatty acids concentration, and shortage of available nutrients are the main factors in the reduction of pathogens during anaerobic mesophilic batch digestion conveyed in laboratory scale digester (Kearney et al. 1993). The pH, volatile fatty acid, and hydraulic retention time as the operational parameters for the determination of pathogen reduction which remained as the subject of discussion among scientists, nevertheless all cited authors agreed on the significance of temperature as the decisive factor during anaerobic digestion for the inactivation of bacterial pathogen (Olsen and Larsen 1987).

Mesophilic anaerobic digestion—it is carried out at 30–35 °C and can inactivate 50–99% of viruses contaminating the sludge. During the anaerobic digestion $NH_3$ production, temperature, virucidal bacteria, and microbial exoenzyme are important factors for the inactivation of virus (Schwartzbrod and Mignotte 1996). $NH_3$ production could inactivate the virus in a pH higher than 6, and poliovirus type 1 was inactivated at 35 °C after 4 days of treatment (Sanders et al. 1979).

The batch anaerobic digestions of beef cattle slurry in laboratory scale digesters demonstrated reduction in the number of *Escherichia coli, Yersinia enterocolitica, Listeria monocytogenes, Salmonella typhimurium,* and *Campylobacter jejuni*, with a thermal reduction time $T_{90}$, i.e., to reduce the initial population by 90% in 12 days, exhibiting better performance than a semi-continuous process where it took 35 days to reduce the number of pathogens. It is noteworthy, that even after the decline in the viable number of other pathogens *C. jejuni* was not affected by either of the treatments and its number was not substantially changed at the end of either process. Therefore, it was concluded that mesophilic treatment does not eliminate the vegetative forms of pathogenic bacteria, also it could not inhibit Salmonella sp. and other enteric bacteria. It is also ineffective against the spores of *Clostridium perfringens* and *Bacillus cereus*, thus, to obtain a sanitation effect the thermophilic treatment should be acquired (Berg and Berman 1980).

## 6 Oily Sludge Toxicity Assessment Requirement

Oily sludge is the semi-solid emulsified wastes generated from the various industrial activities during the exploration, transportation, production, refining and storage of crude oil, which is a major source of energy and feedstock for manufacturing various petrochemical products (ElNawawy et al. 1994). It consists of water, minerals, and petroleum hydrocarbons at varying ratios, and is mostly dependent on the source

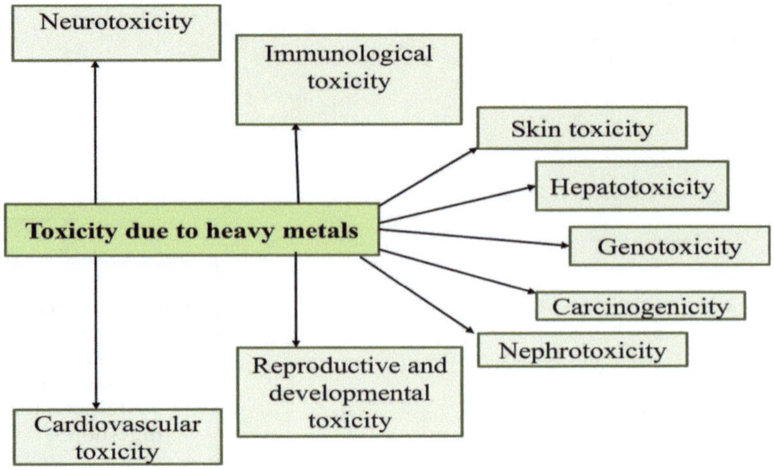


**Fig. 3** Toxicity produced due to heavy metals

of crude oil, the process of generation of oil sludge, and aging time. It contains heavy metals, toxic petroleum hydrocarbons, and radionuclides, which may be toxic to soil microflora, and also to numerous show impact on soil invertebrates, i.e., earthworms, pot worms, mites, and springtails which is an inhabitants of soils and contributes to soil productivity, and as a consequence, have an adverse impact on human health.

Therefore, oily sludge is considered as hazardous material that requires safe handling to minimize contamination and preventing oily sludge from spreading to the surrounding environment by immobilizing via solidification using cementing materials or utilizing storage facilities (Zhang et al. 2018). In addition, detoxification includes the removal of toxic components through chemical or biological degradation or physical separation to reduce their harmful effects (Tahhan et al. 2011), and elimination is intended via high-temperature treatment to destroy the organic components completely. These methods are effective in minimizing environmental hazards to varying degrees but are generally costly. Therefore, cost-effective and innovative technology, along with research on physiochemical properties, together with characterization and toxic assessment of oily sludge is required for the treatment and valorization of the sludge generated by the petroleum industry.

## 7 Heavy Metal Toxicity

Heavy metal toxicity is referred to as the unwanted concentration of heavy metals present naturally on earth but are concentrated as a result of human activities and consequently enter inside animals, plants, and humans via diet, inhalation, and manual handling, and therefore binds and interfere with the functioning of vital cellular components. The toxicity due to heavy metals hampers ecological, environmental, evolutionary, and nutritional significance (Lenntech Water Treatment and Air Purification 2004). The toxicity produced due to heavy metals are presented in (Fig. 3). It is mostly present in aquatic and soil ecosystems, and also small portion is present as particulate vapors in the atmosphere. Some of the heavy metals such as Co, Fe, Ni, Cu, Mn, Mo, Zn, and V in minute quantity are beneficial to organisms, while on the other hand, excessive quantity of these elements could be detrimental to the organisms. Heavy metals such as Hg, Cd, and Pb are considered the "main threats" since they are harmful to both plants and animals and cause air, water, and soil pollution in the environment. Moreover, it does not have any beneficial effect on organisms and may be toxic and cause harm to living things. Metals are accumulated in the ecological food chain where the primary contact site of heavy metal ions are the plant roots which are the primary producer, and it reaches the consumer through consumption. Conversely, in aquatic ecosystems heavy metals are directly absorbed into the leaves as these are deposited on the foliar surfaces. In plants when the concentration of metal exceeds optimal levels, it adversely hampers the plants directly and indirectly which includes inhibition of cytoplasmic enzymes and damage to cell structures due to oxidative stress, moreover, due to the toxicity of heavy metals, there is a replacement of essential nutrients at cation exchange sites of plants (Asati et al. 2016). It also hampers the soil activities and growth of soil microorganisms, which results in the reduction of beneficial soil microbes and leads to less fertility of soil due to a decrease in organic matter decomposition. Furthermore, enzymatic activities which are important for plant metabolism are also hampered due to heavy metal interference, these toxic effects lead to a decrease in plant growth and result in the death of plants (Schaller and Diez 1991).

## 8 Toxicity Assessment of Textile Sludge

The textile industry utilizes a great quantity of water up to 3,000 $m^3$ per day. An industry that consumes 50 $m^3$ of running water per hour can produce 1-10 tonnes of sludge on a wet basis. These operations generate huge amount of wastewater which must be thoroughly treated before being dumped into the environment. The biological treatment of activated sludge does not destroy excess biomass and chemicals. Textile biological sludge comprises a large amount of organic substance, nitrogen, phosphorous, potassium, and micro-nutrients and could be used as manure in soil with poor crop yields or as a fertilizer for the soil of severely polluted areas. Toxicity tests can offer a greater understanding regarding the dangers of those complicated mixes. Due to the wide spectrum of substances present in the waste, the toxicity determination simply by chemical analysis could be difficult (Gomes et al. 2012).

## 9 Toxicity Related to Plants

Phytotoxicity is the intoxication of growing plants by the compounds present in their growth medium which occurs when such compounds are utilized and accumulates in plant tissue. It causes decline in the growth of the aerial portions. Previous studies reported the use of several species for the phytotoxicity analysis namely, *Latuca sativa, Avena sativa, Glycine max, Brassica campestris, Helianthus annuus, Vigna unguiculata, Brassica oleracea, Triticum aestivum,* and *Euruca sativa.* These organisms were employed because of several advantages, including the ability to examine multiple endpoints (such as biomass production, rate of germination and enzymatic activity) for toxicity evaluation in the same test species and at cheap maintenance. In which, most common endpoints for measuring phytotoxicity are plant development and seed germination (Gomes et al. 2012).

For phytotoxicity investigation of textile sludge, the following sample preparation techniques were most widely used. Primarily, the aqueous extraction was employed using leaching and solubilization method followed by dilution of the sample in a control soil. Lastly, *in natura* sample was employed. Aqueous extracts naturally obtained at acidic pH levels were used for aquatic organisms. According to the endpoint associated with the toxicity test, either organic or aqueous extracts were utilized for microorganisms. Earthworm samples were diluted in control soil. In the case of mammals, oral exposure to aqueous extract was utilized (Gomes et al. 2012).

## 10 Microorganisms

The development of bacteria-based tests is aimed to minimize the experiment expense and duration. Several researchers consider that the bio-luminescence assay using *V. fischeri* is the most sensitive for numerous chemicals compared to other bacterial assays that use endpoints such as inhibition of nitrification, luminescence, respiration, and suppression of enzymatic ATP. Several studies reported that the aqueous extract of stabilized sample of sludge found to be non-toxic, however, the fresh sludge aqueous extract found toxic with 50% of the effective concentration values ($EC_{50}$) more than 80%. Similarly, the textile sludge had the highest level of acute toxicity ($EC_{50} = 21.84\%$) (Gomes et al. 2012).

Dehydrogenase enzymes exist in all living organisms and participate in a wide range of energy transfer activities in microbial pathways of metabolism. As dehydrogenase does not function independently of the parent microbial cell in soil as extracellular enzyme, and therefore measuring dehydrogenase activity is an excellent general indicator of microorganisms activity and soil organic matter oxidation. The estimation of substrate-induced dehydrogenase activity has previously been justified as an ecotoxicological bioassay as an indicator of physiologically active biomass (Obbard 2001).

## 11 Mammals

In mammals, chemical exposure can cause chromosomal changes, sperm abnormalities, mutations, premature or postnatal fetal death, birth deformities, lower birth weight, and even cancer. For the toxicity assay, diluted aqueous extracts of the sludge were homogenized accompanied by filtration and distributed to pregnant lady in between 15 and 16th day of their pregnancy with 1:10 and 1:100 dilutions, respectively. At the dosages examined, the samples exhibited maternal toxicity. Toxic signs and symptoms included muscular tremors, convulsions, ataxia, lacrimation, hypersalivation, and restlessness. Additionally, a higher dosage reduced the average mass gain by dams throughout gestation, but there was no dam mortality. External abnormalities such as open eyelids and subcutaneous edema were observed in the fetuses. The free-handed razor section of the brain revealed hydrocephaly, primarily involving the lateral ventricles. The fetal skeletons stained with alzarin showed impaired ossification of skull bones, a decreased count of ribs, and sternebral abnormalities in both dose groups. Hence, it was concluded that at greater sludge concentrations, the most severe consequences were reported. The reported harmful effects may be attributed to the presence of metals, chemical compounds, and even the absence of some critical nutrients in the sludge dilutions, such as magnesium and calcium (Gomes et al. 2012).

## 12 Invertebrates

The invertebrates (earthworms) were subjected to toxicants in polluted soils through direct contact, consumption of liquid and solid particulates, and also through the food chain. The toxicity studies were conducted with *Eisenia foetida* (earthworm) using an adult of minimum two months older. The procedures were reported to be carried out in a culture chamber at constant intensity of light and 20 °C temperature. The organisms were then exposed to a control soil as well as dilution stabilized, and fresh sludge diluted into this soil at concentrations between 6.2 and 50%. Both the fresh and stabilized sludge exhibited to be non-toxic. Since, the stabilized sludge enhanced organism biomass, as previously reported in the algae test. It was then recommended that the stabilized sludge can be employed to restore a non-productive forest soil since it has fertilizer capability (Gomes et al. 2012).

## 13 Sludge Generated from the Water Treatment Plants

The trend of modernization and rapid population growth has increased the demand of portable water, prompting the sources of raw water, as well as the development of supply systems and treatment. Surface water primarily from rivers, transport suspended particulates, colloidal solids, and contaminants and therefore, adequate treatment is necessarily required based on the standards of accessible raw water. During the purification in water treatment plants (WTPs), an enormous amount of leftover or waste is generated which referred as water treatment sludge (WTS). A WTP generates about 100,000 ton/year of sludge, while the daily output of sludge surpasses 10,000 ton. Thus, the environmental researchers and technologists are facing challenges in treating and disposing of such large amounts of sludge generated by WTP. However, the sustainable management and disposal of WTS is a difficult task that must be carefully considered.

In India, most WTPs release their sludge into local drains, which eventually meet the water source on the downstream portion of the intake which in the main cause of contamination. Some WTPs clean their clarifiers once in a year, and the sludge is discharged in nearby open spaces. Due to the presence of chemicals in the product used for the disposal of the WTS which may led to contamination of soil and waterways, hence, this method of disposal is not recommended. Accordingly, the proper treatment of increasing leftover sludge from WTPs in an affordable and environmentally suitable way remains a critical concern (Ahmad et al. 2016).

The consequence of rigorous criteria for wastewater treatment plant effluent, the production of sludge has increased immensely (Forster-Carneiro et al. 2010), thus their accumulation constitutes a heightened environmental problem (Fuentes et al. 2008), as disposal of sludge results in environment pollution, therefore strict guidelines have been enacted for the protection of public health (Liu et al. 2010).

## 14 Physical and Chemical Properties

The term "Water Treatment Sludge (WTS)" refers to all waste generated during water treatment at a water treatment plant (WTP). The quality of WTS is typically defined by the untreated water quality and the type of treatment used. Although, the surface water treatment can cause significant alterations in the production of sludge. WTS is consists of both inorganic and organic particles in liquid, solid, and gaseous phases. The colloidal particles that are present in the unprocessed water are aggregated and settled down with the help of coagulants; the substances that are required to produce a significant portion of sludge. Ferric ion salts ($FeCl_3 \cdot 6H_2O$), aluminium salts ($Al_2(SO_4)_3 \cdot 18H_2O$), and ferrous iron salts ($FeCl_2$, $FeSO_4 \cdot 7H_2O$) were often used as coagulants. The percentage of various oxides in the sludge can differ based on the raw water quality, the type of coagulants employed, the treatment technique used and the overall quality of the water generated (Ahmad et al. 2016). The physiochemical properties of the ferric and alum sludges are tabulated in Table 1.

The introduction of sludge to soil helps to promote the chemical reactions (hydrolysis and redox reactions), physiochemical reactions (precipitation, dissolution, and adsorption), and microbiological mechanisms (metabolite production, microbe multiplication and enzymatic reactions) in the soil system. Recognizing the changes in the physical characteristics of soil caused by the addition of sludge requires a thorough understanding of these interactions including their evolution over time (Metzger and Yaron 1987).

## 15 Ecotoxicological and Toxicity Concerns of Water Treatment Sludge

The literature on possible toxic effects of the WTS is meagre, and the available research is slightly contradictory. Thus, the thorough investigation is required to determine the inorganic contaminants toxicity particularly, metal contaminants along with organic contaminants found in WTS. The toxicity of alum sludge was investigated and revealed its direct effect on growth of algae in receiving streams. The ferric sludge is

**Table 1** Physiochemical characteristics of ferric and alum sludge

| Parameters | Units | Ferric sludge | Alum sludge | References |
|---|---|---|---|---|
| pH | – | 7.0 ± 1.3 | 6.5 ± 0.3 | Ahmad et al. (2016) |
| Total solids | mg/L | 2132–3957 | 2500–52345 | Sotero-Santos et al. (2007) |
| Total dissolved solids | mg/L | Nd | 325.76 ± 12.6 | Barakwan et al. (2019) |
| Total suspended solids | mg/L | 3570 ± 1299 | 12511 ± 235.66 | Sotero-Santos et al. (2007), Barakwan et al. (2019) |
| Total fixed solids | mg/L | 2907.6 ± 1240.4 | 10,962 ± 251.9 | |
| Total volatile solids | mg/L | 813.3 ± 331.3 | 12,696 ± 263.3 | |
| COD | mg/L | 992 ± 1041.2 | 9666.7 ± 942.8 | |
| Metals (mg/kg) | | | | |
| Al | | 105222 ± 52854 | 118700 ± 24260 | Sotero-Santos et al. (2007), Ahmad et al. (2016) |
| Zn | | 61.6 ± 54.8 | 98 ± 31 | |
| Pb | | 59 ± 14.1 | 22 ± 12 | |
| Cd | | Nd | 0.12 ± 0.02 | |
| Ni | | 82.3 ± 6.6 | 28 ± 10 | |
| Fe | | 279166 ± 124749 | 37,000 ± 19,740 | |
| Mn | | 897 ± 524 | 2,998 ± 1,122 | |
| Cu | | 39.6 ± 19.1 | 624 ± 581 | |
| Cr | | 41.3 ± 3.0 | 20 ± 7 | |
| Hg | | Nd | 0.46 | |
| Mg | | Nd | 2,407 ± 572 | |
| Ca | | Nd | 10,360 ± 4,299 | |
| Na | | Nd | 355 ± 142 | |
| K | | Nd | 3,547 ± 582 | |
| S | | Nd | 6,763 ± 2,955 | |

Values are expressed as mean±SD. Nd: Not determined

found to be more toxic as compared to that of alum sludge over prolonged exposure. The toxicological characteristic leaching technique (TCLP) is used to determine the heavy and toxic metals leaching from the WTS. According to the TCLP protocol, heavy metal leaching is significantly beneath the regulatory level for TCLP, and so WTS can be categorized as non-hazardous substance. Despite having differences in raw water quality and sludge characteristics, no association with the amount of toxicity has been developed or validated yet. Most of the concerns about the toxicity and metallic content of WTS have focused on their use in the agricultural sector, horticulture and various other on land applications. It causes heavy metals accumulation in the surrounding environment and possess a risk for higher organisms. In developing nations, the sludge is typically deposited straight into rivers and neighboring streams or else into the water drainage system. It causes an increase in the concentration of aluminium and metallic elements in aquatic species, water and the human body. Aluminium involves in the development of Alzheimer's disease and mental impairment in children. Their absorption into the underground water supply may have a negative impact on the quality of groundwater. Another popular technique of disposal practiced in numerous municipalities is landfilling of WTS followed by dewatering. However, due to the difficulties in identifying landfill sites and the price of maintaining landfills, this disposal approach is no longer viable in many modern cities. The 3'R' idea for sludge, i.e., recovery, recycling, and reuse has been promoted in India (Fig. 4). However, the investigation of secure and affordable methods is still required (Ahmad et al. 2016).

## 16 Sustainable Reuse of the Sludge from Water Treatment

1. **Utilization of WTS in the Removal of Heavy Metals and Contaminants**

Ahmad et al. (2016) reported that the wastewater consists of phosphorous, sulfide, fluorides, glyphosate, arsenate, selenium, hydrogen, boron, perchlorate, mercury, and lead. WTS

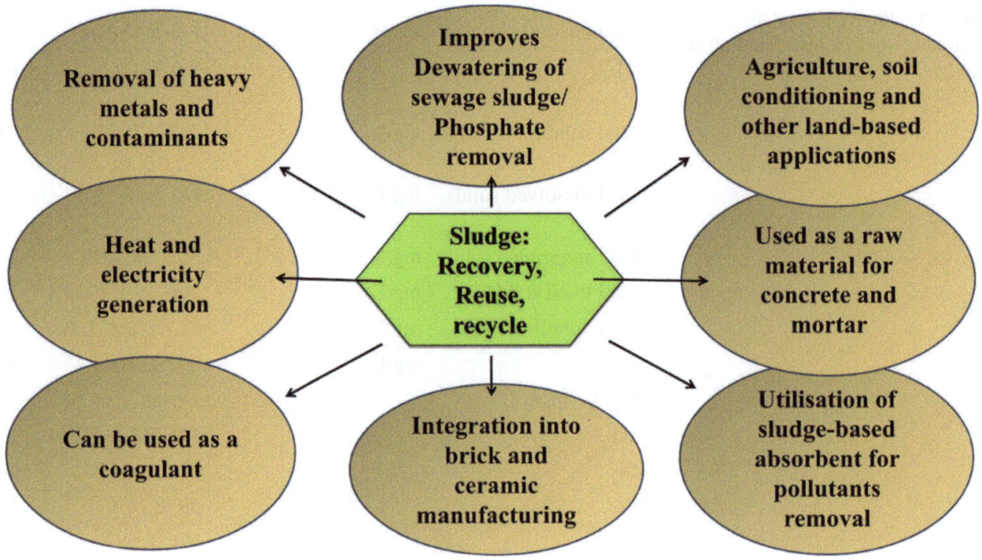


**Fig. 4** Sustainable recovery, reuse, and recycle of the sludge

can serves as an absorbent in adsorption process, through which such contaminants could be removed. Amorphous ferric or aluminium ions, which have higher affinity to anion are considered as a potential adsorbents for the removal of phosphorus present in wastewater. Meanwhile, the adsorption process is significantly depending on the pH level of the solution as the acidic regions having greater adsorption potential. For the removal of all organic and inorganic phosphates, the efficient pH 4–6 is required. The phosphate ions can be selectively adsorbed on alum sludge via an inner-sphere chemical bond that remains unaffected by the presence of sulfate and chloride ions. Further, ferric sludge is found to be a more effective phosphorus adsorbent than alum sludge with higher phosphorus sorption on a molar proportion. Ferric sludge contains more iron compared to that of alum sludge which comprises of unit mass of aluminium. As a result, larger permeability of ferric sludge allows more intra-particle phosphate transport and adsorption over the surface. Reuse and recycling of sludge is considered as an effective and cost-effective method of removing phosphorus from wastewater. Arsenic removal via adsorption onto WTS is preferred among several heavy metals as arsenate compounds are structurally and chemically similar to phosphate molecules. Both the alum and ferric sludge potentially adsorb arsenate but the alum sludge is highly efficient in removing arsenate. Additionally, ferric sludge has a higher affinity as well as sorption capacity for As (III) over As (V). It has been reported that alum sludge has exceptional arsenic absorbance ability and is remain unaffected by solution pH except at low levels of sludge additions. Lead and copper are efficiently removed at a higher pH, whereas chromium sorption is facilitated at low pH. It has also been claimed that by adjusting the pH of the solution, specific heavy metals as well as metalloids can be removed (Ahmad et al. 2016).

2. **Dewatering of Sewage Sludge Through WTS Utilization**

Several studies suggested that the use of WTS in the dewatering and co-conditioning of sewage sludge enhances its treatability. The increasing amount of alum sludge in the blended sludge improves sludge dewaterability and settleability. In addition, the alum sludge act as a skeleton builder making the mixed sludge more malleable to successful dewatering. Alum sludge also enhances the dewatering of anaerobically digested activated sewage and improves the phosphate removal from water. Anaerobic metabolized sludge combined with alum sludge at ratio 2:1 that result in 99 % of phosphorous reduction in the reject water and improve dewaterability of the generated sludge. The ideal polymer dose essential to improve dewaterability is decreased to 15 mg/L from 120 mg/L under this mixed ratio, resulting in a significant savings in polymer addition. Such sludge application and waste co-disposal approaches have been proven to be cost-effective, particularly in terms of improving sludge dewaterability (Ahmad et al. 2016).

3. **Agriculture Reuse of the Anaerobically Digested Sludge**

The sludge is enriched with a concentration of heavy metals which are toxic and nondegradable and could accumulate along the food chain (Dong et al. 2013). Sewage sludge dumped in the anaerobic bioreactor is subjected to an anaerobic stabilization process. However, speciation transformation of heavy metals has been observed when sludge was placed in anaerobic digestion reactors, therefore in a study conducted by Dong et al. (2013) reported that anaerobic digestion of sewage sludge increased the ratios of Ni, Cu, Zn, and

Cr in mobile fractions, while on the other hand, decreased the ratio of Pb. Additionally, sulfate and sulfur compounds were transformed into $S_2$ due to the presence of sulfate-reducing bacteria during the process, and consequently, these $S_2$ formed stable sulfides with heavy metals.

The anaerobically digested sludge is of high quality, meeting the Class A requirements of the US EPA. It is deposited on land in accordance with reduce pathogenic contamination. In this process, the sludge is anaerobically digested for biomass generation and utilized in order to maintain digester at 35 °C. It also used for electricity generation. A polymer is mixed with the digested sludge and this combination is used as a dewatering polymer. It is then passed to a belt filter to lower the water content. Then, the dry matter from the produced product is subsequently re-distributed as an alternative for chemical fertilizers. The amount of organic fertilizer displaced was determined using the amount of nitrogen and phosphorous present in the sludge (Tarpani et al. 2020).

4. **Agricultural and Land-Based Application of WTS**

Sludge applications on agriculture fields include grazing and farmland, forests, botanical nurseries, municipal parks, golf courses, and residential gardens. However, major environmental reuse includes sediment capping and phosphorus immobilization in eutrophic lake recovery, reclamation, and pit filling. This sludge disposal approach is based on the physical, chemical, and biological properties of soil that could make it possible for assimilating the applied sludge without affecting soil quality along with the possibility of soil improvement as lime softening sludge routinely being utilized as a substitute for agricultural limestone. The land-based WTS usage is regarded as an affordable method for the long-term sludge disposal and additionally enhancing soil characteristics or serving for the purpose of the growing medium in agricultural areas. However, previous and current attempts to spread WTS across the soil are being researched primarily to fulfil the subsequent objectives (Ahmad et al. 2016).

5. **Incineration**

The combination of sludge with polymer is employed to improve the dewatering process and also to increase calorific value. Once the dry matter (35%) is centrifuged, sludge is incinerated at 850 °C using fluidized bed combustors, followed by the addition of lime and fuel in order to enhance combustion performance and regulate acid gases. Yet since the United Kingdom has relatively limited energy distribution infrastructure, the energy recovery capabilities were calculated at the existing range of heating district networks in the UK, which is approximately 1% of the population (Tarpani et al. 2020).

6. **Denitrifying Agent (Carbon-Rich Effluent)**

Following the injection of dewatering polymer, the thickened sludge is carried to the reactor to be combined with oxygen at extreme temperatures (235 °C) and pressure (40 bar). The process delivers a carbon-rich output that can be utilized as a methanol alternative for denitrification at sewage treatment facilities (Tarpani et al. 2020).

### Ecotoxicity and Risk Associated with Metal Pollution in Marine Ecosystem

Ecotoxicologists and ecological researchers use a term "heavy metals" to denote metals that can produce adverse effects to the environment. Cadmium, mercury, zinc copper, nickel, chromium, lead, cobalt, titanium, iron, magnesium, silver, and tin are some metals that have received the most attention in recent decades. Metal contamination is caused by a variety of factors including industrial waste disposal and sewage sludge, which results in metal release into the marine environment. Industrial and automobile exhaust provides considerably more Pb to the environment than natural sources. Many metals are required by living organisms for various processes like iron helps in hemoglobin formation, copper contributes to respiratory pigment, zinc and magnesium promotes enzyme activity, cobalt enhances vitamin B12 production. Some metals are essential but have toxic effects at high concentration. Metals like mercury, tin, lead, nickel, arsenic, and chromium are toxic at even lower concentration water (Valavanidis and Vlachogianni 2010).

The bioaccumulation of the metals in aquatic atmosphere caused by facilitated uptake or by passively through the water. The degree of metal bioaccumulation is determined by the overall amount, level of bioavailability of all metals in the environment and the mode of uptake, excretion, and storage. Organisms that are exposed to high metal concentrations in the environment adopt through various mechanism to prevent potential toxicity by minimizing metal intake, increase excretion and sequester metal inside tissue making them non-toxic. For the assessment of the concentration of metals in biological system, certain factors which are taken in account are temperature, concentration of dissolved oxygen, ligands and metals, organic complex and ionic strength competing for pH, site of uptake, general physiological behavior, life history, and life cycle. Metal pollution is also considered biologically toxic due to its carcinogenic and oxidative properties. It has also observed that the metal binds to reacting sites of the nucleic acid of the cellular DNA leading to mutation water (Valavanidis and Vlachogianni 2010).

The heavy metal contamination of freshwater environments has been demonstrated to be common and extensive. Recent studies reported that mercury is a common cause of metal pollution from anthropogenic and natural activities. The estimated value of Hg released from the earth crust is 150.00

tones, while from the industry it is approximately 9 tones. Similarly, lead is also an example of metal pollutants which in recent years got elevated due to the addition of gasoline in vehicles water (Valavanidis and Vlachogianni 2010).

## 17 Ecotoxicological Test of Invertebrates of Marine Environment

The commonly used marine species for toxicity tests are crustaceans copecods (*Acartia tonsa*), mysids (*Mysodopsis bahia*), mussels (*Mytlus edulis*), rotifers (*Brachionus plicatilis*), and brine shrimps (*Artemia salina*). Valavanidis and Vlachogianni (2010) reported that these marine species are easy to handle in laboratory experiments, cost effective, and ecological. The American Society for Testing and Materials (ASTM), the United States Environmental Protection Agency, and the International Organization for Standardization (ISO) also endorse such marine species for ecotoxicological studies water (Valavanidis and Vlachogianni 2010).

## 18 Ecotoxicological Test of Freshwater Invertebrates

Toxicity studies based on different effect demands have been developed using crustaceans, rotifers, amphipods, and insect larvae (mostly dipterans, plecopterans, and ephemeropterans). Acute and chronic life-cycle toxicity tests by using freshwater living organisms which are widely distributed in fresh waterways that are found in a variety of habitats and are an important link in many aquatic food chains, also have a short lifespan and are sensitive to a wide range of aquatic contaminants. Daphnia are commonly used in toxicity testing because they are widely distributed in freshwater and found in a variety of habitats. In addition, they are sensitive to a variety of aquatic contaminants and have a short life cycle. Similarly, protozoans are used to investigate the toxicity of xenobiotics and other chemical contaminants. *Colpidium campylum* is a standard protozoan used for the toxicity test bioassays. Protozoans are unicellular which play a vital role in aquatic ecosystem, particularly in the breakdown of organic materials, rapid generation period and the ability to create an immense quantity of organisms in a small volume. The brine shrimp *Artemia salina*, the rotifers *Brachionus plicatilis*, and *Brachionus calyciflorus* were used in the acute marine bioassays water (Valavanidis and Vlachogianni 2010).

## 19 Ecotoxicological and Toxicology of Fish

The ecotoxicological and toxicological research using fish was first introduced to determine potential hazards in fieldwork and toxicology examinations in 1930s. The importance of ecotoxicological assessment of fish in freshwater is an immediate consequence of their ecological and economic importance. The migration patterns of fish which is linked to their consumption habit and reproductive processes may also have a significant impact on the possibility for exposure to hazardous contaminants. Several international organizations have advocated the use of fish in ecotoxicological investigations. Some of the following species of fishes used for ecotoxicity test:

***Brachydania rerio***—Commonly known as zebrafish and is native to tropical region. The acute, chronic, and subchronic period of exposure is 4, >30 and 250 days, respectively.

***Oncothynchus mykiss***—Also known as rainbow trout and is found in temperate regions. The acute, subchronic, and chronic exposure periods is 7, 250 and >250 days, respectively.

***Pimephales promelas***—Also called fathead minnow and is commonly found in both sub-tropical and temperate regions. The acute, subchronic, and chronic exposure period is 7, 50, and >50 days, respectively.

The primary purpose of toxicology and ecotoxicological research is to ensure that heavy metal contamination caused by anthropogenic pollution does not have negative effects on living creatures. Heavy metal biomonitoring and effect studies on native populations of species must be taken under consideration for pollution-induced community tolerance, which is predicted from communities that have been exposed to certain pollutants for a long time.

## References

Ahmad T, Ahmad K, Alam M (2016) Sustainable management of water treatment sludge through 3 'R' concept. J Cleaner Prod 124:1–13

Asati A, Pichhode M, Nikhil K (2016) Effect of heavy metals on plants: an overview. Int J Appl Innov Eng Manage 5(3):56–66

Barakwan RA, Trihadiningrum Y, Bagastyo AY (2019) Characterization of alum sludge from Surabaya water treatment plant, Indonesia. J Ecol Eng 20(5)

Berg G, Berman D (1980) Destruction by anaerobic mesophilic and thermophilic digestion of viruses and indicator bacteria indigenous to domestic sludges. Appl Environ Microbiol 39(2):361–368

Bruce AM, Davis RD (1983) Utilization of sewage sludge in agriculture. Maximizing benefits and minimizing risks. In: Proceeding of the international symposium: biological reclamation and land utilisation of urban wastes. Naples, pp 121–142
Cao JP, Zhao XY, Morishita K, Wei XY, Takarada T (2010) Fractionation and identification of organic nitrogen species from bio-oil produced by fast pyrolysis of sewage sludge. Bioresource Technol 101(19):7648–7652
Chen M, Li XM, Yang Q, Zeng GM, Zhang Y, Liao DX, Guo L (2008) Total concentrations and speciation of heavy metals in municipal sludge from Changsha, Zhuzhou and Xiangtan in middle-south region of China. J Hazard Mater 160(2–3):324–329
Clark CS, Rylander R, Larsson L (1983) Levels of gram-negative bacteria, *Aspergillus fumigatus*, dust, and endotoxin at compost plants. Appl Environ Microbiol 45(5):1501–1505
Collivignarelli MC, Abbà A, Frattarola A, Carnevale Miino M, Padovani S, Katsoyiannis I, Torretta V (2019) Legislation for the reuse of biosolids on agricultural land in Europe: overview. Sustainability 11(21):6015
Devi P, Saroha AK (2017) Utilization of sludge based adsorbents for the removal of various pollutants: a review. Sci Total Environ 578:16–33
Dong B, Liu X, Dai L, Dai X (2013) Changes of heavy metal speciation during high-solid anaerobic digestion of sewage sludge. Bioresour Technol 131:152–158
ElNawawy AS, ElBagouri IH, Al-Daher R, Khalafawi S (1994) Plant growth and uptake of mineral elements at an oily sludge landfarming site in Kuwait. Resourc Conserv Recycl 11(1–4):111–121
Epstein E, Willson GB, Burge WD, Mullen DC, Enkiri NK (1976) A forced aeration system for composting wastewater sludge. J (Water Pollution Control Federation) 688–694
Forster-Carneiro T, Riau V, Pérez M (2010) Mesophilic anaerobic digestion of sewage sludge to obtain class B biosolids: microbiological methods development. Biomass Bioenergy 34(12):1805–1812
Fuentes A, Lloréns M, Sáez J, Aguilar MI, Ortuño JF, Meseguer VF (2008) Comparative study of six different sludges by sequential speciation of heavy metals. Bioresour Technol 99(3):517–525
Gomes L, Silva F, Barbosa S, Kummrow F (2012) Ecotoxicity of sludges generated by textile industries: a review. Ecotoxicol Environ Contam 7(1)
Havelaar AH, Oosterom H, Noterrnans S, van Knapen F (1983) Hygienic aspects of the application of sewage sludge to land. In: Zucconi F, De Bertoldi M, Coppola S (eds) Biological reclamation and land utilisation of urban wastes. Proceedings of the international symposium, Naples, pp 167–200
Henry DP, Frost AJ, Samuel JL, O'Boyle DA, Thomson RH (1983) Factors affecting the survival of *Salmonella* and *Escherichia coli* in anaerobically fermented pig waste. J Appl Bacteriol 55(1):89–95
Kearney TE, Larkin MJ, Levett PN (1993) The effect of slurry storage and anaerobic digestion on survival of pathogenic bacteria. J Appl Bacteriol 74(1):86–93
Krzyzanowski F Jr, de Souza Lauretto M, Nardocci AC, Sato MIZ, Razzolini MTP (2016) Assessing the probability of infection by Salmonella due to sewage sludge use in agriculture under several exposure scenarios for crops and soil ingestion. Sci Total Environ 568:66–74
Lenntech Water Treatment and Air Purification (2004) Water treatment. Lenntech, Rotterdamseweg, Netherlands. http://www.excelwater.com/thp/filters/Water-Purification.htm
Liu Y, Shi H, Xia L, Shi H, Shen T, Wang Z, Wang Y (2010). Study of operational conditions of simultaneous nitrification and denitrification in a Carrousel oxidation ditch for domestic wastewater treatment. Bioresour Technol 101(3):901–906
Mantis I, Voutsa D, Samara C (2005) Assessment of the environmental hazard from municipal and industrial wastewater treatment sludge by employing chemical and biological methods. Ecotoxicol Environ Saf 62(3):397–407
Mattsby I, Rylander R (1978) Clinical and immunological findings in workers exposed to sewage dust. J Occup Environ Med 20(10):690–692
Metzger L, Yaron B (1987) Influence of sludge organic matter on soil physical properties. Adv Soil Sci 141–163
Millner PD, Marsh PB, Snowden RB, Parr JF (1977) Occurrence of *Aspergillus fumigatus* during composting of sewage sludge. Appl Environ Microbiol 34(6):765–772
Mondala A, Liang K, Toghiani H, Hernandez R, French T (2009) Biodiesel production by in situ transesterification of municipal primary and secondary sludges. Bioresour Technol 100(3):1203–1210
Natal-da-Luz T, Tidona S, Jesus B, Morais PV, Sousa JP (2009) The use of sewage sludge as soil amendment. The need for an ecotoxicological evaluation. J Soils Sediments 9:246–260
Obbard JP (2001) Ecotoxicological assessment of heavy metals in sewage sludge amended soils. Appl Geochem 16(11–12):1405–1411
Olsen JE, Larsen HE (1987) Bacterial decimation times in anaerobic digestions of animal slurries. Biol Wastes 21(3):153–168
Rippon JW (1982) Medical mycology. The W.B. Saunders Company 2:659–660
Samolada MC, Zabaniotou AA (2014) Comparative assessment of municipal sewage sludge incineration, gasification and pyrolysis for a sustainable sludge-to-energy management in Greece. Waste Manage 34(2):411–420
Sanchez ME, Estrada IB, Martinez O, Martin-Villacorta J, Aller A, Moran A (2004) Influence of the application of sewage sludge on the degradation of pesticides in the soil. Chemosphere 57(7):673–679
Sanders DA, Malina JF Jr, Moore BE, Sagik BP, Sorber CA (1979) Fate of poliovirus during anaerobic digestion. J (Water Pollut Control Federation) 51:333–343
Sänger M, Werther J, Ogada T (2001) NOx and $N_2O$ emission characteristics from fluidised bed combustion of semi-dried municipal sewage sludge. Fuel 80(2):167–177
Schaller A, Diez T (1991) Plant specific aspects of heavy metal uptake and comparison with quality standards for food and forage crops. Der Einfluß von festen Abfällen auf Böden, Pflanzen, pp 92–125
Schwartzbrod L, Mignotte B (1996) Virus enterique et boues residuaires urbaines. Bulletin Sociellf Fran9aise Microbiologie 11:11–16
Sommers LE, Nelson DW, Yost KJ (1976) Variable nature of chemical composition of sewage sludges, vol 5, No. 3. American Society of Agronomy, Crop Science Society of America, and Soil Science Society of America, pp 303–306
Sotero-Santos RB, Rocha O, Povinelli J (2007) Toxicity of ferric chloride sludge to aquatic organisms. Chemosphere 68(4):628–636
Stentiford E (1986) Recent development in composting. In: De Bertoldi M, Ferranti MP, L'Hermite P, Zucconi F (eds) Compost: production, quality and use. Elsevier Applied Science, London & New York, pp 52–60
Strauch D (1991) Survival of pathogenic micro-organisms and parasites in excreta, manure and sewage sludge. Revue scientifique et technique (International Office of Epizootics) 10(3):813–846
Tahhan RA, Ammari TG, Goussous SJ, Al-Shdaifat HI (2011) Enhancing the biodegradation of total petroleum hydrocarbons in oily sludge by a modified bioaugmentation strategy. Int Biodeterior Biodegradation 65(1):130–134
Tarpani RRZ, Alfonsin C, Hospido A, Azapagic A (2020) Life cycle environmental impacts of sewage sludge treatment methods for resource recovery considering ecotoxicity of heavy metals and pharmaceutical and personal care products. J Environ Manage 260:109643
Valavanidis A, Vlachogianni T (2010) Metal pollution in ecosystems. Ecotoxicology studies and risk assessment in the marine environment. Dept. of Chemistry, University of Athens University Campus Zografou, p 15784

Wang MJ (1997) Land application of sewage sludge in China. Sci Total Environ 197(1–3):149–160
Wei X, Viadero RC Jr, Bhojappa S (2008) Phosphorus removal by acid mine drainage sludge from secondary effluents of municipal wastewater treatment plants. Water Res 42(13):3275–3284
Yang K, Zhu Y, Shan R, Shao Y, Tian C (2017) Heavy metals in sludge during anaerobic sanitary landfill: speciation transformation and phytotoxicity. J Environ Manage 189:58–66
Yeager JG, Ward RL (1981) Effects of moisture content on long-term survival and regrowth of bacteria in wastewater sludge. Appl Environ Microbiol 41(5):1117–1122
Yuan X, Huang H, Zeng G, Li H, Wang J, Zhou C, Liu Z (2011) Total concentrations and chemical speciation of heavy metals in liquefaction residues of sewage sludge. Bioresour Technol 102(5):4104–4110
Zhang J, Zhang S, Liu B (2020) Degradation technologies and mechanisms of dioxins in municipal solid waste incineration fly ash: a review. J Cleaner Prod 250:119507
Zhang ZY, Li LH, Zhang JS, Ma C, Wu X (2018) Solidification of oily sludge. Petrol Sci Technol 36(4):273–279

# Understanding the Environmental and Health Implications of Sludge

Kundan Samal, Satya Ranjan Samal, and Rajesh Roshan Dash

**Abstract**

Sludge is emerging as major polluting substance as it contains a lot of contaminants, posing both environmental and health challenges. From an environment perspective, untreated sludge can lead to soil and water contamination severely. Soil contamination results from the direct application of untreated sludge in land incorporating heavy metals, organic pollutants, and pathogens. During rainy season runoff transport these pollutants to water bodies polluting aquatic ecosystem and ground water. Incomplete decomposition of sludge emits greenhouse gases (GHGs) polluting atmosphere. From a health perspective, the sludge is a reservoir of pathogenic microorganisms such as bacteria, viruses, and fungi. Exposure can result in various diseases to human. Presence of toxic chemicals and heavy metals like Ni, Pb, and Cd in sludge can lead to significant health issues upon continuous exposure. In bioaccumulation process, contaminants in sludge enter the food chain, impacting human health over time. Addressing these challenges necessitates technological interventions like anaerobic digestion and composting, and stringent regulatory measures to ensure only treated sludge with acceptable contaminant levels finds application. By embracing safe management practices and recognizing sludge's potential for resource recovery, this environmental challenge can be turned into a sustainable opportunity, ensuring both ecological balance and public health safety. This chapter highlights the environmental and health risks associated with sludge and emphasizes the importance of its safe management and potential reuse.

**Keywords**

Sludge · Environmental risks · Bioaccumulation · Health risks · Soil contamination · Greenhouse gases (GHGs)

## 1 Introduction

Sludge is a byproduct of wastewater treatment plant, and has become a prominent focus in the area of waste management recently. It consists of organic matter, pathogens, heavy metals, and various other contaminants (Stefaniuk et al. 2018; Seggiani et al. 2012). Due to urbanization and increase in global population, wastewater treatment plants are treating high volumes of water, leading to increased sludge production. Eventually, the environmental and health implications associated with sludge has become important.

From an environmental perspective, the inappropriate disposal of sludge can lead to the contamination of land, surface waters, and groundwater. The organic compounds present in water reduce oxygen levels in aquatic systems, leading to the death of marine life. Heavy metals leached into water sources, bioaccumulate in aquatic organisms, disrupt the ecosystems, and enter the human food chain. The persistent nature of heavy metals can make areas unsuitable for agriculture or habitation (Samolada and Zabaniotou 2014). Moreover, the release of greenhouse gases from sludge, particularly $CH_4$ during anaerobic decomposition, contributes to global climate change. Health implications associated with sludge are equally concerning. The pathogens including bacteria, viruses, and parasites, can pose direct threats to human health if they infiltrate drinking water sources or contaminate food crops (Samal et al. 2022). Heavy metals like lead, mercury,

K. Samal (✉) · S. R. Samal
School of Civil Engineering, KIIT University Bhubaneswar, Bhubaneswar, Odisha, India
e-mail: kundan.samalfce@kiit.ac.in

R. R. Dash
School of Infrastructure, IIT Bhubaneswar, Bhubaneswar, Odisha, India

© The Author(s), under exclusive license to Springer Nature Switzerland AG 2025

A. Mandpe and M. P. Shah (eds.), *Wastewater Sludge Nexus*, Sustainable Landscape Planning and Natural Resources Management, https://doi.org/10.1007/978-3-031-45758-6_12

and cadmium have severe health impacts, including neurological disorders and developmental issues in children (Li et al. 2018). There's also the potential threat from pharmaceuticals and personal care products that can sometimes be found in sludge.

However, it is essential to recognize that with proper treatment and management, sludge can also be viewed as a resource. Many countries have harnessed its potential as a soil conditioner, improving soil structure and nutrient content. Biogas production from sludge offers a sustainable energy source, turning a waste product into a valuable resource. The chapter includes sludge impact on environment, health, its management and mitigation strategies, and recommendation for minimizing risk.

## 2 Sludge Composition and Properties

Sludge is a semi-solid byproduct that results from the treatment of wastewater in sewage treatment plants and effluent treatment plant. Its composition depends on its source and generally consists of organic materials, pathogens, microorganisms, heavy metals, and chemicals (Smith 2009). The properties of sludge can differ in terms of its moisture content, organic matter content, calorific value, pH, density, and rheological behavior (Table 1). Due to the potential presence of harmful substances, the disposal of sludge requires careful management to protect the environment and human health. Moreover, the characteristics of sludge can also influence its treatability and the efficiency of subsequent processing steps, such as dewatering, digestion, or incineration (Wołejko et al. 2018). Figure 1 represents process of sludge generation and sludge utilization in wastewater treatment plant.

## 3 Environmental Implications of Sludge

### 3.1 Soil Contamination

Sludge can significantly contaminate soil if its mismanagement happens. When the sludge is applied excessively to soil, either as a disposal method or as a fertilizing strategy, contaminants like heavy metals, chemicals, endocrine disruptor, etc. incorporate into the soil matrix (Stefaniuk et al. 2018; Urbaniak et al. 2017). Those toxic substance alter the natural microbial community of the soil, leading to reduced soil fertility and disruptions in the biodegradation of organic matter. Contaminants like heavy metals can persist for extended periods, leading to long-term soil degradation (Collivignarelli et al. 2019). These contaminants gradually accumulate in the plant and obstruct its growth, interfere with nutrient uptake, and subsequently enter the food chain. The leaching or runoff of these contaminants, driven by rainfall or irrigation, can percolate through soil layers, eventually reaching and polluting groundwater reserves (Wołejko et al. 2018). This not only makes groundwater unsafe for human consumption but also

**Table 1** Characteristics of sewage sludge

| Parameters | Value | References |
|---|---|---|
| pH | 5–7 | Samal et al. (2022) |
| Total dry solids (% of TS) | 3–11 | Collivignarelli et al. (2019) |
| Total organic solid (% of TS) | 20–44 | Smith (2009) |
| Total volatile solid (% of TS) | 60–85 | Gao et al. (2020) |
| Organic carbon(% of TS) | 20–50 | Collivignarelli et al. (2019) |
| Total nitrogen (% of TS) | 3–10 | Gao et al. (2020) |
| Total phosphorus (% of TS) | 2–5 | Hoang et al. (2022) |
| Dry matter. DM (%) | 22–32 | Li et al. (2022) |
| Water-holding capacity (%) | 60–90 | Passos et al. (2022) |
| Moisture (%) | 80–98 | Li et al. (2022) |
| Electrical conductivity (mS/cm) | 5–25 | Chen et al. (2020a) |
| C/N | 3–20 | Behera and Samal (2022) |
| Density (g/cm$^3$) | 1–6 | Taki et al. (2020) |
| Heating value (kJ/g) | 16–20 | Hao et al. (2020) |
| Cd | 41–54 (Mg/kg of dry sludge) | Urbaniak et al. (2017) |
| Cr | 102–8810 | Li et al. (2018) |
| Cu | 280–543 | Li et al. (2018) |
| Ni | 192–293 | Chagas et al. (2021) |
| Pb | 91–129 | Chagas et al. (2021) |

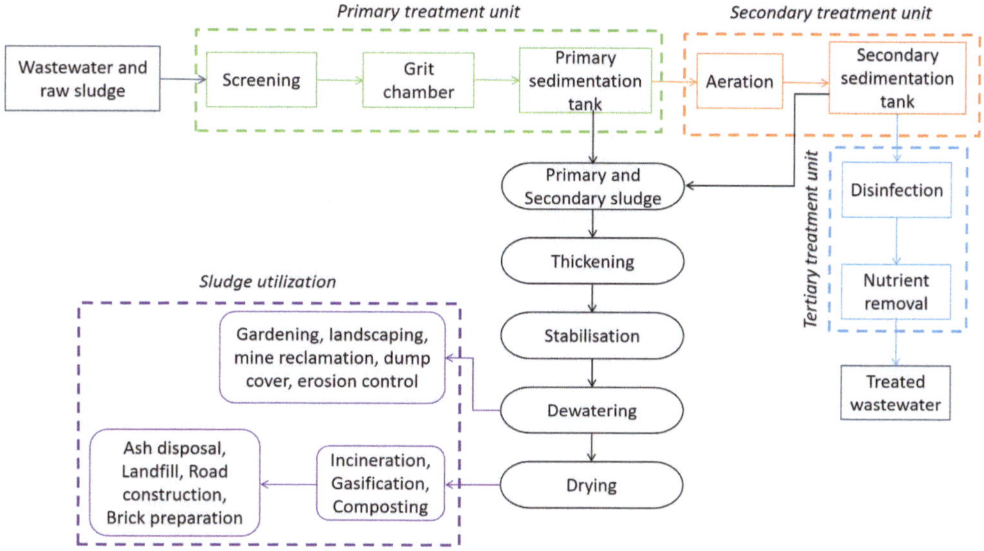


**Fig. 1** Sludge generation and its utilization in wastewater treatment plant

disrupts aquatic ecosystems, harming both flora and fauna in the process. Table 2 shows the impact of sewage sludge on environment.

## 3.2 Water Contamination

Sludge can be a source of water pollution if it directly enters aquatic ecosystems or if its contaminants leach into water sources. Improper disposal practices, such as dumping sludge near water bodies or on land with high runoff potential, can facilitate the migration of pollutants present in sludge into rivers, lakes, or groundwater. Once introduced into water systems, heavy metals can bioaccumulate in aquatic organisms, disrupting aquatic food chains and endangering species at various trophic levels (Turunen et al. 2018; Sabbahi et al. 2018). Organic pollutants, on the other hand, can deplete oxygen levels in water, leading to the death of aerobic organisms and the proliferation of anaerobic ones, thereby disrupting aquatic biodiversity (Ostman et al. 2017). Furthermore, pathogens present in the sludge can increase the risk of waterborne diseases. Contaminated water not only threatens aquatic life but also poses severe health risks for communities relying on these water sources for drinking, irrigation, and other daily activities (Gao et al. 2020).

## 3.3 Air Pollution

Sludge, especially when derived from wastewater treatment processes and industrial activities, can become a source of air contamination in several ways (Fig. 2). The organic components within the sludge often generate gases as they decompose and release odorous compounds like $H_2S$, $NH_3$, and $CH_4$ into the atmosphere (Peng et al. 2018). $CH_4$ for is a potent greenhouse gas with a global warming potential far exceeding that of $CO_2$. Additionally, when sludge is inadequately treated or stored, the disturbance or re-suspension of its particles, often due to wind or mechanical activities, can release particulate matter and pathogens into the air, posing inhalation risks to nearby communities. If the sludge is incinerated without appropriate pollution control mechanisms, hazardous byproducts such as dioxins, furans, and heavy metal aerosols can be emitted. These pollutants not only degrade air quality but also carry significant health implications, ranging from respiratory problems to long-term carcinogenic risks (Ostman et al. 2017). So, managing and treating sludge in an environmentally responsible manner is imperative to prevent the degradation of air quality and the associated health hazards.

## 3.4 Bioaccumulation

Application of sludge to agricultural lands introduces the contaminants into the environment, laying the groundwork for bioaccumulation and biomagnification within food webs (Huang et al. 2018). Plants growing in contaminated soils can uptake and accumulate these contaminants, especially heavy metals like Cd, Pb, and Hg. When herbivores consume these plants, they not only ingest the nutrients but also the accumulated toxins which passed to the next trophic level in food chain. In biomagnification process, animals higher up the food chain, including apex predators and even humans, receive a disproportionately high and potentially hazardous dose of these contaminants. Such concentrations can lead to various health issues, reproductive failures, and

**Table 2** Impact of sewage sludge disposal and monitoring parameters

| Sludge disposal Impact | Pollution indication | Parameters measurement |
|---|---|---|
| Water pollution | ✓ Cloudiness or haziness in water due to suspended particles<br>✓ Unpleasant smells due to organic decay<br>✓ Presence of toxins such as heavy metals, pesticides, and industrial chemicals<br>✓ Excessive foaming, often from detergents or organic decay<br>✓ Sudden growth of algae<br>✓ Fish kills or dies-off of other aquatic organisms | pH, conductivity, hardness, TDS, BOD, COD, $NH_3$, $NO_2^-$, $NO_3^-$ TN, TP, metals and metalloid, indicator organisms |
| Air pollution | ✓ Detectable smells such as sulfur dioxide giving a rotten egg smell<br>✓ Irritated or watering eyes due to smog or particulate matter | PM 10, PM 2.5, $O_3$, $SO_2$, $NO_2$, CO, Heavy metals |
| Soil pollution | ✓ Soil may exhibit unusual colors, such as blackish or oily streaks, indicating the presence of oil, heavy metals, or other pollutants<br>✓ Soil may become excessively acidic or alkaline due to pollutants<br>✓ A decline in the number of soil microorganisms, earthworms, and other beneficial organisms<br>✓ Polluted soil can cause plants to exhibit stress symptoms like stunted growth, yellowing or browning of leaves, or even die-off | Cation Exchange Capacity (CEC), exchangeable cations (Ca, Mg, K, Na), metals, soil texture, pH, carbon content |
| Transmission of diseases | ✓ Pathogens density<br>✓ Vectors (rodents and insects) attractiveness on application site<br>✓ Concentration of toxic compounds | M. *tuberculosis, E. coli, V. cholerae, C. botulinum, S. pneumoniae, S. aureus,* etc |
| Food chain contamination | ✓ Concentration of contaminants in water, soil, and crops<br>✓ Presence of certain contaminants can alter the taste or smell of food items<br>✓ Higher concentrations of pollutants are found in organisms higher up in the food chain compared to those at lower levels<br>✓ Imbalances in predator–prey relationships or the decline of keystone species due to contamination<br>✓ As certain species decline or disappear due to contamination, overall biodiversity may reduce | Bioaccumulation of pollutants and metals, lead, mercury, nitrates, fluorine |
| Aesthetic and social problems | ✓ Improperly treated or stored sludge can produce a foul smell<br>✓ Large piles of sludge or open-air storage can be an eyesore<br>✓ If sludge seeps into nearby water bodies, it can lead to discoloration of water<br>✓ The presence of pathogens in sludge can raise concerns about diseases, leading to anxiety within the community<br>✓ Homes and properties near sludge disposal sites might see a decrease in their value<br>✓ Disposal sites, especially if chosen without community consultation, can lead to conflicts | |
| *Loss of Biodiversity* | ✓ As these toxins bioaccumulate up the trophic levels, apex predators can experience pronounced toxic effects<br>✓ Microbial communities within soil, which play crucial roles in nutrient cycling and organic matter decomposition, can be disrupted by the influx of sludge-based contaminants, altering ecosystem dynamics<br>✓ Pathogens within the sludge can introduce diseases to local wildlife, leading to health issues and potential population declines<br>✓ Aquatic ecosystems, too, can be affected if sludge contaminants leach into water bodies, causing harm to fish, amphibians, and invertebrates | Species richness, genetic diversity, endemic species, ecosystem diversity, carrying capacity |

even death (Fijalkowski et al. 2017; Chagas et al. 2021). The potential for toxins from sludge to infiltrate the food web and culminate in detrimental health effects in top-tier consumers underscores the importance of cautious and informed sludge management practices.

## 3.5 Loss of Biodiversity

Sludge holds a range of substances that can impact biodiversity, both directly and indirectly. When indiscriminately introduced into environments, either through poor disposal

**Fig. 2** Air pollution due to sludge

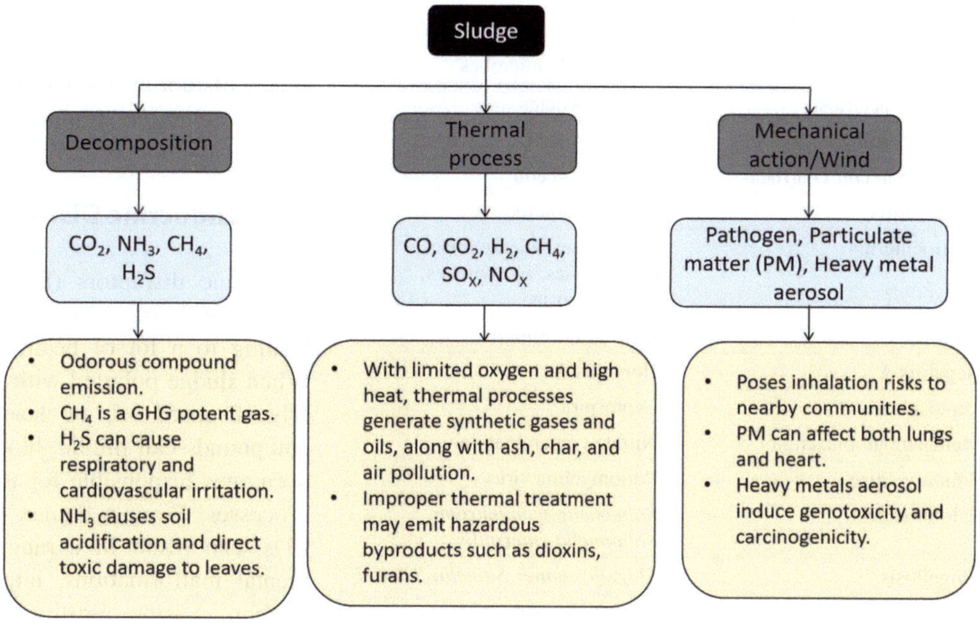


practices or as a perceived soil enhancer, the diverse contaminants within sludge, including heavy metals, organic pollutants, and pathogens, can pose significant threats to local flora and fauna. Plants can absorb and accumulate toxic metals, leading to physiological damage and reduced reproductive success. Such contamination not only impedes plant growth but also risks entering the food chain as herbivores consume these tainted plants. As these toxins bioaccumulate up the trophic levels, apex predators can experience pronounced toxic effects, potentially leading to reduced populations or even local extinctions (Gu et al. 2013; Behera and Samal 2022). Microbial communities within soil, which play crucial roles in nutrient cycling and organic matter decomposition, can be disrupted by the influx of sludge-based contaminants, altering ecosystem dynamics. Additionally, pathogens within the sludge can introduce diseases to local wildlife, leading to health issues and potential population declines. Aquatic ecosystems, too, can be affected if sludge contaminants leach into water bodies, causing harm to fish, amphibians, and invertebrates (Ellis et al. 2018).

## 4 Health Implications of Sludge

### 4.1 Exposure to Pathogens

Sludge contains a diverse range of pathogens including bacteria, viruses, protozoa, and helminths that can pose serious health and environmental risks. Direct exposure to these pathogens can lead to an array of health issues in humans, ranging from gastrointestinal illnesses like diarrhea and vomiting, to more severe diseases such as hepatitis, cholera, and parasitic infections (Table 3). Many of these illnesses can be fatal, leading to prolonged suffering, and in severe cases death of elderly, children, and those with compromised immune systems (Huang et al. 2018). Beyond direct human health impacts, these pathogens, when improperly disposed of or applied to land without adequate treatment, can contaminate surface and groundwater sources. This not only exacerbates the spread of diseases to wider populations but also threatens aquatic ecosystems, with potential disruptions to microbial communities and harm to aquatic fauna (Praspaliauskas and Pedišius 2017). Furthermore, pathogen-laden runoff from sludge-amended soils can enter water bodies, promoting algal blooms and other ecological imbalances.

### 4.2 Toxicity from Heavy Metals

Heavy metals have acquire significant concern due to their persistence, bioaccumulative nature, and potential for generating adverse health and environmental effects. From a health perspective, when sludge containing heavy metals is inadequately treated and then applied to agricultural lands, these toxic elements can be taken up by crops and subsequently enter the human food chain. Consumption of such contaminated food can lead to a range of health issues (Gu et al. 2013). For example, Pb can cause neurological damage, especially in children; Cd can result in renal dysfunction and bone damage; Hg particularly in its methylated form, can impair neurological development; and As can affect multiple organs. Beyond direct human health impacts, heavy metals in sludge pose significant threats to the environment. When introduced into natural ecosystems, these metals can disrupt soil

**Table 3** Diseases caused by sewage sludge

| Diseases | Responsible microbes |
|---|---|
| Campylobacteriosis | Campylobacter |
| Cryptosporidiosis | *Cryptosporidium parvum* |
| *Escherichia coli* Diarrhea | *Escherichia coli* |
| Encephalitis | *Culex* mosquitoes |
| Gastroenteritis | Rotaviruses, adenoviruses, caliciviruses, astroviruses, Norwalk virus |
| Giardiasis | *Giardia intestinalis* |
| Hepatitis A | Hepatitis A |
| Leptospirosis | Leptospira |
| Methaemoglobinaemia | Nitrate contamination |
| Poliomyelitis | Poliomyelitis virus |
| Salmonellosis | *Salmonella typhimurium, Salmonella enteritidis* |
| Shigellosis | *Shigella sonnei, Shigella flexneri* |
| Paratyphoid Fever | *Salmonella paratyphi* |
| Typhoid Fever | *Salmonella typhi* |
| Yersiniosis | *Yersinia enterocolitica* |

microbial communities, diminishing soil health and fertility (Atienza-Martínez et al. 2018). Aquatic organisms exposed to elevated heavy metal levels can experience reduced growth, reproductive failure, and even mortality.

## 4.3 Chemical Contaminants

Chemical contaminants in sludge, constituting a vast amount of substances like pharmaceuticals, endocrine-disrupting compounds, polychlorinated biphenyls (PCBs), and polyaromatic hydrocarbons (PAHs), present a multiple threat to both human health and the environment. From a health perspective, when these chemically contaminated sludges are used without adequate treatment, especially in agricultural lands, that can introduce into crops, entering to human food chain (Stefaniuk et al. 2018; Yıldız and Oran 2018). Many of these chemicals have been linked to severe health issues. For example, endocrine-disrupting compounds can interfere with the hormonal systems of humans and animals, leading to reproductive and developmental problems, cancers, and other endocrine-related disorders. Pharmaceuticals in sludge can lead to antibiotic resistance or unexpected health effects. On the environmental side, these chemical contaminants can have profound effects on ecosystems. They can alter soil microbial communities, affecting soil health and the breakdown of organic matter. In aquatic environments, particularly when contaminated sludge leaches into water bodies, these chemicals can harm aquatic life by causing reproductive issues, altering behavior, and even leading to mortality (Wołejko et al. 2018; Hao et al. 2020). The persistence of such chemicals can result in long-term ecological imbalances, disrupting food webs and posing cascading risks to species at various trophic levels.

## 4.4 Endocrine Disruptors

Endocrine disruptors (EDs) in sludge interfere with the endocrine (hormonal) system of both humans and wildlife, leading to a lot of health and environmental challenges. When sludge polluted with EDs is applied to land, improperly disposed of, or leaches into water bodies, these compounds can mimic, block, or interfere with the natural hormones responsible for the regulation of vital biological processes. From a human health perspective, exposure to EDs, can result in a range of issues including developmental malformations, interference with the reproductive system, causing fertility problems, an increased risk of certain cancers (Bubalo et al. 2021; Taki et al. 2020). Particularly vulnerable are fetuses, infants, and children, where disrupted hormonal activity can lead to lifelong effect like cognitive impairments and developmental delays. On the environmental front, EDs introduced into ecosystems can bring disturbance on wildlife. Aquatic organisms, for instance, are especially at risk, with many fish populations showing skewed sex ratios, reproductive failures, and abnormal development due to exposure. Terrestrial animals too can face hormonal imbalances, reproductive challenges, and altered behaviors upon ED exposure (Makan and Fadili 2020). Given the profound impacts of endocrine disruptors on health and the environment, there's an urgent need for robust strategies to manage and mitigate their presence in sludge and other environmental mediums.

## 4.5 Airborne Diseases

Sludge when undergoes inadequate treatment or mismanagement can be a source of airborne diseases, posing a significant health risk to both local communities and the broader environment. As sludge dries and particulates become airborne, it can release pathogens, volatile organic compounds (VOCs), endotoxins, and other harmful agents into the atmosphere (Ekane et al. 2021). The act of turning, disturbing, or aerating the sludge can enhance this release. Inhaled by humans, these airborne particles can lead to respiratory illnesses, allergic reactions, and the spread of certain diseases. For instance, the aerosolization of certain bacteria or viruses from sludge can result in the transmission of diseases like tuberculosis, influenza, or Q-fever. Additionally, VOCs and other noxious chemicals emitting from sludge can irritate the respiratory system, exacerbate conditions like asthma, or contribute to longer-term respiratory and systemic illnesses (Gwara et al. 2021). Furthermore, in an environmental context, these airborne contaminants can also affect local fauna, impacting their health and altering ecosystem dynamics.

### 4.6 Contamination of Food Sources

Sludge, when inadequately treated and improperly applied, especially to agricultural lands, can act as a major conduit for the contamination of food sources. Rich in nutrients, sludge is often utilized as a soil amendment, but it can also carry a multitude of harmful agents including pathogens, heavy metals, endocrine disruptors, and various organic contaminants. As these agents come into contact with crops, they can be absorbed by plants, accumulate in fruits, vegetables, and grains, or persist in the soil, leading to long-term contamination. When consumed, these contaminated foods introduce these harmful agents into the human body, potentially leading to a spectrum of health issues (Kataki et al. 2016; Saerens et al. 2021; Chen et al. 2022). For instance, heavy metals like cadmium and lead can accumulate in leafy vegetables and cereals, and when ingested, they can cause neurotoxicity, renal dysfunction, and other severe health conditions. Pathogens present in sludge can also contaminate fresh produce, leading to foodborne illnesses. Additionally, the persistent organic pollutants in sludge can enter the food chain and, over time, bioaccumulate in higher trophic levels, posing risks even to consumers of meat and dairy products. This cycle of contamination not only jeopardizes human health but also undermines the safety and trust in the food supply system (Wilk et al. 2021; Gaur et al. 2020). It emphasizes the critical need for rigorous sludge treatment and careful monitoring before its application to agricultural settings.

## 5 Management and Mitigation Strategies

Different mitigation strategies have been adapted to reduce the impact of sludge on health and environment. Figure 3 shows the different criteria for selecting technology for sludge treatment.

### 5.1 Sludge Treatment

Effective sludge treatment processes such as thickening, dewatering, stabilization, digestion, and drying can significantly reduce the volume and its hazard potential. By neutralizing pathogens through techniques like anaerobic digestion or lime stabilization, the risk of disease transmission to humans is drastically reduced (Hoang et al. 2022; Beegle and Borole 2018; Iqbal et al. 2022). Table 4 shows comparison of various sludge pathogens removal technologies and its technical requirement. Similarly, certain treatment methods can immobilize or reduce heavy metals, minimizing the risk of these toxic elements entering the food chain or leaching into water sources. By breaking down or removing organic contaminants, treatment processes ensure reduced environmental contamination and lower the risk of ecosystem disruption (Samal et al. 2022; Paolini et al. 2018). Advanced sludge treatment can also lead to resource recovery, turning waste into valuable products such as

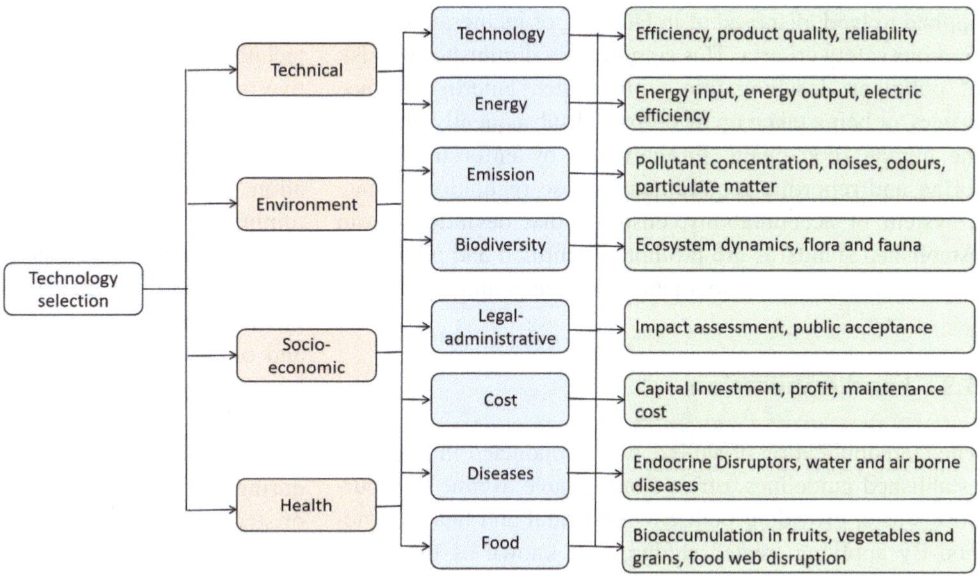


**Fig. 3** Different criteria for selecting sludge treatment technologies

**Table 4** Comparison of various sludge pathogens removal technologies and its technical requirement

| Processes | Bacteria | Viruses | Eggs | Sludge stability | Odor emission |
|---|---|---|---|---|---|
| Windrow composting | +++ | ++/+ | ++ | +++ | ++ |
| In-vessel composting | +++ | +++/++ | ++ | +++ | ++ |
| Aerobic digestion | +++/++ | +++/++ | ++ | ++ | ++ |
| Pasteurization process | +++/++ | +++/++ | +++ | ++ | ++ |
| Lime addition | +++/++ | +++/++ | +++/++ | ++/+ | +++ |
| Thermal drying of sludge | +++ | +++ | +++ | +++ | + |
| Incineration of sludge | +++ | +++ | +++ | +++ | + |

+++: Significant importance, ++: Moderate importance, +: Little importance

biogas or nutrient-rich soil amendments for agriculture (Turunen et al. 2018; Alves et al. 2021). In essence, comprehensive sludge treatment not only diminishes the direct threats posed by untreated sludge to human health and the environment but also promotes sustainable practices, transforming potential pollutants into resources and reducing the ecological footprint of wastewater management.

## 5.2 Regulations

Regulations related to sludge management play an instrumental role in safeguarding both environmental and public health. By establishing stringent standards for the treatment, disposal, and potential reuse of sludge, these regulations act as the first line of defense against the potential sludge hazards. These standards typically address factors such as pathogen reduction, heavy metal concentrations, and limits on organic contaminants (Herbes et al. 2020). By mandating specific treatment processes or setting threshold limits for various contaminants, regulations ensure that any sludge applied to land, disposed of in landfills, or incinerated meets rigorous safety criteria. This significantly diminishes the risk of pollutants leaching into groundwater, entering surface waters, or being taken up by crops and subsequently entering the human food chain. Furthermore, by enforcing monitoring and reporting requirements, these regulations create a system of accountability, ensuring that deviations from established standards are promptly identified and addressed (Sikosana et al. 2017; Boudjabi and Chenchouni 2021).

## 5.3 Land Application

The land application of sludge, when conducted in line with established guidelines, offers a sustainable avenue to repurpose waste, providing both environmental and health benefits. By applying treated sludge, also known as biosolids, to agricultural lands, forests, and reclaimed mining sites, the organic matter and nutrients inherent in the sludge are recycled, enhancing soil fertility and structure. This process helps reduce the reliance on synthetic fertilizers, which, when over-applied, can contribute to nutrient runoff and the resultant eutrophication of water bodies (Boudjabi and Chenchouni 2021; Chagas et al. 2021; Chen et al. 2020a). Additionally, by reintegrating the organic content of sludge into the soil, carbon sequestration is promoted, assisting in mitigating greenhouse gas emissions. From a health standpoint, diverting sludge from landfills reduces the risk of groundwater contamination from leachate and minimizes the production of landfill gases, particularly $CH_4$, a potent greenhouse gas (Peng et al. 2018; Chen et al. 2021). Furthermore, when rigorous treatment standards are met, the land application of sludge reduces the pathogenic load, ensuring that risks of disease transmission through crops or water sources are minimal (Chen et al. 2020b).

## 5.4 Incineration

The incineration of sludge, when executed using advanced technologies and under stringent controls, serves as an effective method for volume reduction and the neutralization of potentially harmful constituents, providing several environmental and health advantages. First and foremost, incineration dramatically reduces the volume of sludge, thereby minimizing the spatial and environmental footprint associated with landfill disposal (Samolada and Zabaniotou 2014; Mahapatra et al. 2022). By reaching high temperatures, incineration ensures the complete destruction of pathogens and organic contaminants, eliminating the risks of disease transmission and the release of pollutants into the environment (Fig. 5.2). From an energy perspective, modern incineration facilities are designed to recover the heat generated during the combustion process, converting it into electricity or steam (Hao et al. 2020). This energy recovery transforms a waste product into a valuable energy resource, reducing dependence on fossil fuels and decreasing greenhouse gas emissions. Additionally, with the integration of advanced air pollution control technologies, modern incinerators can substantially reduce the release of harmful emissions, such as dioxins, furans, and heavy metals, safeguarding air quality and public health (Passos et al. 2022; Yang et al. 2020).

### 5.5 Landfills

Landfilling, a conventional method for sludge disposal, can be designed to minimize the environmental and health impacts of waste. When engineered with modern safeguards, landfills isolate sludge from the environment, effectively preventing the direct release of contaminants. Advanced landfill designs incorporate liners, typically made of clay and synthetic materials, to reduce the leaching of pollutants into the soil and groundwater (Peng et al. 2018; Odirile et al. 2021; Qian et al. 2020). Additionally, collection systems are installed to gather and treat leachate, ensuring that potential contaminants are removed before any discharge. Another critical aspect is the capture and management of landfill gases, like $CH_4$, which are produced as organic matter in the sludge decomposition process (Thomsen et al. 2022). Instead of allowing $CH_4$ to escape into the atmosphere, many modern landfills either burn it off through flaring or harness it for energy generation. From a health perspective, encapsulating the sludge in a landfill diminishes the risk of human exposure to pathogens or other harmful substances present in the waste.

### 5.6 Public Awareness

Public awareness regarding the risks and benefits associated with sludge is essential for informed decision-making, responsible management practices, and fostering trust between communities and wastewater treatment entities. The many-sided nature of sludge: being both a potential pollutant and a resource, underscores the importance of comprehensive public education. On the one hand, untreated or improperly managed sludge can pose significant health threats, including the transmission of diseases, introduction of heavy metals into the food chain, and contamination of water sources. Enlightening the public about these risks empowers communities to demand rigorous treatment standards, ensuring their safety and well-being (Lu et al. 2018; Yuan et al. 2011). On the other hand, when processed appropriately, sludge can be a valuable resource. Treated biosolids can enhance soil fertility, reduce the need for synthetic fertilizers, and even serve as a renewable energy source when utilized for biogas production. Educating the public about these benefits can foster support for sustainable sludge management practices and encourage the responsible integration of sludge into environmental and agricultural systems (Li et al. 2021; Cao et al. 2021).

## 6 Recommendations for Minimizing Risks

These are a list of recommendations for minimizing the health and environmental risks associated with sludge:

- Advanced treatment processes like anaerobic digestion, lime stabilization, and thermal hydrolysis need to be implemented to reduce pathogens and other harmful contaminants in sludge. Frequent testing of sludge for pathogens, heavy metals, and chemical contaminants are need to be conducted compliance with safety standards.
- Public awareness campaigns need to be increased about the risks and benefits of sludge, fostering informed community engagement and feedback. Stringent regulations and guidelines need to be established and enforced concerning the treatment, disposal, and land application of sludge.
- Conversion of treated sludge (biosolids) into biogas or compost needs to be promoted. Industries need to be encouraged to implement cleaner production methods to reduce the introduction of harmful chemicals into wastewater, subsequently decreasing their presence in sludge.
- For landfills, advanced designs with liners and leachate collection systems need to be employed to prevent groundwater contamination. Also gas collection and treatment systems need to be installed to minimize greenhouse gas emissions.
- Collaborations between governments, industries, research institutions, and communities need to be fostered to address sludge-related challenges.

By embracing these recommendations, municipalities and industries can significantly mitigate the health and environmental risks associated with sludge, steering toward more sustainable and safer practices.

## 7 Conclusion

Sludge, a byproduct of wastewater treatment, harbors pathogens, heavy metals, and chemical contaminants that, if not properly managed, can threaten public health and disrupt ecosystems. From the risk of disease transmission to the contamination of food chains and water sources, the consequences of neglect are significant. Conversely, when treated and managed with care, sludge transforms into biosolids that can enhance soil fertility, reduce our dependence on synthetic fertilizers, and even serve as a renewable energy source. As the human being is struggling with various environmental challenges, understanding the use of sludge is important to safeguard the human health and the environment.

## References


Alves O, Calado L, Panizio R, Gonçalves M, Monteiro E, Brito P (2021) Techno-economic study for a gasification plant processing residues of sewage sludge and solid recovered fuels. Waste Manag 131:148–162

Atienza-Martínez M, Ábrego J, Mastral JF, Ceamanos J, Gea G (2018) Energy and exergy analyses of sewage sludge thermochemical treatment. Energ 144:723–735

Beegle JR, Borole AP (2018) Energy production from waste: evaluation of anaerobic digestion and bioelectrochemical systems based on energy efficiency and economic factors. Renew Sustain Energ Rev 96:343–351

Behera S, Samal K (2022) Sustainable approach to manage solid waste through biochar assisted composting. Energy Nexus 7:100121

Boudjabi S, Chenchouni H (2021) On the sustainability of land applications of sewage sludge: how to apply the sewage biosolid in order to improve soil fertility and increase crop yield? Chemosphere 282:131122

Bubalo A, Vouk D, Stirmer N, Nad K (2021) Use of sewage sludge ash in the production of innovative bricks-an example of a circular economy. Sustainability 13:9330

Cao B, Zhang T, Zhang W, Wang D (2021) Enhanced technology based for sewage sludge deep dewatering: a critical review. Water Res 189:116650

Chagas JKM, Figueiredo CC, da Silva J, Paz-Ferreiro J (2021) The residual effect of sewage sludge biochar on soil availability and bioaccumulation of heavy metals: evidence from a three-year field experiment. J Environ Manage 279:111824

Chen YD, Duan X, Zhang C, Wang S, Ren NQ, Ho SH (2020a) Graphitic biochar catalysts from anaerobic digestion sludge for nonradical degradation of micropollutants and disinfection. Chem Eng J 384:123244

Chen Y, Li M, Li Y, Liu Y, Chen Y, Li H, Li L, Xu F, Jiang H, Chen L (2021) Hydroxyapatite modified sludge-based biochar for the adsorption of $Cu^{2+}$ and $Cd^{2+}$: adsorption behavior and mechanisms. Bioresour Technol 321:124413

Chen Y, Yi L, Wei W, Jin H, Guo L (2022) Hydrogen production by sewage sludge gasification in supercritical water with high heating rate batch reactor. Energy 238:121740

Chen YD, Wang R, Duan X, Wang S, Ren NQ, Ho SH (2020b) Production, properties, and catalytic applications of sludge derived biochar for environmental remediation. Water Res 187

Collivignarelli MC, Canato M, Abba A, Carnevale Miino M (2019) Biosolids: what are the different types of reuse? J Cleaner Prod 238:117844

dos Passos JS, Matayeva A, Biller P (2022) Synergies during hydrothermal liquefaction of cow manure and wheat straw. J Environ Chem Eng 10(5):108181

Ekane N, Barquet K, Rosemarin A (2021) Resources and risks: perceptions on the application of sewage sludge on agricultural land in Sweden, a case study. Front Sustain Food Syst 5:647780

Ellis S, Tyrrel S, O'leary E, Richards K, Griffiths B, Ritz K (2018) Proportion of sewage sludge to soil influences the durvival of Salmonella Dublin and *Escherichia coli.* Clean Soil Air Water 46:1800042

Fijalkowski K, Rorat A, Grobelak A, Kacprzak MJ (2017) The presence of contaminations in sewage sludge-the current situation. J Environ Manag 203:1126–1136

Gao N, Kamran K, Quan C, Williams PT (2020) Thermochemical conversion of sewage sludge: a critical review. Prog Energy Combust Sci 79:100843

Gaur RZ, Khoury O, Zohar M, Poverenov E, Darzi R, Laor Y, Posmanik R (2020) Hydrothermal carbonization of sewage sludge coupled with anaerobic digestion: integrated approach for sludge management and energy recycling. Energ Convers Manag 224:113353

Gu C, Bai Y, Tao T, Chen G, Shan Y (2013) Effect of sewage sludge amendment on heavy metal uptake and yield of ryegrass seedling in a mudflat soil. J Environ Qual 42(2):421–428

Gwara S, Wale E, Odindo A, Buckley C (2021) Attitudes and perceptions on the agricultural use of human excreta and human excreta derived materials: a scoping review. Agriculture 11:153

Hao X, Chen Q, van Loosdrecht MCM, Li J, Jiang H (2020) Sustainable disposal of excess sludge: incineration without anaerobic digestion. Water Res 170:115298

Herbes C, Roth U, Wulf S, Dahlin J (2020) Economic assessment of different biogas digestate processing technologies: a scenariobased analysis. J Clean Prod 255:120282

Hoang SA, Bolan N, Madhubashani AMP, Vithanage M, Perera V, Wijesekara H, Wang H, Srivastava P, Kirkham MB, Mickan BS, Rinklebe J, Siddique KHM (2022) Treatment processes to eliminate potential environmental hazards and restore agronomic value of sewage sludge: a review. Environ Pollut 293:118564

Huang K, Mao Y, Zhao F, Zhang XX, Ju F, Ye L, Wang Y, Li B, Ren H, Zhang T (2018) Free-living bacteria and potential bacterial pathogens in sewage treatment plants. Appl Microbiol Biotechnol 102:2455–2464

Iqbal A, Zan F, Siddiqui MA, Nizamuddin S, Chen G (2022) Integrated treatment of food waste with wastewater and sewage sludge: energy and carbon footprint analysis with economic implications. Sci Tot Environ 825:154052

Kataki S, West H, Clarke M, Baruah DC (2016) Phosphorus recovery as struvite from farm, municipal and industrial waste: feedstock suitability, methods and pre-treatments. Waste Manag 49:437–454

Li Z, Deng H, Yang L, Zhang G, Li Y, Ren Y (2018) Influence of potassium hydroxide activation on characteristics and environmental risk of heavy metals in chars derived from municipal sewage sludge. Bioresour Technol 256:216–223

Li E, Wang Y, Zhang D, Fan X, Han Z, Yu F (2021) Siderite/ Pms conditioning- pressurized vertical electro-osmotic dewatering process for activated sludge volume reduction: evolution of protein secondary structure and typical amino acid in EPS. Water Res 201:117352

Li Q, Zhang K, Yang Z, Guo H, Zheng X, Zhang Q, Xiong J, Lu X (2022) Dynamic changes of microbial community and moisture ratio during bio-drying of sludge after electro-dewatering. J Environ Manage 324:116366

Lu J, Liu Z, Zhang Y, Savage PE (2018) Synergistic and antagonistic interactions during hydrothermal liquefaction of soybean oil, soy protein, cellulose, xylose, and lignin. ACS Sustain Chem Eng 6(11):14501–14509

Mahapatra S, Ali MH, Samal K, Moulick S (2022) Diagnostic and treatment technologies for detection and removal of helminth in wastewater and sludge. Energy Nexus 8:100147

Makan A, Fadili A (2020) Sustainability assessment of large-scale composting technologies using PROMETHEE method. J Clean Prod 261:121244

Odirile PT, Marumoloa PM, Manali A, Gikas P (2021) Anaerobic digestion for biogas production from municipal sewage sludge: a comparative study between fine mesh sieved primary sludge and sedimented primary sludge. Water 13(24):3532

Ostman M, Lindberg RH, Fick J, Bj€orn E, Tysklind M (2017) Screening of biocides, metals and antibiotics in Swedish sewage sludge and wastewater. Water Res 115: 318–328

Paolini V, Petracchini F, Segreto M, Tomassetti L, Naja N, Cecinato A (2018) Environmental impact of biogas: a short review of current knowledge. J Environ Sci Health Part A Toxic/hazard Subst Environ Eng 53:899–906

Peng W, Pivato A, Lavagnolo MC, Raga R (2018) Digestate application in landfill bioreactors to remove nitrogen of old landfill leachate. Waste Manag 74:335–346

Praspaliauskas M, Pedišius N (2017) A review of sludge characteristics in Lithuania's wastewater treatment plants and perspectives of its usage in thermal processes. Renew Sustain Energy Rev 67:899–907

Qian L, Wang S, Savage PE (2020) Fast and isothermal hydrothermal liquefaction of sludge at different severities: reaction products, pathways, and kinetics. Appl Energy 260:114312

Sabbahi S, Trad M, Ayed LB, Marzougui N (2018) Occurrence of intestinal parasites in sewage samples and efficiency of wastewater treatment systems in Tunisia. Water Qual Res J 53(2):86–101

Saerens B, Geerts S, Weemaes M (2021) Phosphorus recovery as struvite from digested sludge-experience from the full scale. J Environ Manag 280:111743

Samal K, Moulick S, Mohapatra BG, Samanta S, Sasidharan S, Prakash B, Sarangi S (2022) Design of faecal sludge treatment plant (FSTP) and availability of its treatment technologies. Energy Nexus 7:100091

Samolada M, Zabaniotou A (2014) Comparative assessment of municipal sewage sludge incineration, gasification and pyrolysis for a sustainable sludge-to-energy management in Greece. Waste Manag 34(2):411–420

Seggiani M, Puccini M, Raggio G, Vitolo S (2012) Effect of sewage sludge content on gas quality and solid residues produced by cogasification in an updraft gasifier. Waste Manag 32:1826–1834

Sikosana M, Randall DG, von Blottnitz H (2017) A technological and economic exploration of phosphate recovery from centralised sewage treatment in a transitioning economy context. Water 43:343–353

Smith S (2009) Organic contaminants in sewage sludge (biosolids) and their significance for agricultural recycling. Philos Trans A Math Phys Eng Sci 367:4005–4041

Stefaniuk M, Tsang DC, Ok YS, Oleszczuk P (2018) A field study of bioavailable polycyclic aromatic hydrocarbons (PAHs) in sewage sludge and biochar amended soils. J Hazard Mater 349:27–34

Taki K, Gahlot R, Kumar M (2020) Utilization of fly ash amended sewage sludge as brick for sustainable building material with special emphasis on dimensional effect. J Clean Prod 275:123942

Thomsen LBS, Anastasakis K, Biller P (2022) Wet oxidation of aqueous phase from hydrothermal liquefaction of sewage sludge. Water Res 209:117863

Turunen V, Sorvari J, Mikola A (2018) A decision support tool for selecting the optimal sewage sludge treatment. Chemosph 193:521–529

Urbaniak M, Wyrwicka A, Tołoczk W, Serwecinska L, Zielinski M (2017) The effect of sewage sludge application on soil properties and willow (*Salix* sp.) cultivation. Sci Total Environ 586:66–75

Wilk M, Sliz M, Lubieniecki B (2021) Hydrothermal co-carbonization of sewage sludge and fuel additives: combustion performance of hydrochar. Renew Energ 178:1046–1056

Wołejko E, Wydro U, Jabłonska-Trypuc A, Butarewicz A, Łoboda T (2018) The effect of sewage sludge fertilization on the concentration of PAHs in urban soils. Environ Pollut 232:347–357

Yang J, He QS, Niu H, Astatkie T, Corscadden K, Shi R (2020) Statistical clarification of the hydrothermal co-liquefaction effect and investigation on the influence of process variables on the co-liquefaction effect. Ind Eng Chem Res 59(7):2839–2848

Yıldız S, Oran E (2018) Sewage sludge disintegration by electrocoagulation. Int J Environ Health Res 29:531–543

Yuan H, Zhu N, Song F (2011) Dewaterability characteristics of sludge conditioned with surfactants pretreatment by electrolysis. Bioresource Technol 102(3):2308–2315

# Unveiling Environmental and Health Risks Linked to Sludge

Nitesh Kumar, Stuti Rawat, Neha Thakur, Pratibha, Ruchika Devi, Arti Jamwal Sharma, and Sanjeev Kumar

**Abstract**

This chapter summarises recent researches on the numerous risks that sludge poses during its entire lifecycle, from production to disposal, and potential consequences of improper sludge management for both the ecosystem and human wellness. Sewage sludge is the residual material produced by wastewater treatment. It contains heavy, microplastics metals as well as microplastics, pathogens, organic contaminants, and difficult to dispose of. This kind of garbage cannot be used to improve the soil. The employment of sludge as a potential soil conditioner should be done after proper treatment as the improperly treated sludge may contaminate the soil and decrease yield. The leaching of pollutants and toxins may pollute the groundwater thus affecting nearby water sources and habitats. Studies indicate a possible pattern of chemical irritability apparent in the reports of illnesses and mortality in people in the vicinity of land application sites and are subjected to dust and water runoff from agricultural lands treated with sewage wastewater. The results of risk assessments for human health, including heavy metal concentrations, indicated that there may be concerns, particularly for vulnerable people like children. Pathogens can lead to waterborne diseases and pose risks to workers and nearby communities during handling and application. Sludge treatment through biological, mechanical, and chemical methods should be made mandatory before disposal of sludge and public awareness should arise. Providing PPE and other equipment to workers is also a necessary step.

**Keywords**

Sludge · Wastewater treatment · Environmental risks · Health risks · Contamination · Management strategies

N. Kumar · S. Rawat (✉) · N. Thakur · Pratibha · R. Devi
Department of Biosciences, Himachal Pradesh University, Shimla, India
e-mail: stutirawat011@gmail.com

A. Jamwal Sharma
Department of Botany, Career Point University, Hamirpur, Himachal Pradesh, India

S. Kumar
Department of Basic Sciences, College of Forestry, Dr. Y. S. Parmar University of Horticulture and Forestry, Nauni, Solan, Himachal Pradesh, India

## 1 Introduction

### 1.1 Definition of Sludge

A semi-solid slurry that contains any specific type of solid sedimentary material mixed with any sort of liquid is referred to as sludge. The semi-solid material found in sewage or that which is a by-product of any form of industrial or refining activity is most referred to by this word. It results from several industrial processes, such as wastewater treatment and traditional drinking water treatment, among others. It can also be formed as sewage sludge. After waste solids are removed during the clarification process, which is accomplished by primary sedimentation and dissolved air flotation, the primary sludge is produced. The secondary sludge was produced during secondary treatment, which is often a biological process in which microorganisms use oxygen to consume trash and convert it into carbon dioxide and water (Scott et al. 1995).

To make bricks out of the biological sludge waste (SW) generated by the tannery effluent plant, a study was conducted (Amsayazhi and Mohan 2018). According to (Periyasamy et al. 2022), the biogas produced during the processing of biological sludge can be utilised to create heat

© The Author(s), under exclusive license to Springer Nature Switzerland AG 2025

A. Mandpe and M. P. Shah (eds.), *Wastewater Sludge Nexus*, Sustainable Landscape Planning and Natural Resources Management, https://doi.org/10.1007/978-3-031-45758-6_13

and power or as a natural gas alternative. Sludge improves the soil's capacity to promote plant growth by introducing organic matter and nutrients (Seleiman et al. 2020). Because sludge contains organic matter and plant nutrients (N and P), disposing of it on land can be advantageous. It is a rich source of essential elements and biological matter that can improve agricultural and forest productivity. Because the nutrients in biosolids are inorganic fertilisers, they directly benefit the economy.

### 1.2 Role of Sludge in Wastewater Treatment

According to (Demirbas et al. 2017), sewage sludge, requires additional treatment before it can be disposed of or applied to land (Demirbas et al. 2016). Sewage water therapy in sewage treatment plants yields sewage sludge. Sludge comes in two main forms: primary sludge and secondary sludge, which, in the case of an activated sludge process, is also referred to as activated sludge. Wastewater biosolids is another term for sludge from biological treatment processes. Numerous organic and inorganic chemicals can be found in wastewater sludge (Bharathiraja et al. 2014). To eliminate these impurities and create treated wastewater that is safe for the ecosystem, it involves physical, chemical, and biological procedures.

### 1.3 Purpose and Scope of the Chapter

The primary intention of this chapter is to educate readers about the environmental and health risks associated with sludge, providing a comprehensive understanding of the topic. It aims to raise awareness about the potential dangers of mismanagement or inadequate treatment of sludge, both in terms of environmental impact and public health. The chapter may also delve into the regulatory framework and policies related to sludge management, highlighting gaps and areas for improvement.

A semi-solid slurry that contains any specific type of solid sedimentary material mixed with any sort of liquid is referred to as sludge. It is produced by industrial wastewater, sewage from houses, agricultural waste from farmland, etc. Minimise the progression of chloroplast development. According to a study carried out by (Huerta Lwanga et al. 2017), the presence of microplastics from sludge at concentrations between 0.9 and 2.1 mg/L has been found to have a significant inhibitory effect on the growth of chlorophyll in plants.

### 1.4 Importance of Understanding Environmental and Health Risks

At sewage treatment plants, wastewater is treated to produce sewage sludge. Sludge, dioxins and furans, polychlorinated biphenyls, polycyclic aromatic hydrocarbons are present in sewage (Demirbas et al. 2017). Since exposure to polluted sludge can result in various diseases and detrimental health impacts, it is imperative to understand these hazards to safeguard the public's health. Due to its associations with numerous human-harming infections, including *Ascaris lumbricoides*, *Giardia lamblia*, *Rotavirus*, and *Salmonella* species, sewage sludge poses a risk to human wellness. Applying sewage sludge to soils raises the pH of the soil and enriches vital nutrients for plants. Careless use of biosolids could result in less productive soil. Given that exposure to polluted sludge can cause a variety of illnesses and have an adverse effect on one's health, it is critical to understand these concerns in order to protect the public's health (Shakoor et al. 2017).

## 2 Types and Sources of Sludge

Screening and grit removal are often steps in the preliminary wastewater treatment process. The process of wastewater screening eliminates coarse particles like rags that could cause mechanical equipment malfunctions. Although it does not significantly alter the quality of the wastewater, preliminary treatment helps prepare it for further treatment. Sludge is typically not combined with screens and grit, the by-products of initial wastewater treatment.

### 2.1 Municipal Wastewater Sludge

#### 2.1.1 Primary Sludge

Primary biological sludge is a form of sludge that is produced during the initial phase of wastewater treatment procedures, known as the primary treatment stage. The primary treatment stage constitutes the initial phase of wastewater treatment, wherein the physical separation of solids and other substances from untreated sewage takes place. The formation of primary sludge occurs as a consequence of this particular process. The initial stage of wastewater treatment often entails the use of gravity sedimentation to separate settleable solids from screened and degritted wastewater. The primary sludge undergoes predominantly biological processing using the method of anaerobic digestion. During the process of digestion, anaerobic and facultative bacteria engage in the metabolic breakdown of organic material present in sludge. This metabolic activity leads to a decrease in the volume of sludge that needs to be ultimately inclined of, as well as the transformation of the sludge into a stable state, rendering it non-putrescible. Additionally, this metabolic process enhances the dewatering properties of the sludge. The initial stage of wastewater therapy often entails the use of gravity sedimentation to separate settleable solids from the screened and degritted wastewater. This process typically results in the removal of slightly more than half of the suspended solids present in the wastewater. The portion of biochemical oxygen demand (BOD) that exists in solid form and can be eliminated through the process of sedimentation (usually accounting for approximately one-third of the total BOD) is also eliminated. The primary sludge contains the elements that are removed.

### 2.1.2 Secondary Sludge

The main goal of secondary application is to get rid of all the remaining organic compounds and particles in the fluid by putting the waste water from primary treatment through more treatment steps. In most cases, secondary treatment is often administered after the main treatment. Aerobic biological treatment procedures are employed to eliminate biodegradable, dissolved, and colloidal organic waste and involves the utilisation of oxygen by aerobic microorganisms, primarily bacteria, to metabolise the organic substances present in wastewater. This process results in the generation of additional microorganisms and inorganic by-products, namely carbon dioxide ($CO_2$), ammonia ($NH_3$), and water ($H_2O$). To obtain clarified secondary effluent, it is necessary to employ sedimentation techniques to separate the microorganisms from the treated wastewater. Various aerobic biological procedures are employed for secondary treatment, chiefly differing in the method of oxygen supply to microorganisms and the rate of organic matter metabolism by these organisms. Microorganisms present in suspension, such as in the "activated sludge" process, or adhering to media, as observed in a "trickling filter" or its variants, as well as in ponds or other similar processes, are employed for the purpose of eliminating biodegradable organic matter. The remaining organic material serves as a source of energy and essential resources to sustain the microorganism community. Microorganisms undergo biological flocculation, resulting in the formation of settleable particles. This sludge is also known by various names, such as "biological sludge," "waste-activated sludge," or "trickling filter humus" (Council 1996).

### 2.1.3 Tertiary Treatment

Secondary wastewater treatment doesn't always produce effluent that meets the standards for receiving water or is good enough for other reasons. This is when tertiary treatment is used in municipal wastewater treatment plants. The most prevalent form of tertiary treatment involves disinfection to control harmful bacteria and viruses. Using a coagulant along with filtration can effectively lower the amounts of suspended particles and the biochemical oxygen demand (BOD) that goes along with them in treated effluent. The process of adsorption, often employing activated carbon as the adsorbent, has the potential to effectively eliminate some biological compounds and toxic elements. Nitrification has the potential to decrease the ammonia concentration in secondary effluent. The implementation of tertiary treatment methods aimed at reducing nitrogen and phosphorus levels to mitigate the enrichment of surface waters with nutrients is widespread. Typically, nitrification and denitrification are the two steps that remove nitrogen, while microbial uptake or chemical precipitation remove phosphorus (Council 1996).

## 2.2 Industrial Wastewater Sludge

Industrial wastewater sludge, sometimes referred to as industrial biological sludge, is a by-product of sewage water treatment derived from industrial activities (Demirbas et al. 2017). There are notable distinctions between industrial and municipal wastewater sludge, mostly due to toxins, pollutants, and elements unique to the industry from which the sludge originates. Municipal and industrial wastewater treatment plants (WWTP) are not suitable for disposal in any natural or regulated setting. Due to this rationale, it is important to subject the material to processing procedures such as stabilisation and drying in order to facilitate its subsequent utilisation as a fertiliser, construction component, found in concrete plants (Demirbas et al. 2017). The utilisation of heat and electricity in the drying of sludge is characterised by a significant consumption of energy and lacks economic viability. In addition to concerns regarding energy efficiency, a significant challenge associated with sludge drying is the inherent diversity in the quality of sludge as a raw material. If the desiccated sediment comprises coliform bacteria, viruses, helminth eggs, or limited amounts of heavy metals, its application as a fertiliser is precluded; nonetheless, it remains viable as a fuel source.

Industrial sludge often exhibits elevated levels of heavy metals, and improper management of such sludge may result in the potential release of these heavy metals through leaching processes. As depicted, the utilisation of wastewater for the cultivation of food crops or other reuse purposes, such as replenishing groundwater or irrigating urban landscapes, often takes place subsequent to the implementation of secondary wastewater treatment and may require supplementary treatment measures. The process of generating reclaimed water frequently involves the implementation of coagulation, filtration, and disinfection techniques in conjunction with secondary treatment.

The pretreatment of industrial wastewaters serves as a method for effectively addressing the presence of hazardous pollutants in both treated wastewater effluents and sludge residuals. According to the National Research Council (1977), industrial wastewater treatment refers to the process of eliminating harmful substances from the industrial facility prior to its discharge into the municipal sewer system. Pretreatment programmes have proven to be efficient in mitigating the concentrations of various toxic materials in wastewater due to the significant contribution of harmful chemicals from industrial activities in densely populated urban regions (Council 1996). Various treatment methods are employed to address the issue of industrial sludge, encompassing chemical precipitation, biological treatment, physical separation, and other modern treatment processes that are customised to effectively target the specific toxins found in the sludge.

## 2.3 Agricultural and Livestock Waste Sludge

The inadequate handling of agricultural wastewater, generated by diverse farming practices such as animal feeding operations and agricultural product processing, possesses the capacity to pollute both surface and groundwater reservoirs. Agricultural wastewater comprises various sources, including manure, wash water from milking centres, runoff from barnyards and feedlots, runoff associated with composting, among other sources (Khan et al. 2022). Furthermore, the discharge of water from agricultural fields can introduce sediment, fertilisers, and pesticides into bodies of water.

The process of sludge dewatering poses challenges due to the specific properties of sludge found in the animal breeding business, namely the presence of long fibres and high sludge output. These qualities contribute to the increased likelihood of clogging during the dewatering process. Agricultural irrigation is the primary application for recovered water, as it offers enhanced water value through the inclusion of fertilisers, nutrients, and macroelements (Almeida et al. 2017). It is imperative to subject wastewater to appropriate treatment measures in order to effectively eliminate a diverse range of contaminants, including pesticides, medicines, and metals, prior to their reuse. When considering cattle, the primary contaminants that raise concern are metals and medicines. For example, the utilisation of Zn and Cu in elevated quantities has been observed to augment the immune systems of animal.

According to (Santos et al. 2019), the treatment of livestock sewage water is generally more challenging compared to urban wastewater. This is primarily attributed to the complex composition and elevated levels of organic matter present in livestock wastewaters. The discharge of wastewater from livestock and poultry operations is a significant threat to environmental integrity, as it contributes to the escalation of disease transmission and early mortality rates. The water quality is distinguished by elevated levels of chemical oxygen demand, suspended solids, heavy metals, and pathogens, among several other pollutants. The presence of these contaminants has been found to have adverse effects on the overall quality of soil, and air, hence posing a possible risk to human health (Santos et al. 2019). Livestock and poultry wastewater presents a significant threat to ecosystem contamination as it contributes to the acceleration of disease transmission and premature mortality (Vaishnav et al. 2023). Numerous sources of sludge are shown in Fig. 1.

**Fig. 1** Shows different sources of sludges

## 2.4 Emerging Sources of Sludge

### 2.4.1 Pharmaceutical Wastewater

The global population growth in recent years has necessitated the provision of reliable access to high-quality freshwater to support sustainable livelihoods. Conversely, the water bodies have been affected by unregulated discharge of effluents because of growing development. The emission originating from pharmaceutical enterprises has become a subject of growing concern due to the potentially significant impact of pharmaceutical wastewater. Pharmaceutical chemicals provide potential hazards due to their uncertain implications and uncharacterised toxicity levels. The resolution of this matter can be achieved by the identification of origins and the implementation of diverse treatment methodologies. Water is a crucial primary resource that is utilised in significant quantities within the pharmaceutical sector for the purposes of manufacturing, processing, and serving as a coolant. The phrase "pharmaceutical wastewater" refers primarily to the effluents generated by pharmaceutical enterprises. However, the disposal of pharmaceuticals from households and hospitals, as well as their presence in municipal sewage water, also contribute to the overall volume of pharmaceutical waste (Khan et al. 2023). It is imperative to accurately ascertain the specific sources of pharmaceutical waste to effectively address the treatment of wastewater streams and facilitate the recovery of valuable components. Multiple strategies are suggested with the purpose of identifying a resolution to this issue. Various biological, physical, and chemical methodologies can be employed for the treatment of pharmaceutical wastewater. Several biological approaches can be employed to handle pharmaceutical wastes, including constructed wastelands, bioaugmentation, activated sludge processing, and biotransformation. The potential dangers connected with pharmaceutical wastewater are a subject of concern. However, the implementation of contemporary and traditional approaches for wastewater treatment holds promise as a potential solution (Khan et al. 2023). Notwithstanding this, there exist several opportunities for the investigation and advancement of sophisticated interventions aimed at safeguarding the environment from the release of pharmaceutical substances.

### 2.4.2 Microplastic Wastewater

Microplastics, which are plastic particles measuring less than 5 mm in size, can exist in many forms, such as fragments, fibres, pellets, foam, and films (Arthur et al. 2008). Microplastics have recently emerged as contaminants that are widely distributed around the planet. Although microplastic particles are small, they have demonstrated detrimental effects on plants, animals, humans, and the whole ecosystem. Water serves as a crucial medium for the transportation of microplastics across various locations (Amrutha and Warrier 2020). According to sewage and wastewater treatment plants are currently unable to effectively eliminate microplastics, resulting in the discharge of significant amounts of microplastics into the effluent. Previous research has indicated that microplastics have the potential to induce genotoxicity, carcinogenesis, fibrosis, an increased inflammatory response, oxidative stress, and other adverse effects (Deng et al. 2017). In the process of transferring energy and nutrients via successive trophic levels in an ecological community, Microplastics have the potential to undergo trophic transfer, thus entering the human food chain. Hepatic damage refers to the impairment or injury to the liver, which is a vital organ responsible for several metabolic functions in liver damage has been observed in certain organisms as a result of metabolic stress induced by microplastics (MPs) and the accumulation of contaminants on their surface.

Microplastics are characterised as plastic particles with dimensions less than 5 mm. According to (Arthur et al. 2008), these particles have the ability to exist in diverse states, including fragments, threads, pellets, foam, and films. In recent times, microplastics have emerged as pervasive contaminants that have been detected in many geographical locations across the globe. Despite their small size, microplastic particles have been shown to have negative impacts on plants, animals, humans, and the whole ecosystem. Water has a pivotal role as a conduit for the transmission of microplastics across diverse geographical regions. Moreover, it is important to acknowledge that water plays a crucial role as a conduit for the ingestion of microplastics by people, resulting in many health difficulties like cancer, mutagenic effects, and teratogenic abnormalities. According to a study conducted by (Huerta Lwanga et al. 2017), the presence of microplastics from sludge at concentrations between 0.9 and 2.1 mg/L has been found to have a significant inhibitory effect on the growth of chlorophyll in plants.

Amrutha and Warrier (2020) asserts that the existing sewage and wastewater treatment facilities are presently incapable of efficiently eradicating microplastics, releasing substantial quantities of microplastics in the effluent. Preliminary studies have suggested that microplastics possess the capacity to elicit genotoxicity, carcinogenesis, fibrosis, an augmented inflammatory response, oxidative stress, and other associated consequences (Evanylo 1999).

# 3 Environmental Risks Associated with Sludge

## 3.1 Heavy Metals and Toxins

Even minute quantities of heavy metals present in sludge can pose a significant risk, whereas elevated levels of arsenic found in food and water have the potential to cause fatalities (Wong and Selvam 2006). Cadmium has been associated with adverse health outcomes such as renal impairments, spontaneous pregnancy loss, and foetal demise. Copper, nickel, and zinc have been identified as factors that can induce growth-related issues in agricultural crops. Lead exposure has been associated with the development of behavioural and learning difficulties in children. Exposure to mercury during critical periods of foetal development has been found to be associated with the manifestation of learning impairments and neurological diseases. Molybdenum exhibits bioaccumulation in livestock that primarily consume grass, and its excessive ingestion can lead to adverse health effects such as anaemia, diarrhoea, and growth impairments (Wong and Selvam 2006).

### 3.1.1 Nutrient Overload

The processes of urbanisation, industrialization, and population growth result in the generation of a significant volume of waste. Improper treatment of industrial waste and untreated sewage, which are used for irrigation purposes, contributes to the accumulation of harmful substances such as trace elements, micro plastics, and high salt content in the soil. Consequently, this accumulation adversely affects the overall soil quality. Likewise, the discharges emitted by industrial sectors such as distilleries, tanneries, and textile factories also encompass various origins of trace elements, sodium, salts, and organic compounds (Bahuguna et al. 2022). Nitrogen (N) and phosphorus (P) are essential components for the sustenance of both plants and animals. However, they can also have detrimental effects on the environment if they are released into the atmosphere or hydrosphere. Additionally, incorrect agricultural management practices can lead to soil acidification, with nitrogen and phosphorus being contributing factors (Oenema et al. 2003). Currently, the biogeochemical cycling of nitrogen (N) and phosphorus (P) because of anthropogenic activities has surpassed the established threshold for environmental sustainability, leading to unforeseen repercussions. To mitigate the occurrence of eutrophication in natural bodies of water, it is imperative to decrease the levels of nitrogen and phosphorus originating from industrial wastewater as well as agricultural and household practices. The process of rapid urbanisation, industrialization, and population growth results in the generation of a substantial quantity of waste. When this waste, originating from both industrial activities and untreated sewage, is utilised for irrigation purposes without adequate treatment, it leads to the accumulation of harmful substances in the soil. Consequently, the overall quality of the soil is diminished.

## 3.2 Water Pollution

### 3.2.1 Release of Pathogens

Microbial infections provide significant health concerns in relation to water and sewage water. Common markers of faecal contamination encompass total coliforms, *Escherichia coli*, and *Clostridium perfringens*. The monitoring of faecal markers is conducted to gather data on their progression during the various stages of wastewater treatment procedures. Helminth eggs exhibit a notable capacity for prolonged survival in the environment and possess a considerable propensity for transmission through water, hence establishing their efficacy as dependable markers of contamination. Sewage harbours a rich and diverse assortment of microorganisms, which originate primarily from human domestic waste, including faecal matter, laundry water, bathing water, urine, and perspiration. The bacteria in question are introduced into the human body and can be found in many locations, such as the skin, respiratory tract, and gastrointestinal system. Wastewater is recognised as a significant reservoir of various pathogens, encompassing Faecal Coliform, *E. coli, Salmonella, Shigella, Vibrio cholera*, parasitic eggs and cysts, viruses, and fungi. Additionally, it harbours intestinal nematodes such as *Ancylostoma duodenale*, *Ascaris lumbricoides*, and *Trichuris* (Ajonina et al. 2015).

### 3.2.2 Eutrophication of Water Bodies

In various bodies of water, where the issue of eutrophication has been observed, the concentration of total inorganic phosphate has been observed to rise significantly by a factor of 20–25 (permissible concentration range 0.03–0.04 mg $l^{-1}$) over the course of the past ten to twenty years, particularly in urban areas and industrial regions. Approximately 60% of the phosphate found in the waterways of the United States is derived from residential sewage and is additionally derived via mining activities, the runoff of fertilisers, and the discharge of domestic sewage that contains a significant proportion of phosphate. Animal excreta are containing nitrates and phosphates. Agricultural runoff is the primary contributor to phosphorus in riverine sediments, serving as a significant supply for benthic algae and rooted plants to utilise. Eutrophication has emerged as a significant issue in both industrialised and emerging nations, particularly in densely populated countries like India, Bangladesh, Indonesia, China and Pakistan. The contribution of phosphorus input from diffuse sources to the eutrophication of rivers may have little importance, as the timing of these transfers typically does not coincide with the period of highest biological demand. The symptoms associated with eutrophication predominantly occur during the period of plant growth, particularly in the seasons of spring and summer. These symptoms are observed due to factors such as reduced water flow, extended water residence durations, elevated amounts of sunshine, and higher water temperatures, all of which contribute to the rapid proliferation of algae.

## 3.3 Air Pollution

### 3.3.1 Odour Emissions

Particles are emitted during the aeration phase of the activated sludge process, as well as from the distributor nozzles on percolating filters. The areoles found in sewage treatment plants have the potential to harbour many harmful bacteria, viruses, and endotoxins commonly associated with sewage. These contaminants can cause infections in animals and people through inhalation. The release of odorous substances into the atmosphere (Sharma et al. 2023).

The implementation of the SS biological treatment method elicits several environmental and social concerns that are inherent and cannot be circumvented. These concerns encompass the emission of several gaseous compounds that contribute to both olfactory discomfort and air pollution. The gaseous emissions from SS composting consist primarily of ammonia, hydrogen sulphide, and various volatile organic compounds (VOCs) (Lebrero et al. 2013). The implementation of the SS biological treatment method elicits several environmental and social concerns that are inherent and cannot be circumvented. These concerns encompass the emission of several gaseous compounds that contribute to both olfactory discomfort and air pollution.

### 3.3.2 Greenhouse Gases

Greenhouse gases (GHGs) encompass carbon dioxide ($CO_2$), including bio-based sources, nitrous oxide ($N_2O$). These gases are quantified in terms of carbon dioxide equivalents ($CO_2$eq). At present, the primary environmental concern associated with biodegradable waste, which includes sludge, is the production of methane. Furthermore, it should be noted that the unregulated anaerobic breakdown of biodegradable substances, results in significant emissions of $CO_2$ and $CH_4$ (Piippo et al. 2018). Sludge possesses a substantial quantity of decomposable organic matter, rendering it a noteworthy contributor to the release of greenhouse gas (GHG) emissions. The practice of disposing of sludge in landfills without any form of treatment is frequently employed; nevertheless, this approach is seen as unfavourable due to the substantial quantities of methane ($CH_4$) emissions associated with such a practice. According to a prior research, it was observed that the greenhouse gas (GHG) emissions resulting from the combination of sludge heat drying and incineration were significantly higher compared to the emissions connected with land use following composting or digestion (Sharma et al. 2023). Nevertheless, this therapeutic procedure is accompanied by other issues. There exists a significant level of public apprehension surrounding the issue of exhaust gas emissions and their impact on atmospheric pollution. The severity of haze and pollution in China is getting worse, which makes these worries even worse. Notwithstanding these difficulties, incineration continues to be a recognised form of waste management, mostly due to its comparative advantages over alternative ways of sludge disposal. Furthermore, numerous water bodies exhibit eutrophic or nutrient-rich states, hence exacerbating the potential for discharged wastewater to contribute to elevated greenhouse gas (GHG) emissions (Sharma et al. 2023). Nevertheless, the consideration of greenhouse gas (GHG) emissions originating from receiving waters is infrequently undertaken, mostly because of the scarcity of available data pertaining to the aquatic health of the recipient body of water and the subsequent paths of downstream discharge.

D. Impact on Aquatic Ecosystems

### 3.3.3 Disruption of Aquatic Life

Over the course of recent decades, ecosystems worldwide have had detrimental effects on their health because of unregulated sewage disposal. The protection of freshwater ecosystems from the adverse effects of untreated sewage is progressively emerging as a formidable task. Aquatic ecosystems that receive untreated or partially treated sewage have been observed to experience a range of detrimental effects, including diversity reduction, physiological and behavioural alterations in species, shifts in community composition, and increased fish mortality. Sewage contains a significant amount of nutrients, including nitrogen and phosphorus, which are major contributors to the process of eutrophication in aquatic ecosystems. Numerous freshwater ecosystems across the globe have experienced the detrimental effects of eutrophication as a result of inadequate treatment of sewage discharge, resulting in alterations to their trophic status. The absorption of pharmaceutical substances by fish can transpire through two primary routes: the dermal and gill surfaces for pharmaceuticals present in water or sediment, and orally through the consumption of contaminated food. Additionally, medicines can also be transferred to fish offspring through the lipid reserves of eggs, known as maternal transfer. The growth of fish, including their larvae, can be impeded by the presence of toxic materials. Additionally, heavy metals can diminish the size of fish populations and pose a significant threat to the overall fish population, particularly when found in high concentrations. The presence of a significant quantity of aluminium can lead to the occurrence of osmoregulatory dysfunction in aquatic organisms, such as fish (Vosylienė and Jankaitė 2006).

Physical Changes Due to Change in Temperature

The phenomenon of sewage discharge is frequently linked to observable alterations in aquatic environments. Aquatic ecosystems rely on a specific temperature range for their sustenance. The alteration in water temperature can have a significant impact on several forms of aquatic organisms, including bacteria, invertebrates, algae, and fish. An elevation in temperature leads to a decrease in the solubility of oxygen in water, thus resulting in an augmented oxygen demand necessary for the bacterial breakdown of waste materials. Tissue hypoxia may manifest at elevated, lethal temperatures in aquatic organisms. The physicochemical conditions, such as oxygen concentrations, and energy processes related to primary generation and litter decomposition are influenced by temperature.

Environmental Changes

Minimise the progression of chloroplast development. According to a study conducted by (Huerta Lwanga et al. 2017), the presence of microplastics at concentrations between 0.9 and 2.1 mg/L has been found to have a significant inhibitory effect on the growth of chlorophyll in plants. In the process of transferring energy and nutrients via successive trophic levels in an ecological community, Members of Parliament (MPs) have the potential to undergo trophic transfer, thus entering the human food chain. Hepatic damage refers to the impairment or injury to the liver, which is a vital organ responsible for several metabolic functions in liver damage has been observed in certain organisms because of metabolic stress induced by microplastics (MPs) and the accumulation of contaminants on their surface.

# 4 Health Risks Linked to Sludge

## 4.1 Occupational Exposure

### 4.1.1 Health Risks for Wastewater Treatment Workers and Farmers

Exposure to sewage sludge presents considerable health hazards, resulting in the development of several infectious diseases among individuals. The main factors contributing to this issue encompass two fundamental causes. Firstly, it involves the direct exposure of farmers to sewage sludge in agricultural areas. Secondly, it involves the eating of fruits, grains, and vegetables cultivated in these regions where sewage has been utilised. Employees working at sewage treatment facilities face significant risks as a result of their direct contact with untreated wastewater, hence classifying their career as very perilous. Moreover, the transmission of infectious diseases can occur via several means, including (i) individuals who reside in proximity to sewage treatment plants, (ii) youngsters who engage in activities near ponds or lakes that are contaminated, and (iii) individuals who live in close proximity to areas where waste is dumped. The transmission of infections is further facilitated by the improper management of sludge and insufficient adherence to hand hygiene and sanitation practices. This is particularly evident among farmers who are directly exposed to sludge during its application in agricultural fields. Farmers that engage in the practice of applying sludge in their fields, as well as those employed in sewage treatment plants, may encounter extended periods of exposure to infectious diseases as a result of direct interaction with sewage sludge. Individuals engaged in the processes of dewatering and digested sludge settlement are exposed to the most significant health risks. While the dilution and wastewater treatment methods employed in handling sewage sludge effectively decrease the presence of numerous pathogens, it is important to note that individuals working in these occupations still face a considerable risk of exposure to infectious diseases. According to a study, exposure to sewage sludge presents a potential hazard of illnesses caused by diverse pathogens in human beings, with the degree of risk being contingent upon the magnitude of their exposure. The potential impact on human health is more pronounced when individuals are exposed to viruses, even if it occurs only once, in comparison to encounters with protozoa and bacteria. This phenomenon can be attributed to the relatively diminished presence of viruses in sewage sludge. Furthermore, it has been noted that non-viral microorganisms exhibit limited potential for infecting human health, whereas viruses are responsible for causing significant illnesses, particularly among people who have direct contact with sewage sludge. The authors also observed that the consumption of vegetables cultivated in soil treated with sewage sludge poses a relatively lower risk of infection to human health compared to direct exposure to sewage sludge. Another study has indicated that those employed in wastewater treatment facilities are at risk of experiencing gastrointestinal disorders, headaches, weariness, respiratory ailments, as well as skin and eye allergies. A research investigation was carried out in Nam Dinh, a region located in North Vietnam, with the objective of examining the prevalence of skin illnesses among farmers who had been exposed to wastewater. The study centred on regions where wastewater was employed for agricultural irrigation and observed variations in the incidence of dermatological conditions among different communities.

### 4.1.2 Safety Measures and Protective Equipment

Personal Protective Equipment (PPE)

Provide appropriate personal protective equipment (PPE), training on its use, and hand washing facilities to workers handling sewage or human waste. After taking off their PPE, workers should immediately wash their hands with soap and water. When handling sewage or human waste, the following personal protective equipment is advised: Goggles: to shield the eyes from sewage or human waste splashes. Splash-proof face shields or protective face masks are used to guard the mouth and nose from sewage or human waste splashes. Other such PPE required:

(A) Coveralls that repel liquid: to prevent sewage or human waste from getting on clothes. Workers should wear appropriate coveralls, aprons, or uniforms to prevent skin contact with sludge.
(B) Gloves: Chemical-resistant gloves should be worn to protect the hands when handling sludge.
(c) To avoid contact with sewage or human waste, wear waterproof gloves.
(D) Rubber boots: to shield against sewage and human waste (Gautam et al. 2021). Safety Boots: Steel-toed, waterproof boots with non-slip soles are recommended for sludge.
(E) Respiratory Safety: Relying on the environment of the sludge, workers may need respirators to protect against inhalation of airborne contaminants.

## 4.2 Community Exposure

### 4.2.1 Airborne Pollutants

The cost-effectiveness of sludge combustion is widely acknowledged since it allows for the recovery of energy that can be used for other purposes. However, it is important to note that this process may result in the emission of significant quantities of air pollutants, including certain harmful toxins. As an illustration, in the process of sludge combustion, (Wang et al. 2002) reported a mean total mass rate of 9440 mg/h for polycyclic aromatic hydrocarbons (PAHs). Furthermore, this process may result in significant emissions of zinc (Zn), copper (Cu), manganese (Mn), chromium (Cr), nickel (Ni), and lead (Pb). The application of TSS to agricultural land involves the utilisation of either a liquid or dewatered (semi-solid) cake form, which is often carried out by the use of tractors equipped with sprayers, injectors, or manure spreaders (Harrison et al. 1999). Residents residing in close proximity to areas where TSS (thermal soil sterilisation) is administered have reported the presence of unpleasant odours, an abundance of flies, irritation of mucous membranes, headaches, and the manifestation of acute respiratory. The range of values is 4 to 8. Animal wastes are subjected to a far lower level of treatment in terms of pathogen removal and vector control measures. Inhabitants residing in these areas have reported heightened instances of irritation in the mucous membranes, acute respiratory symptoms, elevated stress levels, increased anxiety, and disruption in their daily activities during periods of acute exposure in comparison to periods of non-exposure. Additionally, these individuals exhibit higher blood pressure levels. There is a positive correlation between the proximity of children's residences or schools to industrial swine operations and elevated occurrences of wheezing and medically diagnosed asthma. Both TSS (total suspended solids) and animal dung have the potential to be put in the same fields or farms in close proximity (Pavilonis et al. 2013).

### 4.2.2 Contaminated-Groundwater and Drinking Water

The consumption of arsenic-contaminated groundwater leading to chronic arsenic toxicity, commonly known as arsenicosis, poses a significant environmental health risk on a global scale, especially in India. Prolonged exposure to arsenic present in drinking water has been found to be associated with the development of cancer in various organs, including the skin, lungs, bladder, and kidneys. Additionally, it has the potential to induce various alterations in the skin, including thickness and pigmentation. Black foot disease (BFD), a variant of peripheral vascular disease, has been identified as a significant consequence of chronic arsenic toxicity (CAT) in Taiwan. Inhalation of arsenic in occupational settings has been identified as a potential contributor to the development of lung cancer. There is an established association between non-malignant skin modifications, including keratosis as well as hypo- and hyper-pigmentation, and the presence of arsenic. According to the World Health Organisation (2004), it has been estimated that around 1.8 million individuals succumb to diarrheal diseases annually on a global scale. Studies reported that approximately 10% of the population residing in developing nations experiences severe infection with intestinal worms as a result of inadequate waste disposal and faecal sludge management. Microbiological pathogens, such as bacteria, viruses, fungi, and protozoa, present in water contaminated with faecal matter might pose significant health risks to individuals engaging in bathing activities. The entry of these contaminants into the human body can occur through several routes, including ingestion, inhalation, or contact with water that has been contaminated. Insufficient sanitation and substandard hygiene practices contribute to the proliferation of waterborne illnesses. Upon entering the human body, these pathogens have the potential to induce various ailments, including gastroenteritis, cholera, hepatitis, and parasitic infections.

### 4.3 Food Chain Contamination

#### 4.3.1 Uptake of Sludge-Borne Contaminants by Crops

The utilisation of organic materials, such as compost, waste, urine, human waste, wastewater, and biosolids, in agricultural practices has been extensively documented in existing scholarly literature. The objective of this endeavour is to enhance the fertility of the soil and augment the provision of nutrients to agricultural crops. Nevertheless, it has been shown that certain organic substances harbour harmful contaminants, and the accumulation of these contaminants in agricultural soil could potentially jeopardise the ecological processes of soil, hinder plant development, and constitute a substantial threat to human well-being (Sun et al. 2018). Human consumption of contaminated food.

When considering cadmium, it is important to note that the kidney is the primary organ of concern. Studies have indicated that a cadmium concentration of 200 mg $g^{-1}$ in the renal cortex can lead to renal tubular dysfunction. Research conducted on males who are occupationally exposed to cadmium has revealed that approximately 10% of the exposed individuals, who have a renal cortex concentration of cadmium of approximately 200 mg $kg^{-1}$, exhibit an elevated excretion of the protein in question. Contaminants in sludge can enter the food chain through crops that are consumed by livestock, and when humans consume meat or dairy products from these animals, they can be exposed to a higher concentration of contaminants. This is known as bioaccumulation and can increase health risks over time. The use of vegetables that are contaminated with heavy metals has been found to result in significant adverse effects on human health. These effects include the development of gastrointestinal cancer, compromised immune function, and malnutrition (Dickin et al. 2016). Leafy vegetables like spinach, lettuce, and kale are more prone to contamination as they have large surface areas that can trap contaminants from sludge. These plants are particularly vulnerable to pathogens like *E. coli* (Truschi et al. 2023). Sludge contamination may carry the hepatitis A virus, which can infect food crops and, when consumed, can lead to hepatitis A, a liver infection. Symptoms include jaundice, fatigue, and flu-like symptoms. In the context of cadmium, the kidney is identified as the essential organ. The initial indication of this dysfunction is an elevation in the excretion of low-molecular-weight proteins, such as ß2-microglobulin. Research conducted on men who are occupationally exposed has revealed that around 10% of individuals within the exposed group, who possess a renal cortex concentration of cadmium at approximately 200 mg $kg^{-1}$, have an elevated excretion of such protein. The presence of toxins inside sludge has the potential to infiltrate the food chain via crops that are then ingested by livestock. Consequently, individuals who consume meat or dairy products derived from these animals may be exposed to an elevated level of contaminants. This phenomenon is commonly referred to as bioaccumulation, which has the potential to escalate health hazards gradually. According to (Dickin et al. 2016), the consumption of contaminated vegetables can lead to adverse effects like the development of gastrointestinal cancer, compromised immune function, impaired mental growth, and malnutrition. Leafy crops such as spinach, lettuce, and kale exhibit a higher susceptibility to contamination due to their expansive surface areas, which possess the capacity to accumulate toxins originating from sludge. These plant species exhibit a heightened susceptibility to pathogenic microorganisms such as *E. coli* (Truschi et al. 2023). The presence of sludge contamination has the potential to harbour the hepatitis A virus, which has the ability to contaminate food crops. Ingestion of these contaminated crops can result in the development of hepatitis A, a viral infection that mostly affects the liver. Common signs of this condition are jaundice, exhaustion, and symptoms like those of influenza. Figure 2 shows sources, applications, and effects of sludge on environment and human bodies.

## 5 Regulatory Framework and Guidelines

In 1993, the US Environmental Protection Agency (EPA) implemented regulations setting essential standards for the land application of sludges (Chen et al. 2012). These rules specify concentration limits for nine metals and pathogens, as well as measures to reduce attraction to vectors like flies and rodents (Gao et al. 2020). The regulations categorise sludges into Class A, where pathogens have been mostly eliminated, and Class B, where pathogens have been reduced but are still present (Smol et al. 2015). While there are specific restrictions for Class B use under federal 503 rules, individual site permits are unnecessary for its application (Hu et al. 2013). The government regulations additionally delineate criteria for nine pollutants. The aforementioned standards encompass criteria commonly referred to as "excellent quality," in addition to stipulations pertaining to the mitigation of pathogens and vectors. In relation to the levels of metal concentrations, it is possible to utilise sludges and their by-products that do not meet one or more of the prescribed pollutant concentrations known as "EQ," as long as they remain below a higher threshold concentration. Nevertheless, it is imperative for the applicator to diligently oversee the overall quantity of each metal being administered and cease the treatment process upon reaching a prescribed threshold for cumulative pollutant loading, as stipulated by regulatory guidelines (Smith 1904). Sludge products that do not satisfy the pollutant concentrations specified by the "EQ" standards but remain below the maximum allowable concentration may still

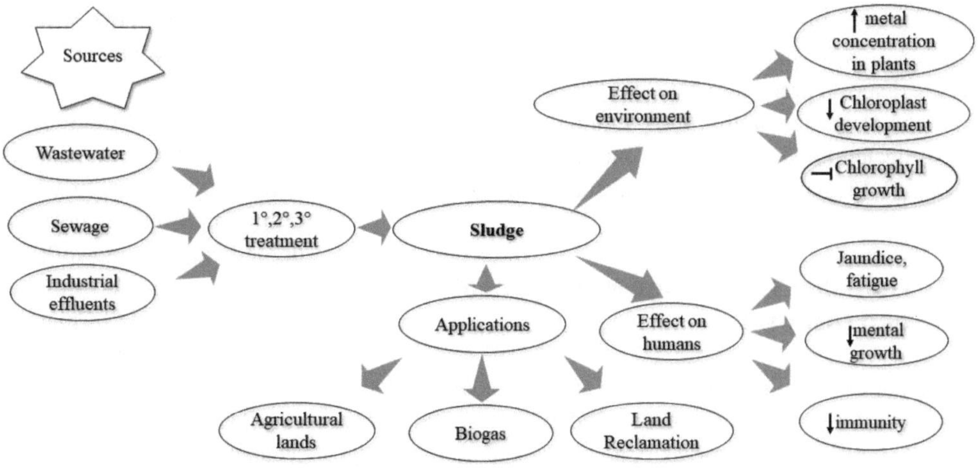


**Fig. 2** Applications, sources, and effect of sludge on humans and environment

be supplied to households or packaged in bags, as long as the user is provided with information regarding the allowed annual pollutant loading rate (APLR).

Various states in the US have the flexibility to impose regulations that are stricter than federal standards. New York State (NYS) has historically enforced stricter regulations in several aspects. However, NYS, along with many other states, is contemplating revising or aligning their regulations more closely with the federal Part 503 rules. When more stringent state rules are in place, land applications must adhere to these stricter regulations (Bressman et al. 2023).

## 5.1 National and International Regulations

The legal framework for Sludge Management (FSSM) incorporates environmental laws, regulations prohibiting "manual scavenging," and institutional laws defining the creation, authority, and responsibilities of local governing bodies (Mehta et al. 2019).

The initial classification encompasses legislative acts such as the Environment (Protection) Act of 1986 and the Water (Prevention and Control of Pollution) Act of 1974. These acts lay the foundation for the establishment of a regulatory framework aimed at governing the release of effluent, wastewater, and septage. Furthermore, the Municipal Solid Waste (MSW) Rules of 2016, which fall under the jurisdiction of the Environment (Protection) Act, regulate the appropriate and secure management of treated residual faecal sludge and septage (Verma et al. 2020). The primary aim of these laws is to mitigate the risk of pollution in groundwater, surface water, and ambient air. In addition, the 2016 MSW Rules encompass regulations pertaining to the management of faecal sludge and septage, encompassing their disposal and treatment processes, both pre- and post-processing, whether at landfills or for utilisation as compost.

The Employment of Physical Scavengers and Construction of Dry Latrines (Prohibition) Act of 1993 enforced a ban on the construction of dry latrines, which are characterised by the absence of water-seal or flushing devices, as well as the engagement of humans in the physical transportation of human excreta. The Prohibition of Employment as Manual Scavengers and Their Rehabilitation Act of 2013 further solidified this notion by expanding the prohibition to encompass the act of engaging in "hazardous cleaning" of sewers and septic tanks. According to current legislation, the process of manually cleaning sewers and septic tanks, when deemed appropriate, must be conducted in a meticulously regulated environment. This entails implementing comprehensive safety measures and adhering carefully to predetermined regulations and norms.

It is mandatory for all individuals employed in both the public and private sectors to adhere to the safety protocols specified in the Sewage Treatment, a publication authored by the Ministry of Urban Development. Furthermore, it is imperative for individuals to strictly comply with the protective measures outlined in the Prohibition of Employment as Manual Scavengers and their Rehabilitation Act of 2013. In addition, they must also adhere to any supplementary regulations established by the Urban Local Body (ULB) in accordance with its own governing principles. In order to manage the disposal of septage, the ULB is required to comply with the regulations specified in the Environment (Protection) Act of 1986 and the Municipal Solid Waste (MSW) Rules of 2016. The specific guidelines to be followed depend on the selected method of disposal (Bassi et al. 2023).

It is recommended that local authorities adopt protocols and guidelines for the strategic organisation and implementation of Faecal Sludge and Septage Management (FSSM) procedures. In order to enhance the effectiveness of this endeavour, it is imperative to undertake a comprehensive review of building laws to ensure the inclusion of provisions that mandate the construction of appropriate site facilities capable of accommodating anticipated loads and facilitating secure disposal practices (Akadiri et al. 2012). Prior approval from the competent authority is required for the operation of sites selected for sludge application, whether by Urban Local Bodies (ULBs) or other entities such as residential layouts.

### 5.2 Guidelines for Sludge Management and Disposal

The Sludge Guidelines promote the advantageous utilisation of sludge, aiming for sustainable sludge management practices. According to (Kacprzak et al. 2017), not all the sludge generated in South Africa meets the necessary quality standards for beneficial use. Consequently, disposal remains the only viable option for such sludge.

Sludge constitutes a waste product unique to certain industries, possessing distinct characteristics and properties. The Minimum Requirements primarily address the management of domestic and industrial solid waste, with only a limited section dedicated to the disposal and co-disposal of sludge (Watson-Craik and Sinclair 2020). The policy concerning the disposal of land-derived water containing waste in the marine environment essentially prohibits the disposal of sludge into marine environments as per its definition. Moreover, there are practices like dedicated land disposal (DLD) that are not covered in these guiding documents.

The distinctive nature and features of sludge, distinct from other types of waste, underline the necessity for industry-specific guidelines to ensure its responsible management. In practice, many wastewater treatment plants (WWTPs) resort to on-site waste piles and lagoons as sludge management and disposal methods. However, these facilities are not specifically designed for waste disposal and often do not meet the minimum requirements. This volume strives to act as a "bridge" between the waste management industry and the wastewater industry, addressing the unique challenges posed by sludge management.

Regular monitoring of sludge is crucial to ensuring its quality remains within the necessary limits for co-disposal in landfills. The same sampling analyses and frequencies should be applied for on-site and off-site land disposal. The sampling frequency for monitoring purposes depends on the quantity of sludge produced (Cofie et al. 2009). However, the landfill operator might require additional monitoring, particularly for pollutant-class B and c sludge that needs treatment before it can be delisted for co-disposal.

### 5.3 Monitoring and Enforcement Mechanisms

The Greater Noida Industrial Development Authority (GNIDA) plans to establish a monitoring and evaluation (M&E) framework to assess performance. Additionally, they will design data collection and reporting systems utilising an indicator framework. GNIDA will develop a database specific to on-site sanitation systems and create a comprehensive reporting format to ensure households and establishments comply with outcomes and process standards (Saroha et al. 2013). To oversee faecal sludge management operations, GNIDA will establish a dedicated cell. Furthermore, a Management Information System (MIS) will be developed to monitor and track the progress of these initiatives.

It is advisable to conduct sludge monitoring to assess whether sludge quality has improved or deteriorated over time. If the quality of sludge increases, it may be necessary to re-evaluate the available sludge management options.

Ensuring the quality of sludge is paramount for landfill site management. The landfill operator must have confidence in the sludge quality, particularly when the sludge needs treatment (such as liming) before it can be delisted. The landfill operator should consistently receive the results of sludge analyses and/or monitoring information (Adar et al. 2017). In cases where small wastewater treatment plants (WWTPs) are only mandated to monitor on an annual or quarterly basis, this frequency might not align with the permit conditions required by the landfill operator. Therefore, additional sludge quality data might be necessary for landfill operators to make informed decisions.

## 6 Case Studies

### 6.1 Examples of Environmental Incidents Related to Sludge Mismanagement

One notable environmental disaster related to sludge mismanagement occurred in 2008 in Kingston, Tennessee. A massive containment pond, holding coal ash sludge generated by the Tennessee Valley Authority's Kingston Fossil Plant, ruptured, releasing over a billion gallons of toxic sludge into the Emory River and surrounding areas. The spill covered hundreds of acres, damaging homes, aquatic habitats, and farmland. The sludge contained hazardous substances like arsenic, lead, and mercury, posing severe risks to human health and the ecosystem. This incident highlighted the catastrophic consequences of inadequate sludge containment and disposal methods. It prompted widespread public outrage, leading to increased awareness and advocacy for stricter regulations and better management practices within the coal and power industries. The Kingston disaster stands as a stark example of the environmental havoc that can result from sludge mismanagement, emphasising the urgent need for responsible handling and disposal of industrial waste materials (Luther 2010).

### 6.2 Health Impact Assessments in Communities Near Sludge Treatment Facilities

Risks arising from the land application of biological sludge are chronic and might only become apparent after prolonged exposure. As a result, identifying, quantifying, and documenting harmful effects directly linked to sludge are challenging tasks (Harder et al. 2017).

Residents residing near sludge land application sites in different parts of the US have reported illnesses, particularly those associated with Class B sewage sludge. These complaints have varied from headaches and respiratory issues to severe cases resulting in fatalities. Recently, there have been increasing concerns about the risk of food contamination with antibiotic-resistant bacteria (ARB) from sludge, especially in light of the COVID-19 pandemic (Agoro et al. 2020).

In the context of Greece, many methods have been employed for the disposal of sewage sludge, which encompass agricultural utilisation, incineration (accounting for 27% of cases), landfilling (14%), and other miscellaneous applications (17%). On the other hand, it is worth noting that in Germany, the primary method of waste management is thermal disposal, which involves incineration for the purpose of energy production. This method accounts for 55% of waste management practices in the country. Following thermal disposal, land application is the second most prevalent choice, constituting 42% of waste management practices. Material recovery, on the other hand, represents a mere 3% of waste management practices in Germany. It is important to highlight that landfilling is not employed as a waste management option in Germany. The aforementioned assortment of methodologies underscores the imperative for customised resolutions that take into account the distinct conditions prevailing in each geographical area (Smith 1904; Delibacak et al. 2020).

## 7 Mitigation and Best Practices

The concept of sustainability in sludge management entails the fulfilment of current societal needs while simultaneously safeguarding the ability of future generations to fulfil their own requirements. The management of sludge encompasses the appropriate processing and disposal of by-products produced during the treatment of wastewater. If not successfully managed, these residues can present substantial environmental risks. Achieving a harmonious equilibrium between present and future requirements in the realm of sludge management entails the implementation of methodologies that are both ecologically sustainable, socially acceptable, and economically feasible. A recommended approach for achieving sustainable sludge management entails conceptualising sludge as an asset rather than a mere by-product of waste. This particular viewpoint involves the conversion of sludge into biogas, which can be utilised as a sustainable energy resource, or the application of sludge as a fertiliser in agricultural practices. By using these strategies, it is possible to mitigate the environmental consequences associated with sludge management while also providing economic advantages.

Indeed, another vital facet of sustainable sludge management is ensuring that the practices are socially acceptable. This entails involving stakeholders, including local communities, in decision-making processes and addressing their concerns. By establishing trust and gaining support for sludge management practices through this participatory approach, it becomes crucial for the long-term sustainability of such initiatives. In conclusion, the attainment of sustainable sludge management necessitates a comprehensive approach that incorporates environmental, social, and economic factors. Sustainable sludge management aims to achieve a harmonious equilibrium between the demands of present and future generations, ensuring the realisation of the benefits associated with wastewater treatment while safeguarding the capacity of future generations to meet their own requirements.

### 7.1 Sustainable Sludge Management Options

Traditional methods for treating and disposing of sludge can be expensive, energy-consuming, and harmful to the environment. Yet, recent times have seen a rise in innovative approaches to sewage sludge treatment and reuse. These methods focus on sustainability, resource recovery, and minimising environmental impact. This emphasis on eco-friendly solutions has led to the development of effective ways to manage sewage sludge.

The practice of crop water and sludge reuse entails the treatment of wastewater and sewage sludge with the objective of maximising water extraction and minimising waste generation. The processed water and sludge can thereafter be employed as a source of fertiliser or for the purpose of irrigation in agricultural practices. This approach presents the opportunity to optimise water utilisation, minimise wastage, and maximise agricultural yield. Current scholarly investigations highlight the several benefits associated with the utilisation of reclaimed water and sludge in agricultural practices. These advantages encompass cost reduction, the establishment of a dependable local water supply, and the promotion of a circular economy. Through the adoption of cutting-edge treatment technologies, the agricultural industry has the potential to make substantial contributions to the practice of sustainable resource management. This, in turn, can facilitate the development of a more streamlined and ecologically conscious approach to the management of water and sludge.

The method of stabilising sewage sludge using clay minerals and biochar entails adding these substances at the same time. This approach may reduce the number of microorganisms present, control heavy metal levels, and promote effective nutrient recycling. Research suggests that stabilised sewage sludge can be used as a fertiliser and soil conditioner without risk. The efficacy of this technique in enhancing soil characteristics and its potential as a safe fertiliser and soil conditioner are being investigated in an ongoing study. Utilising these combined methods can make treating and using sewage sludge more ecologically friendly and sustainable.

Indeed, hydrodynamic cavitation (HC) represents a promising technique in the realm of sewage sludge management. By utilising cavitation bubbles to break down activated sludge and dissolve organic matter, HC effectively reduces the size of suspended sludge particles. This process enhances the stabilization of sewage sludge in both anaerobic and aerobic digestion units.

Additionally, the environmentally friendly and cost-efficient nature of HC makes it a valuable pretreatment method, particularly in the context of agricultural applications. By enhancing the stabilization of sewage sludge, HC contributes to the development of sustainable practices, ensuring the safe and efficient reuse of this valuable resource in agriculture. Ongoing research and application of HC in sewage sludge management hold promise for improving overall efficiency and environmental impact in wastewater treatment processes.

Absolutely, sustainable sewage sludge management is paramount for ensuring environmental protection, public health, and efficient use of resources. The emphasis on innovative techniques that prioritise sustainability, resource recovery, and environmental impact is crucial. These approaches not only address the challenge of managing sewage sludge but also contribute to the development of eco-friendly solutions. By integrating sustainable practices, societies can effectively manage sewage sludge, reduce environmental pollution, and promote circular economy principles, ensuring a cleaner and more sustainable future (Hubbard and Evans 2010).

### 7.2 Risk-Reduction Strategies

A risk assessment is a simplified representation of real-world situations, relying on assumptions and subjective judgements. Its accuracy depends on the quality of the underlying data. As systems become more complex, these assessments are more prone to errors due to the disparities between the model and reality. Ecosystems are challenging to assess comprehensively, often missing the combined impacts of multiple factors. These limitations underscore the importance of careful interpretation and awareness of broader ecological contexts when using risk assessments to guide environmental decisions.

Creating regulations for a diverse country like the US is incredibly challenging. The Part 503 risk assessment assumes that models designed for a ranch in west Texas can be applied to a vegetable farm in New York, which might need to be revised due to the diverse conditions (Haimes 2005). Models have inherent limitations, so their results should acknowledge uncertainty, either as a range of values or by applying a safety factor. Surprisingly, the US EPA risk assessment needs to incorporate such recognition of uncertainties, which goes against more recent practices (Haimes 2005).

Despite significant efforts by the US EPA and other organizations in conducting the risk assessment for the Part 503 regulations, this study reveals fundamental errors in the assessment structure, questionable assumptions, and crucial omissions, leading to regulations that lack adequate protection. Given the complexities of this risk assessment, the potential long-term effects on agricultural productivity, and the challenges of remediating soil contamination impacts, adopting a precautionary approach, as practiced by several other nations, seems more appropriate. Many of the concerning pollutants do not degrade and can persist in the soil for extended periods, although their impacts might be limited due to limited bioavailability (Aven et al. 2013).

## 8 Future Directions

### 8.1 Advances in Sludge Management Research

The problems involved in sludge management research include various considerations, including the potential for leaching of contaminants, emissions of greenhouse gases, and the presence of disagreeable odours. Additionally, the treatment of such biosolids incurs significant expenditures. The sustainable management of sludges, which are classified as solid wastes, necessitates the exploration and implementation of novel technologies that optimise waste utilisation.

In the context of Greece, many methods have been employed for the disposal of sewage sludge, including agricultural utilisation, incineration (accounting for 27% of cases), landfilling (14%), and other miscellaneous applications (17%). In contrast, Germany exhibits a distinct waste management pattern, with thermal disposal, specifically burning for energy production, emerging as the primary method (55%). This is followed by land application (42%) and material recovery (3%), while landfilling is notably absent. The aforementioned assortment of approaches underscores the necessity for customised solutions that take into account the distinctive characteristics of each respective place.

#### 8.1.1 Landfilling

Sludge disposal in landfills usually entails combining dried or concentrated biological sludge with other solid sewage wastes at open-air disposal sites. However, as sludge landfilling is bad for the environment, the contaminated water treatment industry has been forced to adopt more sustainable management practices due to rigorous legislation and higher landfill costs. A lot of nations have passed laws to reduce the amount of sludge dumped in landfills. For example, to prevent groundwater pollution and lower greenhouse gas emissions, Germany legislated in 2005 that the organic content of landfill wastes be less than 5%. In the European Union (EU), there was a significant reduction in sludge landfilling between 2002 and 2013. Countries like Norway, France, and Germany witnessed reductions of 85.8%, 88.4%, 94.2%, 96.4%, and 100%, respectively. These reductions were achieved through measures aimed at curbing environmental impact and promoting more sustainable waste management practices.

#### 8.1.2 Incineration

Completely desiccated sewage sludge exhibits promising potential as a valuable renewable energy resource, with a calorific value ranging from around 11 to 13 MJ/kg, a level comparable to that of lignite. The incineration of sewage sludge results in the emission of greenhouse gases at a notably reduced level in comparison to the combustion of fossil fuels. In terms of emissions reduction, sewage sludge demonstrates a comparative advantage by emitting 58% and 80% fewer emissions than natural gas respectively, while generating equivalent energy output. Nevertheless, the cost of the incineration procedure might be considerably high, particularly in cases where forced drying, co-combustion, and the utilisation of supplemental fuel become imperative. Despite the associated expenses, the burning of sludge continues to be a prevalent method of waste management in various nations, such as Japan, France, and Germany. The aforementioned nations, characterised by their densely populated urban areas and sophisticated technological infrastructure, persist in employing sludge incineration as a means of waste disposal.

### 8.2 Role of Public Awareness and Advocacy

Public awareness and advocacy are pivotal in promoting responsible sludge management practices. By educating communities about the sources, risks, and sustainable disposal methods of sludge, awareness campaigns empower citizens to demand proper treatment and disposal measures. Advocacy efforts drive the formulation and implementation of stringent regulations, ensuring industries and municipalities adhere to environmentally sound sludge management practices. Informed and engaged citizens also hold authorities accountable, encouraging transparency and responsible actions. Moreover, public pressure stimulates innovation, prompting research into eco-friendly sludge management technologies. Through these collective efforts, public awareness and advocacy create a culture of accountability, fostering the adoption of safe and sustainable sludge management practices that protect both public health and the ecosystem (Elgarahy et al. 2023).

### 8.3 Policy Recommendations and Future Outlook

The practice of land application for sludge is currently seeing a surge in popularity, driven by robust scientific research, cleaner sludge materials, and proactive public awareness efforts. This upward trend is expected to continue, especially with ongoing research emphasising the advantages of utilising sludge as an organic fertiliser. Additionally, the focus on resource recovery from sludge is growing in significance due to the rising demand for phosphorus and nitrogen.

The utilisation of sewage sludge as fertiliser in agricultural practices has become a pressing concern due to the growing need for sustainable development and the escalating costs of chemical fertilisers. However, it is imperative to conduct a meticulous analysis of sewage sludge properties before application to prevent any detrimental effects on soil quality and human health (Chojnacka et al. 2020; Chojnacka et al. 2022).

In summary, the future of sludge management will centre on resource recovery, energy utilisation, and safe agricultural application. These advancements will necessitate the meticulous assessment of the environmental consequences associated with sludge management choices. It is crucial to choose a sludge management approach through a thorough evaluation of the project's unique requirements rather than following prevailing trends.

## 9 Conclusion

The management of sludge presents a significant challenge to sewage water therapy plants because of its high-water content, poor dewaterability, and stringent disposal or reuse regulations. This chapter unveils the problems with different kinds of sludge, their improper treatment and disposal which pollutes land and waterbodies further affecting human and other living beings. They disturb the soil microbiota also. Even minute quantities of heavy metals present in sludge can pose a significant risk, whereas elevated levels of arsenic found in food and water have the potential to cause fatalities.

Sludge can contain a variety of pollutants, including toxic materials and pathogens. Improper disposal or management can lead to the pollutant of soil, groundwater, and surface water, which can have severe ecological and human health consequences. Responsible management minimises environmental risks. Responsible management involves recovering and utilising these resources, contributing to sustainability, and reducing the need for external fertilisers and energy sources. This reduces the reliance on finite resources, conserves energy, and lessens the environmental impact of resource extraction and processing.

It is possible to implement sustainable methods that enhance sludge management, water quality, hygiene, and sanitation through the introduction of community-led comprehensive sanitation participation (Sigler et al. 2015). The common waste management hierarchy, which places the highest priority on recovery, recycling, reuse, and prevention with landfilling and dumping being illegal, is a requirement for all countries to follow. The majority of research on sewage sludge looks at greenhouse gas emissions, carbon storage, nutrient leaching, metal introduction into soils, fuel requirements (transporting sludge over agricultural land), and other environmental effects associated with sludge application.

Sludge management is shifting from mere disposal to resource recovery. Advances in technology are allowing for the extraction of valuable resources from sludge, including energy, nutrients (e.g., phosphorus and nitrogen), and bio-based chemicals. This approach helps reduce waste and create a circular economy. The future of sludge management is closely linked to broader environmental and sustainability goals. As awareness of resource scarcity and environmental impacts continues to grow, innovative solutions that treat biological garbage sludge as a valuable resource rather than a waste product will be at the forefront of the industry. Research, technology, and international collaboration will play pivotal roles in shaping the future of sludge management.

## References


Adar E, Ince M, Karatop B, Bilgili MS (2017) The risk analysis by failure mode and effect analysis (FMEA) and fuzzy-FMEA of supercritical water gasification system used in the sewage sludge treatment. J Environ Chem Eng 5(1):1261–1268

Agoro MA, Adeniji AO, Adefisoye MA, Okoh OO (2020) Heavy metals in wastewater and sewage sludge from selected municipal treatment plants in Eastern Cape Province, South Africa. Water 12(10):2746

Ajonina C, Buzie C, Rubiandini RH, Otterpohl R (2015) Microbial pathogens in wastewater treatment plants (WWTP) in Hamburg. J Toxicol Environ Health A 78(6):381–387

Akadiri PO, Chinyio EA, Olomolaiye PO (2012) Design of a sustainable building: a conceptual framework for implementing sustainability in the building sector. Buildings 2(2):126–152

Almeida CMR, Santos F, Ferreira ACF, Lourinha I, Basto MCP, Mucha AP (2017) Can veterinary antibiotics affect constructed wetlands performance during treatment of livestock wastewater? Ecol Eng 102:583–588

Amrutha K, Warrier AK (2020) The first report on the source-to-sink characterization of microplastic pollution from a riverine environment in tropical India. Sci Total Environ 739:140377

Amsayazhi P, Mohan K (2018) Use of sludge waste as ingredient in making of brick. Int J Eng Technol 7:419

Arthur C, Baker J, Bamford H (eds) (2008) International research workshop on the occurrence, effects, and fate of microplastic marine debris. Conference proceedings

Aven T, Baraldi P, Flage R, Zio E (2013) Uncertainty in risk assessment: the representation and treatment of uncertainties by probabilistic and non-probabilistic methods. John Wiley & Sons

Bahuguna A, Singh SK, Sharma S, Arvind, Pandey A, Dadarwal BK et al (2022) Soil pollution due to sewage sludge and industrial effluents. In: Shit PK, Adhikary PP, Bhunia GS, Sengupta D (eds) Soil health and environmental sustainability: application of geospatial technology. Springer International Publishing, Cham, pp 345–367

Bassi N, Gupta S, Chaturvedi K (2023) Reuse of treated wastewater in India

Bharathiraja B, Yogendran D, Ranjith Kumar R, Chakravarthy M, Palani S (2014) Biofuels from sewage sludge-a review. Int J ChemTech Res 6(9):4417–4427

Bressman LS, Rubin EL, Stack KM (2023) The regulatory state. Aspen Publishing

Chen H, Yan S, Ye Z, Meng H, Zhu Y (2012) Utilization of urban sewage sludge: Chinese perspectives. Environ Sci Pollut Res 19:1454–1463

Chojnacka K, Moustakas K, Witek-Krowiak A (2020) Bio-based fertilizers: a practical approach towards circular economy. Biores Technol 295:122223

Chojnacka K, Mikula K, Skrzypczak D, Izydorczyk G, Gorazda K, Kulczycka J et al (2022) Practical aspects of biowastes conversion to fertilizers. Biomass Convers Biorefinery 1–19

Cofie O, Kone D, Rothenberger S, Moser D, Zubruegg C (2009) Co-composting of faecal sludge and organic solid waste for agriculture: process dynamics. Water Res 43(18):4665–4675

Council NR (1996) Use of reclaimed water and sludge in food crop production. National Academies Press

Delibacak S, Voronina L, Morachevskaya E (2020) Use of sewage sludge in agricultural soils: useful or harmful. Eurasian J Soil Sci 9(2):126–139

Demirbas A, Taylan O, Kaya D (2016) Biogas production from municipal sewage sludge (MSS). Energy Sourc Part A: Recov Util Environ Eff 38(20):3027–3033

Demirbas A, Edris G, Alalayah WM (2017) Sludge production from municipal wastewater treatment in sewage treatment plant. Energy Sourc Part A: Recov Util Environ Eff 39(10):999–1006

Deng Y, Zhang Y, Lemos B, Ren H (2017) Tissue accumulation of microplastics in mice and biomarker responses suggest widespread health risks of exposure. Sci Rep 7(1):46687

Dickin SK, Schuster-Wallace CJ, Qadir M, Pizzacalla K (2016) A review of health risks and pathways for exposure to wastewater use in agriculture. Environ Health Perspect 124(7):900–909

Elgarahy AM, Maged A, Eloffy M, Zahran M, Kharbish S, Elwakeel KZ et al (2023) Geopolymers as sustainable eco-friendly materials: Classification, synthesis routes, and applications in wastewater treatment. Sep Purif Technol 124631

Evanylo GK (1999) Land application of biosolids for agricultural purposes in Virginia

Gao N, Kamran K, Quan C, Williams PT (2020) Thermochemical conversion of sewage sludge: a critical review. Prog Energy Combust Sci 79:100843

Gautam M, Wankhade K, Sarangan G, Sudhakar S (2021) Framework for addressing occupational safety of de-sludging operators: a study in two Indian cities. J Environ Manage 289:112243

Haimes YY (2005) Risk modeling, assessment, and management. John Wiley & Sons

Harder R, Peters GM, Svanström M, Khan SJ, Molander S (2017) Estimating human toxicity potential of land application of sewage sludge: the effect of modelling choices. Int J Life Cycle Assess 22:731–743

Harrison EZ, McBride MB, Bouldin DR (1999) Land application of sewage sludges: an appraisal of the US regulations. Int J Environ Pollut 11(1):1–36

Hu G, Li J, Zeng G (2013) Recent development in the treatment of oily sludge from petroleum industry: a review. J Hazard Mater 261:470–490

Hubbard D, Evans D (2010) Problems with scoring methods and ordinal scales in risk assessment. IBM J Res Develop 54(3):2:1–2: 10

Huerta Lwanga E, Mendoza Vega J, Ku Quej V, Chi JLA, Sanchez Del Cid L, Chi C et al (2017) Field evidence for transfer of plastic debris along a terrestrial food chain. Sci Rep 7(1):14071

Kacprzak M, Neczaj E, Fijałkowski K, Grobelak A, Grosser A, Worwag M et al (2017) Sewage sludge disposal strategies for sustainable development. Environ Res 156:39–46

Khan MM, Siddiqi SA, Farooque AA, Iqbal Q, Shahid SA, Akram MT et al (2022) Towards sustainable application of wastewater in agriculture: a review on reusability and risk assessment. Agronomy 12(6)

Khan AH, Khan NA, Naushad M, Aziz HA (2023) The treatment of pharmaceutical wastewater: innovative technologies and the adaptation of treatment systems. Elsevier

Lebrero R, Rangel MGL, Muñoz R (2013) Characterization and biofiltration of a real odorous emission from wastewater treatment plant sludge. J Environ Manage 116:50–57

Luther L (ed) (2010) Regulating coal combustion waste disposal: issues for congress. Library of congress, congressional research service

Mehta M, Mehta D, Yadav U (2019) Citywide inclusive sanitation through scheduled desludging services: emerging experience from India. Front Environ Sci 7:188

Oenema O, Kros H, de Vries W (2003) Approaches and uncertainties in nutrient budgets: implications for nutrient management and environmental policies. Eur J Agron 20(1):3–16

Pavilonis BT, Anthony TR, O'Shaughnessy PT, Humann MJ, Merchant JA, Moore G et al (2013) Indoor and outdoor particulate matter and endotoxin concentrations in an intensely agricultural county. J Eposure Sci Environ Epidemiol 23(3):299–305

Periyasamy S, Kavitha S, Temesgen T (2022) Chapter 22—Biogas recovery from sludge. In: An A, Tyagi V, Kumar M, Cetecioglu Z (eds) Clean energy and resource recovery. Elsevier, pp 381–394

Piippo S, Lauronen M, Postila H (2018) Greenhouse gas emissions from different sewage sludge treatment methods in north. J Clean Prod 177:483–492

Santos F, Almeida CMR, Ribeiro I, Mucha AP (2019) Potential of constructed wetland for the removal of antibiotics and antibiotic resistant bacteria from livestock wastewater. Ecol Eng 129:45–53

Saroha P, Yadav S, Meena R (2013) Investment in real estate sector in NCR-A high pay master. Int J Trade Commerce IIARTC

Scott GM, Smith A (eds) (1995) Sludge characteristics and disposal alternatives for the pulp and paper industry. Tappi International environmental conference. Tappi Press

Seleiman MF, Santanen A, Mäkelä PSA (2020) Recycling sludge on cropland as fertilizer—advantages and risks. Resour Conserv Recycl 155:104647

Shakoor MB, Nawaz R, Hussain F, Raza M, Ali S, Rizwan M et al (2017) Human health implications, risk assessment and remediation of As-contaminated water: a critical review. Sci Total Environ 601:756–769

Sharma A, Bhardwaj SK, Aggarwal RK, Sharma R, Agrawal G (2023) Greenhouse gas emission potential of sewage treatment plants in Himachal Pradesh. Sci Rep 13(1):9675

Sigler R, Mahmoudi L, Graham JP (2015) Analysis of behavioral change techniques in community-led total sanitation programs. Health Promot Int 30(1):16–28

Smith S (1904) Organic contaminants in sewage sludge (biosolids) and their significance for agricultural recycling. Philos Trans Roy Soc A: Math Phys Eng Sci 2009(367):4005–4041

Smol M, Kulczycka J, Henclik A, Gorazda K, Wzorek Z (2015) The possible use of sewage sludge ash (SSA) in the construction industry as a way towards a circular economy. J Clean Prod 95:45–54

Sun S, Sidhu V, Rong Y, Zheng Y (2018) Pesticide pollution in agricultural soils and sustainable remediation methods: a review. Curr Pollut Rep 4:240–250

Truschi S, Baldi A, Bruschi P, Cacciari I, Marvasi M, Lenzi A (2023) Foliar roughness and water content impact on Escherichia coli attachment in baby leafy greens. Biology 12(1):102

Vaishnav S, Saini T, Chauhan A, Gaur GK, Tiwari R, Dutt T et al (2023) Livestock and poultry farm wastewater treatment and its valorization for generating value-added products: Recent updates and way forward. Bioresour Technol 382:129170

Verma R, Sengupta S, Anand S (2020) Toolkit: managing faecal sludge in rural areas. Centre for Science and Environment, New Delhi

Vosylienė MZ, Jankaitė A (2006) Effect of heavy metal model mixture on rainbow trout biological parameters. Ekologija 4:12–17

Wang L-C, Lee W-J, Tsai P-J, Chen S-J (2002) Potential method for reducing emissions of polycyclic aromatic hydrocarbons from the incineration of biological sludge for the terephthalic acid manufacturing industry. Environ Sci Technol 36(15):3420–3425

Watson-Craik IA, Sinclair KJ (2020) Co-disposal of industrial wastewaters and sludges. Microbiology of landfill sites. CRC Press, pp 91–130

Wong JW, Selvam A (2006) Speciation of heavy metals during co-composting of sewage sludge with lime. Chemosphere 63(6):980–986

# Sludge Management Strategies for Developing Nations

Humira Assad and Ashish Kumar

**Abstract**

The rising demand for clean water around the world has led to an increase in the creation of sludge. Since the removal of sludge from various industrial activities is acknowledged as an ecological burden on society, sludge management has grown to be a major concern. The efficient management of sludge is a complicated problem since developing countries frequently have limited resources, poor infrastructure, and low technological capabilities. Thus, handling the sludge left over after wastewater treatment is a serious problem for the environment and human health, especially in poor countries. However, in many developing countries, many sludge management studies have been evaluated to identify the most up-to-date scientific and technological knowledge encompassing issues of sludge generation, evaluation, management, reuse of construction and agricultural materials, and final disposal. This is necessary to effectively manage sludge adequately. In this regard, this chapter elucidates the diverse sludge management techniques used in various developing countries, emphasizing economical and environmentally friendly options. The main objective is to evaluate the state of the art, pinpoint issues, and suggest novel strategies to tackle the distinct socio-economic and infrastructure limitations encountered by developing countries.

H. Assad
Department of Chemistry, School of Chemical Engineering and Physical Sciences, Lovely Professional University, Punjab, India

A. Kumar (✉)
Science, Technology and Technical Education Department, Government of Bihar, Nalanda College of Engineering, Bihar Engineering University, Patna, India
e-mail: drashishchemlpu@gmail.com

**Keywords**

Sludge · Wastewater treatment · Public health · Environment · Sustainable

## 1 Introduction

Although biological methods are an efficient means of treating wastewater from civilization and guaranteeing that there is no negative impact on the aquatic environment, they have one major flaw: they generate sludge. In general, a process produces more sludge the more successful it is in eliminating contaminants. Schematics that illustrated the liquid treatment plan in detail, complete with all necessary unit operations and an arrow that concluded with the words "sludge to disposal," were frequently seen in the past (Campbell 2000). The fact that just a small portion of the engineering design money went toward sludge management contributed to the belief that the sludge would miraculously vanish. Sadly, this strategy does not reflect reality, and the wastewater treatment industry is well aware that managing sludge or bio-solids, as it is now called represents serious technical difficulties and contributes significantly to the initial and ongoing costs of any facility (Cieślik at al. 2018). Sludge management is a major problem in the field of wastewater treatment, especially in emerging countries where population growth, industry, and urbanization are often occurring at a faster rate than the infrastructure that is needed (Świerczek et al. 2018). To minimize environmental concerns and optimize resource recovery, sludge—the semi-solid residual material produced during wastewater treatment (see Fig. 1)—contains a complex mixture of organic and inorganic chemicals that need to be handled carefully (Mansor et al. 2022).

Unprecedented rates of urbanization in developing countries are increasing the amount of wastewater produced. The

© The Author(s), under exclusive license to Springer Nature Switzerland AG 2025

A. Mandpe and M. P. Shah (eds.), *Wastewater Sludge Nexus*, Sustainable Landscape Planning and Natural Resources Management, https://doi.org/10.1007/978-3-031-45758-6_14

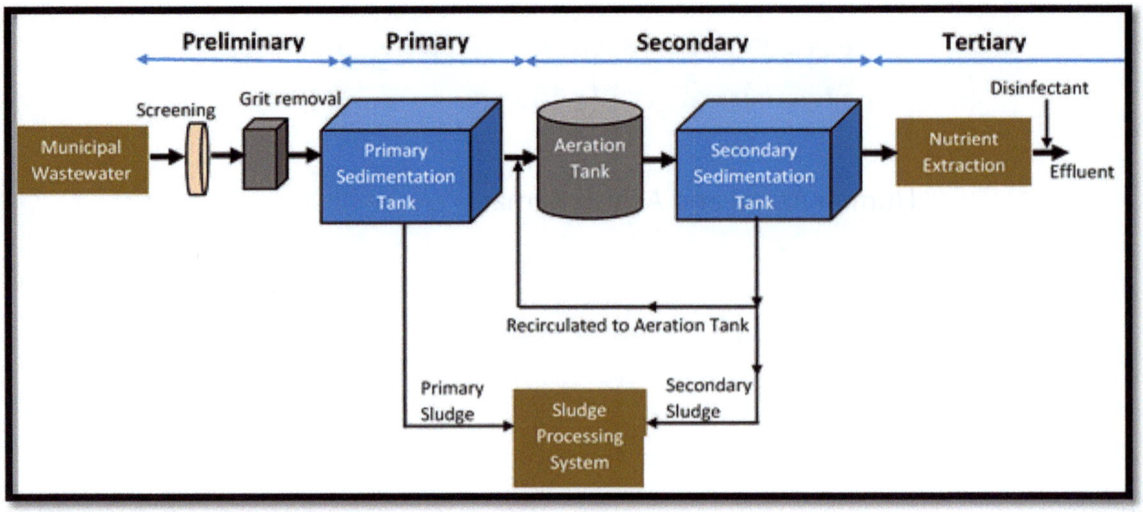


**Fig. 1** Schematic illustration of sewage sludge production from wastewater treatment (Enebe et al. 2023)

problem is made worse by the concurrent expansion of industrial activity, which produces larger volumes of sludge that need to be properly treated and disposed of (Ahmad et al. 2016). The issue is frequently made worse by inadequate infrastructure and a lack of funding, which calls for creative and economical sludge management solutions (Świerczek et al. 2018). Furthermore, there are serious threats to the environment and public health associated with poor sludge management. Sludge that has been left untreated or managed incorrectly can pollute land, water sources, and the atmosphere, which can spread pollutants and diseases. Communities' health and welfare are put in danger as a result, underscoring the need of swiftly putting into practice efficient sludge management techniques to protect the environment and the general public's health.

Sludge is not just a waste product, despite the difficulties; it has significant resources that can be extracted and put to new uses. Innovative techniques and technologies can be used to harness elements including energy, nutrients, and organic matter, which will reduce waste while optimizing resources (Babatunde and Zhao 2007). Developing nations can greatly benefit from sludge management systems that give priority to resource recovery. Sludge management has seen a boom in best practices and technology advancements targeted at increasing sustainability and efficiency (Cieślik at al. 2018). Developing countries have access to a variety of instruments that can be customized to their unique requirements and limitations, ranging from sophisticated anaerobic digestion and thermal treatment procedures to creative dewatering and drying methods (Christodoulou and Stamatelatou 2016). Strong institutional and policy frameworks must be developed and put into place for sludge management in developing countries to be successful. To promote the adoption of sustainable sludge management procedures, governments and regulatory organizations must provide policies, norms, and incentives (Cieślik at al. 2018). To foster a climate that is conducive to investment and innovation, cooperation between the public and private sectors is vital.

There is an urgent need for comprehensive and context-specific policies that strike a balance between environmental sustainability, public health concerns, and resource recovery as developing countries struggle with sludge management. The need for effective and sustainable sludge management methods is a challenge for developing countries. This chapter provides a deeper look at the many factors that must be taken into account in order to address the complexities of sludge management in developing countries.

## 1.1 Historical Development

Over the past 50 years, sludge management has changed from a strategy that is probably best described as an "afterthought" to a high-priority issue that now accounts for up to 50% of the cost of treating wastewater and is, in many ways, a far more sensitive topic with the public than the liquid treatment train (Ahmad et al. 2016). While there is an incredible range of particular management approaches that have been used, the majority of them can be broadly classified into three categories: land application, landfill disposal, and incineration/thermal technology.

Sludge has long been stored on land, frequently with no restrictions on the use of the space. Sludge was sometimes put at very high loading rates without any thought to the agronomic requirements for nutrients; in fact, there were sometimes no attempts made to even grow a crop. In these instances, the area was not utilized for sludge utilization, but rather as a disposal site (Egan 2013). The emphasis shifted to "sludge utilization" in the early 1970s, and a vigorous phase of research was launched to examine the long-term effects of heavy metals in sludge (Campbell 2000). Concurrently, a greater effort was made to create and implement industrial sewer ordinances to prevent pollutants from entering the sewage and create cleaner sludges. Guidelines, codes of conduct, and laws for the use of sewage sludge (bio-solids) on agricultural land were developed as a result of this time. These policies, rules, etc., made it possible to exploit sludge' nutritional value for agriculture while posing little damage to the environment or public health.

Composting has been a popular land-based substitute as well. A range of bulking materials, such as wood chips, bark, municipal solid waste, etc., have been used in the composting of sludge, utilizing various processes (Gomes et al. 2019). Composting was first hailed as a highly desirable "green" technique by the environmental community. In reality, the NIMBY syndrome is still a factor in plant location, and there has been a significant failure rate for large-scale projects. Numerous factories have been shut down because of community concerns about the difficulty in controlling odors in both maturation/storage regions and processing plants. The premise that a sizable sum of money would be generated from the sale of the completed compost without needing to invest the capital required to build the market has frequently formed the basis of sludge composting facility economics. Without it, a lot of facilities didn't make the kind of money they had hoped to and in other cases didn't have anywhere to put their finished compost (Campbell 2000; Goldbold et al. 2003).

Sludge management in municipalities has always been thought to be the easiest and potentially least expensive when it comes to disposing of sludge in a sanitary landfill. Why not use the landfill for sludge as well? After all, they needed it to handle municipal solid trash. The challenges associated with locating new landfills turned into a political nightmare in the 1970s and 1980s, which led to major questions about this method. As the NIMBY syndrome developed, several neighborhoods became vocally opposed to the location of new landfills in their area. Sludge dumping in landfills, which would shorten their useful lives, started to be actively questioned, especially when there were alternatives. In addition to public opinion, there were operational problems with sludge disposal in landfills. There was worry that the sludge's high organic content would overwhelm any leachate treatment system, especially in the landfill's early years. Over time, the raw sludge became intolerable, and treatment was needed to reduce the attraction of vectors. Landfill owners started turning away liquid sludge because of issues with compacting the sludge/garbage mixture. It became standard practice to dewater sludge to a certain minimal solids concentration. Due to its lack of considerable benefits for sludge reuse, landfilling has been deemed as one of the least desired solutions in numerous evaluations conducted in recent years (Christodoulou and Stamatelatou 2016).

Incineration became a popular disposal method for large communities in the 1960s and 1970s. Air pollution was not a top concern, energy was cheap and plentiful, and incineration produced a minimum volume of waste for final disposal. The years that followed saw a significant change in this circumstance. The substantial increase in energy prices made it clear that should there be a scarcity of energy, sludge incineration would not be given high priority. Better dewatering facilities were put in place to raise the solids content of sludge cakes and lower energy consumption. Numerous efforts were undertaken to enhance the energy balance surrounding incinerators; nevertheless, their effectiveness was limited as the issue was not tackled from a comprehensive systems perspective. More lately, incineration has come under fire from neighborhood interest organizations and has become one of the century's most contentious environmental discussions in many places. The public's perception of incineration reached a record low, even though many of the problems, like the release of dioxins and furans, were mostly associated with burning municipal waste and had little to do with burning sludge. No matter what kind of material was being burned, the public viewed the release of pollutants into the air as an unavoidable health risk that was intolerable, regardless of the degree of danger that was calculated using mathematical risk assessment techniques (Mansor et al. 2022). To handle organics, nitrogen and sulfur compounds, heavy metals, and particulates, new emission standards were implemented. Multiple hearth furnaces were mostly replaced by fluidized bed combustors, which provided higher energy efficiency and improved emission control. While airborne emission control technologies kept advancing, public perception of incineration did not.

## 2 Sludge-Related Issues

Economics is the primary consideration for all decision-makers when assessing projects on a global scale. Sludge management is not a part of the treatment plans of many wastewater treatment plants, and sanitation initiatives have only lately begun to take it into account as part of their overall programs. Sludge is frequently applied, disposed of, or discharged without any kind of treatment; occasionally, it is even left to air dry (Fungaro and Silva 2014). Sludge management can account for as much as 50% of the overall cost of wastewater treatment, which is one of the limitations that needs to be considered (Campbell 2000).

Although several developing nations have been receiving funding from abroad to promote sanitation initiatives, the combined amount of national and foreign investment is frequently insufficient to meet the objectives. Even though a significant amount has been invested, only 25% of that has been allocated to water supply. This illustrates how supplying drinking water to people is prioritized, downplaying the significance of sanitation and how it directly affects public health (Jiménez et al. 2004). Sludge management costs pale in comparison with those associated with sanitation. Furthermore, most nations invest very little in national research, which makes them more reliant on information and innovations from other countries. Environmental costs are typically understood to be the costs associated with ecological harm that has already occurred; however, the focus should be on lowering these costs through the prevention of ecological damage. Additionally, while estimating the environmental cost of sludge management, a few other elements are overlooked (Jiménez et al. 2004). Among these is the possibility of saving money on public health, especially when treating waterborne illnesses; a reduction in the use of chemical fertilizers for crop production, which is limited when it comes to mineral phosphorus; and a decrease in land degradation, which accounts for five to ten percent of agricultural output in developing nations (Young 2000). In this context, it is important to remember that soil degradation causes the loss of 5–7 million hectares of arable land annually on a global scale (Rhoades 1997). By halting the spread of diseases and parasites, supplying nutrients for soil fertility, and enhancing soil properties (framework, organic matter content, and water-holding capacity), proper sludge management may have a positive effect on those factors. Furthermore, using sewage in food production could boost some crops' yields and assist lower-income countries in meeting their demands while also lowering the price of inorganic fertilizers.

The quality of the sludge is the most crucial factor to take into account in this topic since it will dictate the stabilization procedure that is used. As was previously indicated, the majority of parasite infections have been documented in these countries, and sludge and wastewater may be potential exposure points. Furthermore, the microbiological quality of sludge from underdeveloped nations is poorly understood, frequently as a result of inadequate financial and analytical resources. However, several nations are working hard to characterize the sludge's microbiological makeup. About 83% of the eggs in the Mexican sludge sample are consistent with *Ascaris* spp. This particular helminth exhibits the highest resistance to traditional stabilization methods. It can only be rendered inactive through the use of severe pH conditions (>12 units) or temperatures exceeding 50 °C (Jiménez et al. 2004). Therefore, to lower their densities, procedures like as composting, lime stabilization, or thermophilic aerobic or anaerobic digestion should be employed, but only under particular conditions. For example, the extended retention times needed for composting can be doable for tiny plants but prohibitively expensive in major cities due to the amount of acreage needed. Conversely, thermophilic digestion might be a possibility if the plant is run by qualified staff and there is enough room to erect the reactors. Ultimately, the quick retention period required for sludge stabilization makes lime treatment an appealing procedure; yet, the increased bulk of solids may result in higher transportation and disposal costs. As of right now, a few alternative methods have proven to be quite effective in fighting helminths, but their suitability for use in developing nations should be assessed economically. For instance, the concentrations of helminth eggs in raw and treated sludge were lowered from 112 to 5.5 ova/g TS by treating the sludge with peracetic acid (Barrios et al. 2001). Furthermore, employing sludge to produce bricks could seem like a good idea, particularly in nations where housing is a major concern. Foreign technologies are occasionally employed for stabilization; however, their effectiveness is constrained by the lack of knowledge on the quality of the sludge that needs to be treated. Consequently, there is still a significant chance of infection, which puts at risk the adoption of sludge for advantageous uses. Characterizing the sludge is a crucial step in the development of the treatment operations. If sludge is not being produced, it is still necessary to conduct a thorough microbiological analysis of the wastewater to approximate the density of pathogens and parasites in the sludge that would be generated (Jiménez et al. 2004). To choose technologies on an individual basis, it is crucial that local groups, such as educational institutions, research organizations, and professional associations, be involved in this process.

## 3 Sludge Processing and Characterization

Sewage sludge treatment and disposal are thought to account for half of the operational costs of secondary sewage treatment plants in Europe. Therefore, the final appearance of sewage sludge is influenced not only by the grade of the raw sewage sludge but also by several procedures implemented inside sewage sludge treatment at wastewater treatment plants (WWTPs) as shown in Fig. 2.

The following is a summary of a typical procedure: (Metcalf et al. 2013; Wójtowicz et al. 2013)

- **Preliminary treatment**: This includes screening and comminuting.
- **Primary thickening**: This includes gravity, flotation, drainage, belt, and centrifuges.

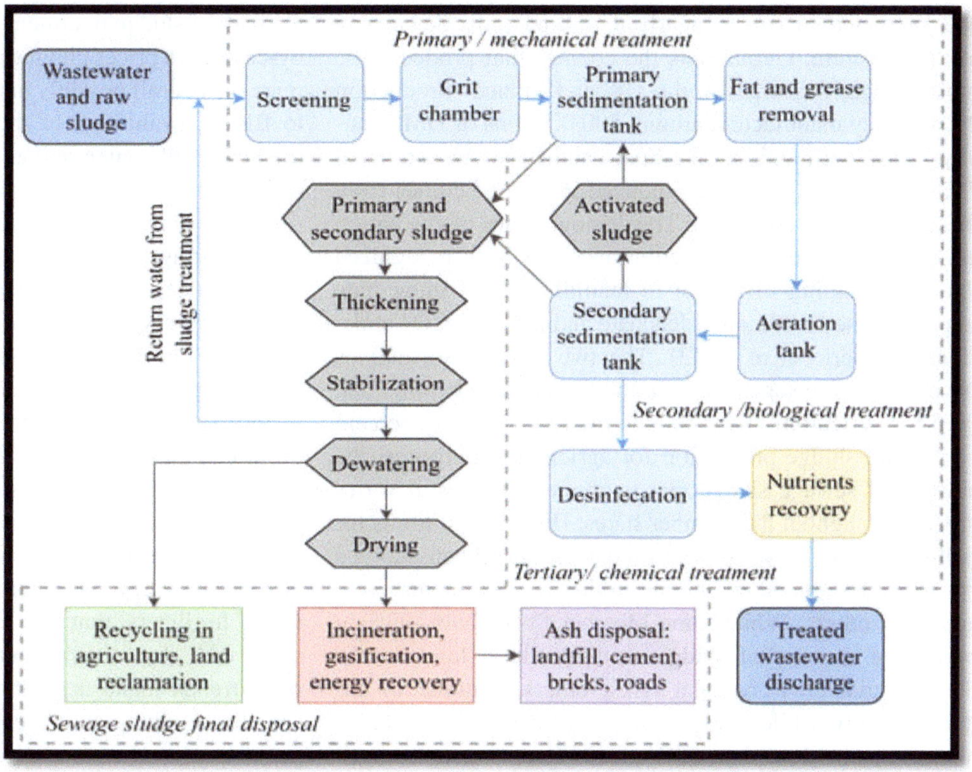


**Fig. 2** Overview of sewage sludge and wastewater treatment processes (Đurđević et al. 2022)

- **Liquid sludge stabilization**: This includes anaerobic digestion, aerobic digestion, and lime addition.
- **Secondary thickening**: This includes gravity, flotation, drainage, belt, and centrifuges.
- **Conditioning**: This includes elutriation, chemical, and thermal procedures
- **Dewatering**: This includes plate press, belt press, centrifuge, and drying bed.
- **Final treatment**: This includes composting, drying, line addition, incineration, wet oxidation, pyrolysis, and disinfection.
- **Storage**: This includes liquid sludge, dry sludge, compost, and ash.
- **Transportation**: This includes road, pipeline, and sea.
- **Final destination**: This includes landfill, agriculture/ horticulture, forest, land reclamation, land building, and other uses.

Depending on stabilization procedures, dewatered sewage sludge (dry) typically comprises 50–70% organic matter, 30–50% mineral content (including 1–4 % C), 3.4–4.0% N, 0.5–2.5% P, and a sizable amount of different nutrients, including micronutrients (Tyagi and Lo 2013; Samolada and Zabaniotou 2014; Fytili and Zabaniotou 2008). Because sewage sludge has very little lignin or cellulose, organic matter from the sludge mineralizes quickly. This rapid deterioration may cause the concentration of nitrate and other pollutants in the soil to peak. Sewage sludge can be used as a fertilizer for plants because it also has high levels of phosphate and nitrogen. However, it also contains a variety of contaminants, including both organic and inorganic substances, including pesticides, surfactants, hormones, medications, heavy metals, and many more (Siebielska 2014). Furthermore, the coexistence of many harmful species, such as bacteria, viruses, and protozoa, in addition to other parasitic helminths, might pose a risk to the health of people, animals, and plants (Fijalkowski et al. 2014; Kacprzak and Stańczyk-Mazanek 2003). By properly treating the sludge (lime treatment), the amount of pathogenic and parasitic microbes in it can be greatly decreased before its application to the land, hence lowering the dangers to human health (Farzadkia and Bazrafshan 2014; Olszewski et al. 2013).

## 4 Sludge Management Strategies

A report published by the European Commission states that in 2008, 26 EU Member States generated over 10 million tons of dry solids (DS) of sewage sludge; of which, about 36% (3.7 million tons DS) were recycled into agricultural applications (Kacprzak and Stańczyk-Mazanek 2003). For instance, the sewage sludge produced annually in the nations bordering the Baltic Sea watershed was estimated to be around 3.5 million tons of dry solids; by 2020, that amount was predicted to rise to nearly 4 million tons. According to the literature, Germany is the country that produces the most sludge, followed by the UK and France (Production 2021). They manufacture around 500,000 tons of DM annually, along with nations like Spain and Italy. An estimated 75% of the sewage sludge produced in Europe comes from these 5 countries combined. The demographics of any nation are often reflected in this circumstance. It is well-recognized that the dumping of sludge in landfills has an immediate impact on the formation of leachate and airborne $CO_2$ emissions (Barberio et al. 2013). The two primary techniques for managing sewage sludge in the EU are still burning and used for agriculture. Even though the EU uses forty percent of its total sludge production for agricultural purposes, the amount of sewage sludge that is applied to the soil varies greatly throughout the member states. Higher limit limits for pollutants than those found in the SSD have been adopted by some EU members. Some additional pollutants were added by others. While some Member States have outright prohibited the use of sludge on agricultural land due to the potential environmental risks, others continue to use it extensively while also making improvements to sludge management. More than 50% of the sludge used for agriculture in Belgium, Denmark, Spain, France, Ireland, and the United Kingdom was utilized in that year (Kacprzak and Stańczyk-Mazanek 2003). However, less than 5% is utilized for agriculture in some nations, such as Belgium and Finland. Greece, the Netherlands, Romania, Slovenia, and Slovakia do not employ sludge in their agricultural practices. Sewage sludge landfilling has been gradually declining in Poland, while thermal sludge conversion has increased. Increased amounts of sludge are used in agriculture in non-EU nations like Norway and Switzerland. As per Directive (1999), Norway's official goal of recycling 60% of bio-solids to farmland was already accomplished; in 2008, up to 80% of bio-solids were recycled to farmland or green spaces. In contrast, Switzerland forbade the application of bio-solids on agricultural land back in 2005.

The average cost of treating various wastewater and disposing of untreated sludge is estimated to be between 160 and 210 EUR per ton of dry matter. Costs rise to roughly 210-300 EUR/ton DM when dewatered sludge is used in forestry or agriculture, notably when it is burned with domestic trash or when degraded lands are reclaimed as per information on the average prices of various sewage sludge management systems throughout all European nations (Jakobsson 2012). These days, recycling of organic materials and energy is the main method of handling sewage sludge, albeit costs could differ between countries as shown in Fig. 3 (Wiechmann et al. 2013). According to Smith (Wiechmann et al. 2013), one strategy that can be used to decrease pollution and damage to the environment while also improving its quality is the Best Pragmatic Environmental Option (BPEO). Using this technique, the path of sludge treatment and disposal can be chosen case-by-case, with prices playing a secondary role, in order to minimize the overall possibly harmful environmental impacts. As to BPEO, evaluation of the effect on every environmental medium (soil, water, and air) is feasible. However, it usually calls for numerous other contextual considerations in addition to objective explanations of significance. Using sewage sludge for agricultural purposes is typically regarded as the best practice (BPEO), but it necessitates a detailed examination of the sludge and is dependent on numerous outside variables. For instance, a farmer's cooperation can only come from his goodwill and acceptance. Due to the possible risk of infections and contamination spread, many governments prohibit the use of sewage sludge in food production. On the other hand, where sewage sludge quality forbids its usage on land, as is the situation with major metropolitan WWTPs, sewage sludge must be burned.

Sewage sludge and its byproducts can be recycled to make fertilizers that are meant for commercial sale, as discovered by several nations, including Norway. Two main trends make up that type of approach. To start, two specific products with the potential for sustainable energy generation are heat and electricity (produced from biogas). Approaching bio-solids as products where sewage sludge plays a significant role is the second trend. These products have two potential uses: They can be sold on the open market or used on-site, which has been a long-standing practice at wastewater treatment plants. Two key components of sustainable development of sewage sludge procedures are the following:

- Creating cutting-edge technology and management frameworks that are appropriate for the regional environmental factors, including the climate, geo-economics, etc. and
- Maximizing the benefits of sludge recovery and recycling using frameworks that rely on renewable energy.

No single technological advancement can solve local issues or promote sustainable growth (Barberio et al. 2013). Therefore, every nation should perform a thorough life cycle assessment (LCA) that considers the production, processing, and ultimate disposal procedures for the full stream of sewage sludge generated in the nation or the region. The preferred sewage sludge treatment techniques are shown in Fig. 4 and are mostly determined by the size of the WWTP and the amount of sewage sludge generated. The technology selection factors were separated as follows by other authors (Wójtowicz et al. 2013; Tyagi and Lo 2013):

- **Social factors**: (protect public health and safety, minimize community disruption, land requirement),

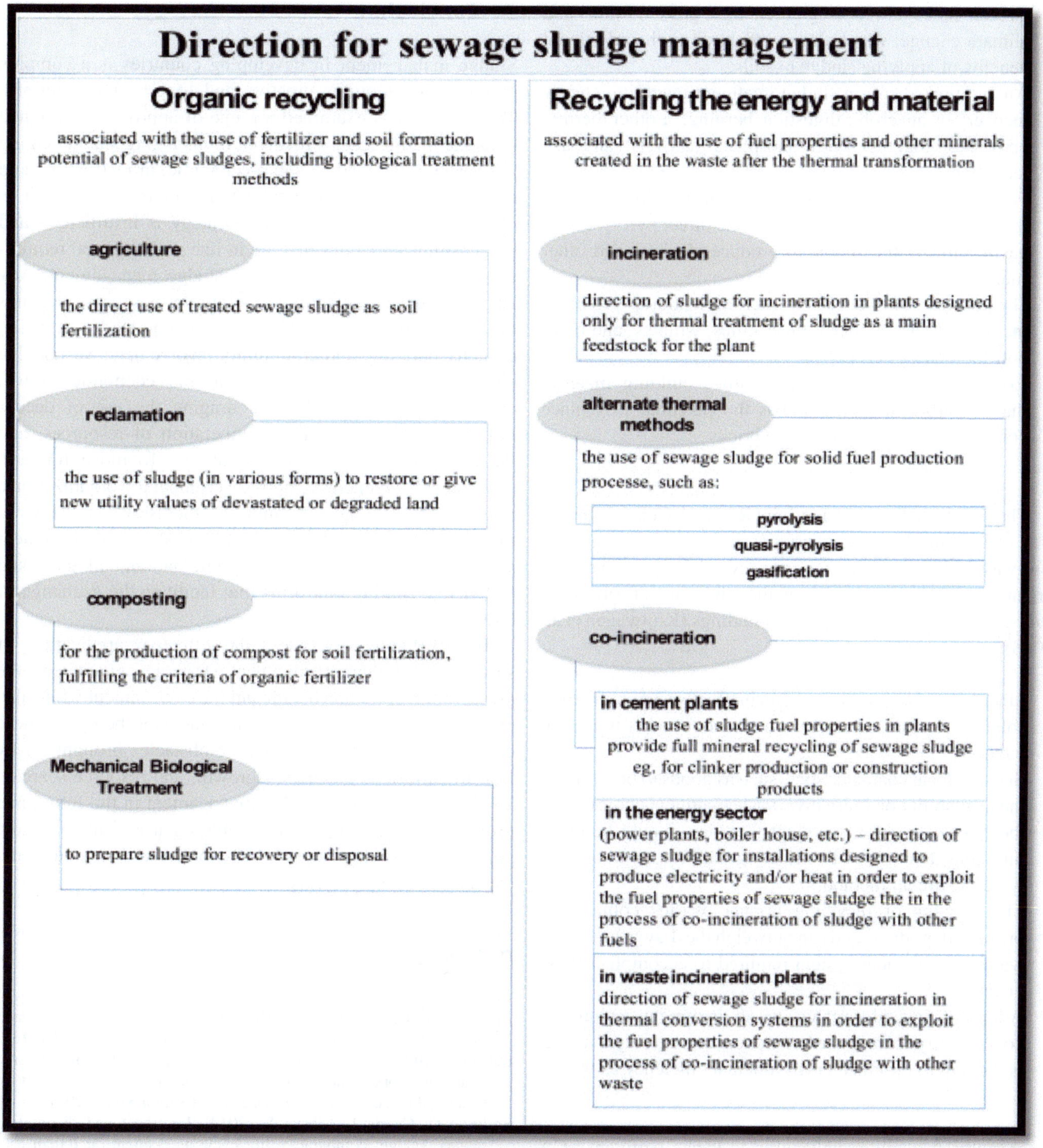


**Fig. 3** Overview of fundamental sludge disposal strategies (Kacprzak and Stańczyk-Mazanek 2003)

- **Economic factors**: (technology status, affordability),
- **Environmental factors**: (greater energy efficiency input–output, upstream and on-site emission, secondary waste production).

Sludge management, however, can also be influenced by a wide range of other issues, including advancements in renewable energy and climate change policies (Kacprzak and Stańczyk-Mazanek 2003):

- Greater focus on reducing greenhouse gas emissions and climate change, which draws emphasis to the additional benefits of applying sludge to soils;
- An anticipated rise in sludge treatment involving energy recovery by anaerobic digestion, burning, or other thermal processes, together with phosphate recovery and ash recycling;
- Potential rise in the generation and consumption of biogas from sewage sludge, as well as some direct pyrolysis and gasification of the sludge to produce alcohols and other fuels;
- Greater use of sludge on fuel agricultural products, including hybrid poplars, miscanthus, willows, and other non-food energy crops;
- Conversion of sewage sludge into a building material that is carbon neutral and has the potential to replace conventional clay and concrete bricks.

The European Commission Joint Research Center's Inventory of Existing Research Applying Life Cycle Thinking to Bio-waste Management examines several LCA research on organic waste management, even though it isn't a formal case study. It offers the following initial findings (Krutwagen et al. 2008):

- Since bio-waste landfilling is an ineffective environmental solution, it should be eschewed in favor of other treatment methods;
- Specific local characteristics, such as goods to be avoided, the availability of alternative waste treatment choices, and the effectiveness of treatment facilities, should be taken into consideration when selecting the preferable organic waste treatment method;
- The benefits of recycling organic wastes come from prevented products and can outweigh the drawbacks of the waste management system required to accomplish these goals;
- Adopting several organic waste treatment techniques concurrently, as opposed to concentrating only on one, is thought to optimize environmental efficiency as well.

Regretfully, it is quite challenging to locate pertinent LCA research on this subject. Many studies are limited in scope, do not include all organic wastes, concentrate on certain system activities, or make very crude predictions.

## 5 Conclusion

Sludge management in developing countries is a complex problem that requires creative and long-lasting solutions. This chapter has examined a range of approaches designed to tackle the intricacies involved in managing sludge, taking into account the distinct socio-economic, infrastructure, and environmental circumstances of developing nations. It is obvious that a one-size-fits-all strategy is insufficient, and that customized tactics that are in line with regional requirements are required for effective sludge management. Any sludge management initiative's long-term success depends on highlighting the significance of knowledge transfer, capacity building, and community involvement. Sustainable development of the environment and economic growth can also be achieved by encouraging the use of decentralized treatment systems, recuperation of resources, and the incorporation of natural solutions. To move forward, government agencies, non-governmental organizations, researchers, and the communities themselves must work together. The introduction of proven sludge management solutions can be greatly aided by international alliances and knowledge-sharing platforms that facilitate the exchange of best practices.

In developing countries, the effective application of sustainable sludge management strategies guarantees the preservation of human health and the environment while also enhancing the general prosperity and well-being of these communities, even in the face of obstacles brought on by fast urbanization, population growth, and resource scarcity. To ensure responsible sludge management in the future and provide a cleaner, more sustainable environment for future generations, we must keep funding research, innovation, and capacity-building projects.

## References


Ahmad T, Ahmad K, Alam M (2016) Sustainable management of water treatment sludge through 3 'R' concept. J Cleaner Prod 124:1–13

Babatunde AO, Zhao Y (2007) Constructive approaches toward water treatment works sludge management: an international review of beneficial reuses. Crit Rev Environ Sci Technol 37(2):129–164

Barberio G, Cutaia L, Librici V (2013) Treatment and disposal of sewage sludge: comparative life cycle assessment on Italian case study extended abstract. Environ Eng Manag J 12:7–10

Barrios J et al (2001) Application of peracetic acid to physicochemical sludge to reduce its microbial content. In: 6th European and organic residuals conference, workshop and exhibition

Campbell H (2000) Sludge management–future issues and trends. Water Sci Technol 41(8):1–8

Christodoulou A, Stamatelatou K (2016) Overview of legislation on sewage sludge management in developed countries worldwide. Water Sci Technol 73(3):453–462

Cieślik BM, Świerczek L, Konieczka P (2018) Analytical and legislative challenges of sewage sludge processing and management. Monatshefte für Chemie-Chem Monthly 149:1635–1645

Đurđević D, Žiković S, Blecich P (2022) Sustainable sewage sludge management technologies selection based on techno-economic-environmental criteria: case study of Croatia. Energies 15(11):3941

Egan M (2013) Biosolids management strategies: an evaluation of energy production as an alternative to land application. Environ Sci Pollut Res 20:4299–4310

Enebe NL et al (2023) Biogas and syngas production from sewage sludge: a sustainable source of energy generation. Methane 2(2):192–217

Farzadkia M, Bazrafshan E (2014) Lime stabilization of waste activated sludge

Fijalkowski K, Kacprzak M, Rorat A (2014) Occurrence changes of Escherichia coli (including O157: H7 serotype) in wastewater and sewage sludge by quantitation method of (EMA) real time—PCR. Desalination Water Treatment 52(19–21):3965–3972

Fungaro DA, Silva M (2014) Utilization of water treatment plant sludge and coal fly ash in brick manufacturing. Am J Environ Prot 2(5):83–88

Fytili D, Zabaniotou A (2008) Utilization of sewage sludge in EU application of old and new methods—a review. Renew Sustain Energy Rev 12(1):116–140

Goldbold P et al (2003) The potential reuse of water utility products as secondary commercial materials. WRc Report No. UC, p 6081

Gomes SDC et al (2019) Progress in manufacture and properties of construction materials incorporating water treatment sludge: a review. Resourc Conserv Recycl 145:148–159

Jakobsson C (2012) Sustainable agriculture. Baltic University Press

Jiménez B et al (2004) Sustainable sludge management in developing countries. Water Sci Technol 49(10):251–258

Kacprzak M, Stańczyk-Mazanek E (2003) Changes in the structure of fungal communities of soil treated with sewage sludge. Biol Fertility Soils 38:89–95

Krutwagen B, Kortman J, Verbist K (2008) Inventory of existing studies applying life cycle thinking to biowaste management

Mansor MI, Fatehah MO (2022) Sludge waste management techniques and challenges in water resources supply. In: Handbook of solid waste management: sustainability through circular economy. Springer, p 877–900

Metcalf E et al (2013) Wastewater engineering: treatment and resource recovery. McGraw-Hill, New York, NY, USA

Olszewski JM et al (2013) The effect of liming on antibacterial and hormone levels in wastewater biosolids. J Environ Sci Health Part A 48(8):862–870

Production SS (2021) Disposal from urban wastewater. Eurostat

Rhoades J (1997) Sustainability of irrigation: an overview of salinity problems and control strategies. US Salinity Laboratory, Riverside, California

Samolada M, Zabaniotou A (2014) Comparative assessment of municipal sewage sludge incineration, gasification and pyrolysis for a sustainable sludge-to-energy management in Greece. Waste Manage 34(2):411–420

Siebielska I (2014) Comparison of changes in selected polycyclic aromatic hydrocarbons concentrations during the composting and anaerobic digestion processes of municipal waste and sewage sludge mixtures. Water Sci Technol 70(10):1617–1624

Świerczek L, Cieślik BM, Konieczka P (2018) The potential of raw sewage sludge in construction industry–a review. J Cleaner Prod 200:342–356

Tyagi VK, Lo S-L (2013) Sludge: a waste or renewable source for energy and resources recovery? Renew Sustain Energy Rev 25:708–728

Wiechmann B et al (2013) Sewage sludge management in Germany. Umweltbundesamt

Wójtowicz A et al (2013) Modelowe rozwiązania w gospodarce osadowej. Izba Gospodarcza Wodociągi Polskie

Young A (2000) Land resources: now and for the future. Cambridge University Press

# Design, Estimation and Implementation Strategy of Wastewater Treatment Plant—A Case Study at Centurion University of Technology and Management, Odisha

Prafulla Kumar Panda, Rajib Kumar Majhi, and Laxmidhar Behera

**Abstract**

Water scarcity nowadays not only causes problems for humans, but it also poses serious challenges in the case of eco-system growth. As a result, there is a serious need to investigate any alternative uses for used water in both domestic and agricultural activities. The present investigation used a robust approach to build a wastewater treatment plant at Centurion University of Technology and Management (CUTM), Paralakhemundi campus in Odisha and use it for agricultural purposes. The study area is located in the Paralakhemundi block of the Gajapati district of Odisha, where water is scarcer than in other parts of the district. It has been noticed that at the time of summer, all water bodies dry up, and the people of the district depend largely on this contaminated water which causes several health problems. The current study was carried out by designing and constructing a 750 $m^2$ area wastewater treatment plant. The campus includes three staff quarters, five student's hostels along with three hostel messes, with a total population estimated to be more than 1500. Wastewater from the aforementioned structures is collected, and the plant treats it mostly for agricultural use. A bar and screen chamber, an oil and grease tank, a collecting tank, a sedimentation tank, a filter tank, and a control panel make up the treatment plant. The daily water supply is 279 cum, and the average garbage generated per hour is 9.29 cum. The design has been done as the highest flow capacity is to be 27.89 cum/hr when the peak factor is 3. The anticipated cost amounts to be Rs. 25,75,539. Successfully treated water meets the appropriate standards meant for agricultural and gardening practises. A total of 10 hectares of field crops, horticultural crops, plantation crops, an herbal garden, grass, and a playground are watered by this treatment facility.

**Keyword**

Wastewater treatment · Bar and screen chamber · Oil and grease tank · Collection tank · Sedimentation tank · Filtration tank · Control panel

## 1 Introduction

Removal of contaminants from the waste water in the process of primary treatments like physical, chemical and biological process from household wastes is known as sewage treatment. The process is environment-friendly treated waste water which is also known as treated effluent. The by-product released from the sewage treatment is mainly waste or slurry type which is known as sewage slurry that has to undergo for further process to make it suitable for the disposal purposes for different land applications. Removing pollutants from wastewater requires the use of energy, chemicals, and the treatment of surplus sludge, which is running more cost that, cannot be countered by economically productive. Therefore the waste water treatment plant (WWTP) has some environmental effects like direct or indirect emission of greenhouse gases (GHG) released from biological and energy consumption processes respectively (Dhote et al. 2012; Sonaje 2015). Moreover, the disposal of solid wastes or sludge contains some toxic chemicals which may also affect the land (Meneses et al. 2015; Meneses et al. 2016; Revollar et al. 2020).

In sewage treatment plants (STP), the waste water released to the drain before releasing it into the environment (Rangwala et al. 2005). The waste water treatment plants are mainly focused on the treatment of contaminated water including solid wastes. The process of clearing the water from the solid materials, screening and settlement is

P. K. Panda (✉) · R. K. Majhi · L. Behera
Centurion University of Technology and Management, Bhubaneswar, Odisha, India
e-mail: prafullapanda@cutm.ac.in

© The Author(s), under exclusive license to Springer Nature Switzerland AG 2025

A. Mandpe and M. P. Shah (eds.), *Wastewater Sludge Nexus*, Sustainable Landscape Planning and Natural Resources Management, https://doi.org/10.1007/978-3-031-45758-6_15

known as primary treatment (Meneses et al. 2015; Meneses et al. 2016; Rangwala et al. 2005) which takes place in the first stage of the treatment to remove larger debris particles. After the removal of large items still the waste water contains organic material which can contaminate the water bodies by decreasing the oxygen availability if comes in contact directly without the proper treatment (Drolka et al. 2001; Ibeje and Okoro 2013; Rai 2005). A framework has been presented by considering the performance of the treatment phases like primary, secondary, and tertiary which are based upon some proposed equations and the condition index for each phase was calculated (Qasem et al. 2009; Kaddoura et al. 2020).

By using the process of wastewater treatment contaminated water can be cleaned and used in different sectors like agriculture, vegetation, plantation, gardening etc. as well as the treated water also may be used for toilet and various domestic uses (Asiwal et al. 2016). By treating the polluted water the cost of transportation can be minimised and the treated water can be stored in tanks for uses at time of water scarcity. The treatment process also can be carried out and recycled for the ground water, surface runoff and polluted for the wellbeing of society (Das et al. 2007; Hoak 1961). The site is located at Centurion University of Technology and Management which is at Alluri Nagar, Paralakhemundi that covers the study area of about 10 ha. There are five multi-storied buildings situated on the campus with proper road alignment and a well-planned drainage system is also connected to each building through pipe lines in a proper drainage process which further connects to the main drain for the purpose of outlet. New buildings are under construction which may lead to increase in the amount of waste water released in the nearby future. In the present work of study a large of waste deposition tank, gardens, a big forest and hill area nearby and a big multipurpose playground are situated on site. Wastewater sample are also collected from the bathrooms of student hostels and staff quarters as well as from kitchen effluents.

## 1.1 Objective

The bellows given are the objectives of the study:

- To identify the physical, chemical and biological characteristics of the domestic wastewater from hostels of Centurion University, Paralakhemundi Campus
- Comparative analysis with respect to the prescribed standards
- To design the sewage treatment plant and cost estimation
- To demonstrate its implementation in agricultural activities

## 1.2 Location

The wastewater treatment facility is situated within the premises of Centurion University of Technology and Management, Paralakhemundi Campus, Gajapati district, Odisha, India. This facility has been designed to cater to the

**Fig. 1** Site location for waste water treatment plant

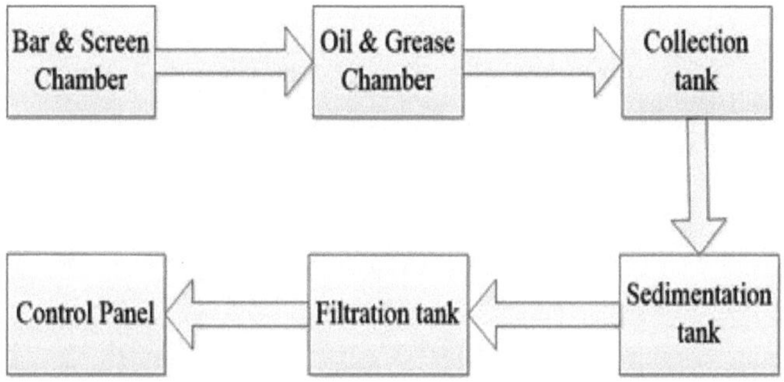


**Fig. 2** Schematic representation of the model

needs of a population of approximately 1550 individuals. The treatment plant occupies an area of 750 $m^3$ (Fig. 1).

## 2 Methodology

In order to remove larger-size solids, the wastewater is initially permitted to pass through the bar and screen chamber, which has 21 numbers of 20 mm thick bars and have a width of 1700 mm. For the removal of oil and grease from the wastewater a grease trap is provided in the tank. Following that, the water and oil that are free of coarse materials are collected in a collection tank that measures 12 m × 4.5 m × 4 m. The collected water is once more allowed to flow into similar size sedimentation tank for a four-hour holding period (Fig. 2).

After the process of separating coarse particles and sedimentation is finished, the water is filtered using a slow sand filter in a filtration tank with a size of 11.12 m × 5.5 m × 4 m. After the aforementioned processes are finished, the water is made ready for use by an electric motor that is located in the control panel chamber of similar size as that of filtration tank.

## 3 Experimental Investigation

This section briefly explains about the experimental investigations of the model which has been carried out.

### 3.1 Development of Contour Maps

Contour lines have been derived using Google Earth Pro Platform, TCX converter and finally represented in quick grid system. Contour maps assist in the selection of a location for the wastewater treatment plant and determining the slope of the land for the placement of wastewater treatment units such as sedimentation, filtration, and aeration at the proper level (Fig. 3).

### 3.2 Tests on Raw Water

Different tests of raw sewage water (Total hardness, BOD, pH, COD, O&G, TDS, & TSS) have been tested by considering BIS (IS: 4733-1972) and the corresponding results articulated in Table 1.

### 3.3 Design of Waste Water Treatment Plant

**Plant capacity**

Maximum daily demand per head = 180 lit/day

Average water supply per day = 180 × 1550 = 279000 lit = 279 cum

Average sewage produced/ day = 85% of supplied water = 0.85 × 279 = 223 $m^3$

Average sewage generated per hour = 223/24 = 9.29 cum/hr

Peak factor = 3

Max. design flow = 9.29 × 3 = 27.85 cum/hr

**Design of screen chamber**

Max. discharge = $30m^3/h$

Average discharge = $10m^3/h$

Average velocity = 0.3 × 40 = $12m/h/m^2$

Required C/S area = Max. flow/Avg. velocity = 30/12 = 2.5 $m^2$

Required depth of liquid = 1 m

Velocity at max. flow through screen = 1.6m/sec

Clear area = 2.5/1.6 = $1.3m^2$

Number of clear spacing = 1.3/0.02 = 65 nos.

Channel width = (65 × 20) + (67 × 6) = 1702 mm

Screen width = 1700mm

**Dimensioning of collection pit**

Retention time = 8 h

Design flow = 27.87 $m^3/h$

Collection sump capacity = 8 × 27.87 = 222.96 $m^3$

Assumed depth of liquid = 4 m

**Fig. 3** Contour map of the study area

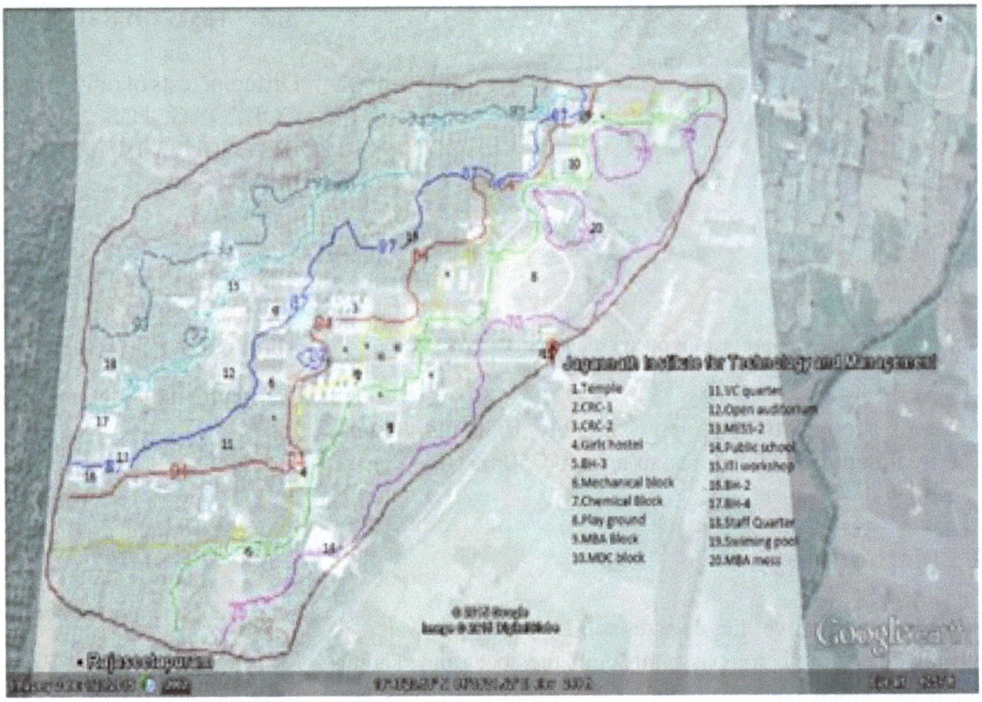


**Table 1** Tests on wastewater

| Locations | Total hardness (ppm) | BOD | pH | COD | Oil and grease (mg/l) | TDS | TSS |
|---|---|---|---|---|---|---|---|
| Hostel-2,4 and Mess-2 | 265.3 | 5.75 | 7.72 | 513 | 5.62 | 2851 | 222 |
| Hostel-5 and Mess-1 | 432.3 | 5.75 | 7.02 | 507 | 7.54 | 2538 | 201 |
| Hostel-3 | 256.8 | 11.41 | 7.80 | 502 | 6.83 | 2369 | 213 |
| Hostel-1 | 260.5 | 5.75 | 7.61 | 521 | 6.50 | 2808 | 224 |
| Girls hostel (B.Tech) | 346.7 | 5.75 | 7.06 | 568 | 7.42 | 3272 | 198 |
| Staff quarter-1 | 271.22 | 11.5 | 7.21 | 533 | 6.12 | 3058 | 206 |
| Staff quarter-2 | 321 | 11.5 | 6.52 | 544 | 8.06 | 3184 | 245 |

Collection pit area = 222.96/4 = 55.74 $m^2$

Let it be a rectangle tank

$L : B = 2.5 : 1$

$L = 12$ m & $B = 4.7$ m

Provide tank size is 12 m × 4.7 m × 4 m

**Sizing calculation of Sedimentation Tank**

Quantity of water to be treated in 24 h = 223000 L = 22.3 × $10^4$ L

Quantity of water to be treated during the detention time period of 4 h

$= \frac{223000}{24} \times 8 = 37166.6 \text{ L} = 38000 \text{ L}$

The capacity of tank required = 222.96 cum

Length of tank = 13 m

Width of tank = 4.3 m

**Sizing calculation of Slow Sand Filter**

Maximum Daily Demand = 223 cu.m

Rate of filtration per day = (150 × 24) L/$m^2$/day

Total Surface Area of Filter required $= \frac{\text{maximum daily demand}}{\text{rate of filtration per day}} = \frac{223 \times 10^3}{150 \times 24} = 62$

Assume Length, $L = 2B$

$2B^2 = 62$

or, $B = 5.56$ m

& $L = 2B = 2 \times 5.56 = 11.12$ m

## 3.4 Waste Water Treatment Flow Diagram

A flow diagram is shown in Fig. 4 showing the process of treatment in the plant in an appropriate way.

**Process Description**

The activated sludge principle is used by the wastewater treatment facility's impounding process (also known as the SBR process). In this way, biomass is created by floating microorganisms (activated sludge) that remove contaminants from wastewater. The pre-treatment chamber is where the wastewater enters. The sedimentation of coarse material requires the presence of this chamber. Mechanical pretreatment is another name for it. With the aid of a natural gradient, coarse materials fall to the bottom and the pre-treated effluent flows into the following chamber. The buffer tank, which is the second chamber, serves as a holding area until the effluent is injected into the reactor.

**Reactor**

Three cycles of eight hours each day are used in the reactor to remediate wastewater. Phases of a cycle are as follows:

1. Filling/Loading Phase
2. Reaction
3. Settling phase
4. Decanting
5. Sludge Removal

With the aid of the air lift pumps, a specific volume of water will first be added to the reactor. Microorganisms are needed to break down the organic waste in the effluent, which is how

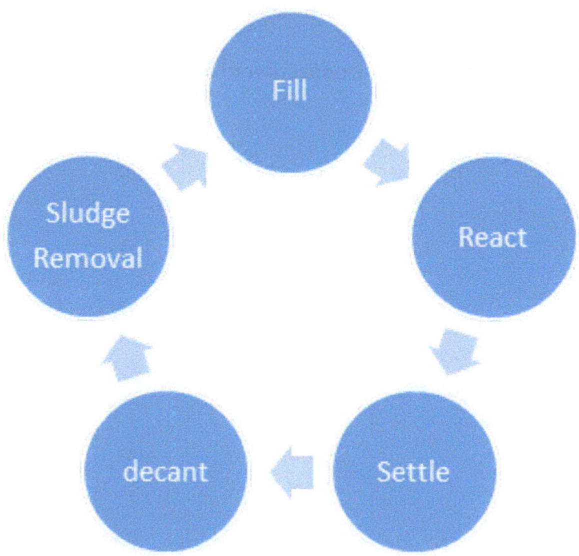


**Fig. 4** Treatment flow diagram

the effluent is treated. The controlled infusion of air causes the microorganisms to proliferate, which quickly reduces the amount of organic materials in the effluent. Based on the SBR-timing theory, the Control Unit regulates four magnetic valves and a compressor (Blower) to operate the plant. The air lift principle is used in the loading, sludge return, and treated water disposal pipes. The matching valve opens to start the water transport or aeration process. The blower starts after two seconds of waiting. In the opposite situation, the valve will close after a two-second wait after the blower has shut down.

**Control Cabinet**

To control all the performance of the treatment unit electrically a control cabinet is established which contains many controlling units such as blower, PLC, magnetic valves, and pressure hose. The whole control system is mounted on a strong and solid concrete base. All control cabinets are manufactured from CRCA sheet and powder coated of adequate gauge confirming to IP 5 protection. The detailed view of the control unit is shown in Fig. 5.

**Solenoid Valves**

The size of valves has been use ¾″ to 2″

The technical data of the magnetic valves are shown in Fig. 6.

Working Pressure: 0–12 bar, Medium Temperature: −10° to +80°, Environment temperature: −10° to +40°, Viscosity for hot and cold water, Oil and air: 20 mm/Sec

**Membrane Disk Diffuser**

For the aeration in the reactor membrane disk diffusers have been used. The support construction of the disk diffusor is an EPDM disk coated with PTEF. The membrane disk diffusers have been installed in pipe and the diameters are free member diameter of 300 mm (Fig. 8). Empty membrane disk diffusers are resistant to soiling and produce fine bubbles with a diameter of >2 mm. The big advantage of disk diffusers is that entire diffuser surface is on the level, which ensures that, the whole area is evenly aerated. The disk diffusers have a control range of 0–10 N/m$^3$/hr (Figs. 7 and 9).

## 3.5 AutoCAD Plan and the Reinforcement Details of the Model

The detailed plan and reinforcement detail of the proposed model is prepared using AutoCAD drafting (Fig. 10). It shows different components of the model. Similarly, the design details of the different chambers in the waste water treatment plant are provided in Table 2.

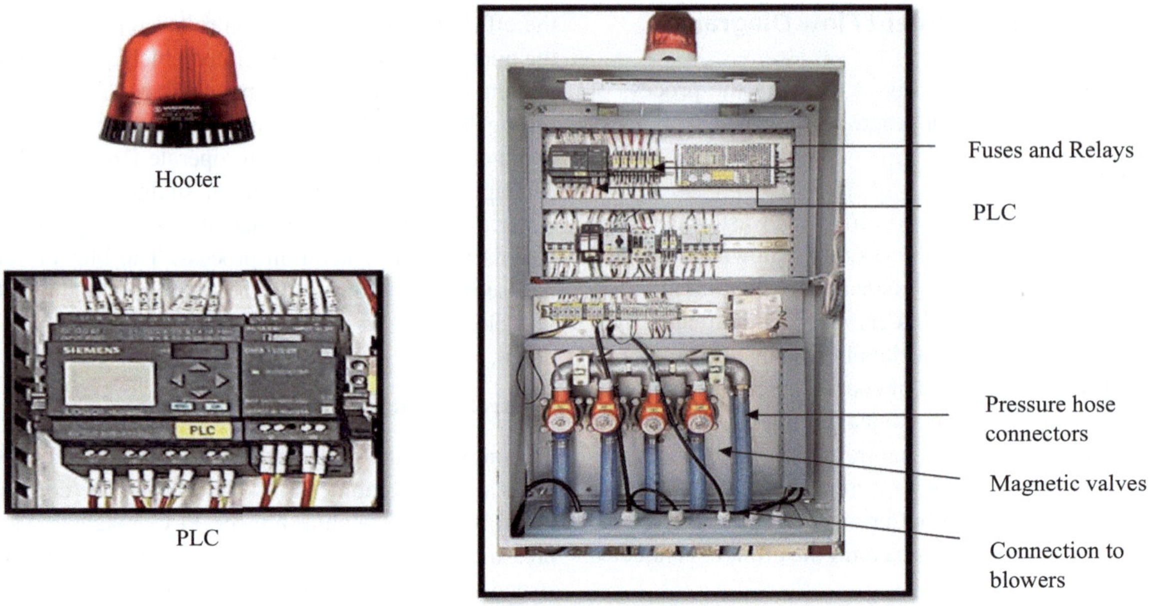


**Fig. 5** Detailed view of the control unit

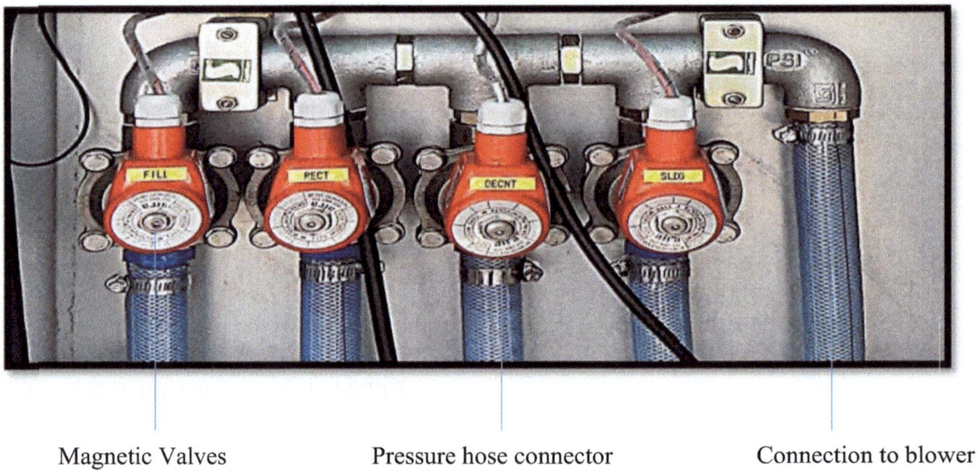


**Fig. 6** Magnetic valves

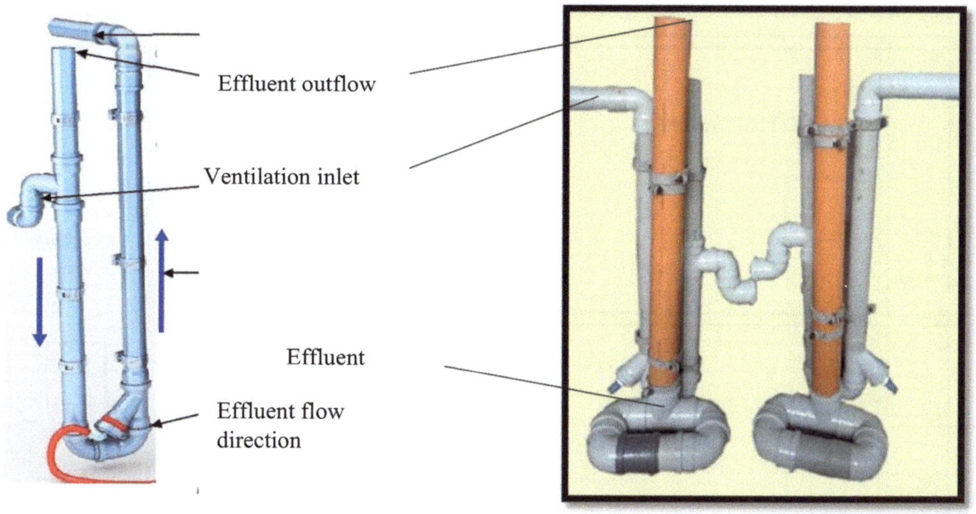


**Fig. 7** Air lift technology

**Fig. 8** Diffuser

**Fig. 9** Air blowers

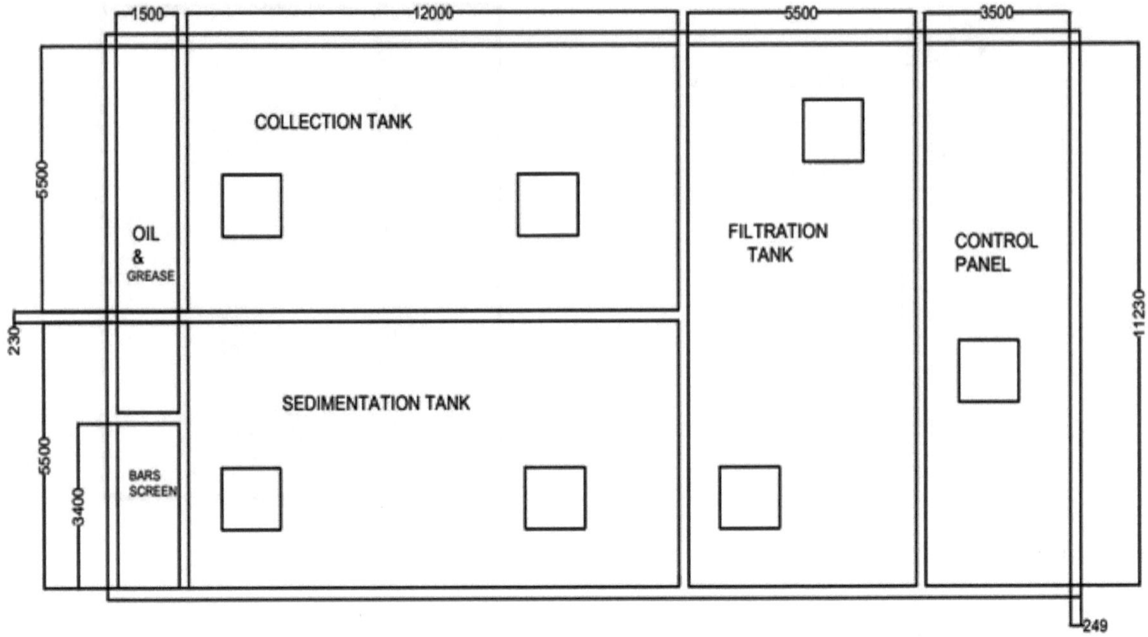


**Fig. 10** AutoCAD plan of the model

### 3.6 Model Estimation

The detailed description of approximated cost of Rs. 25,75,539.20 as estimated for the construction of the model is abstracted in Table 3.

### 3.7 Test of the Treated Effluent

For agricultural uses treated water has been tested and comparative analysis has been done as per standard of BIS (IS: 4733-1972) presented in Table 4.

## 4 Implementation Strategy and Utilisation

After treatment of effluent the results have been found as BOD 7.67 mg/l, COD 89 ppm, O&G 4.32, Chlorides 598.86, pH 7.85, TDS 2801.23 ppm, and TSS 192 ppm satisfying the standards IS: 4733-1972 for agricultural practices. Ten hectares of land, including field crops, horticultural crops, plantation crops, a herbal garden, a lawn, and a playground, are supplied with treated water as shown in Figs. 11, 12, 13, 14, 15, 16, 17.

## 5 Conclusion

The current study illustrates that the construction model of wastewater treatment plant can be used to treat the ground water for drinking purposes and to store it for future purposes. The irrigation problem can be mitigated by using the treated waste water and it can prevent the loss of water. On the other hand, the treatment process can also help the human and aquatic life in a better manner to reuse the treated water. Model of the treatment plant is less maintenance and operational cost-oriented and also serves an eco-friendly environment. The model also focuses on the challenges of low level of groundwater and the wasted surface runoff. The wastewater is collected from different sources on the campus like hostel buildings, staff quarters, toilets, and kitchen and used in different agricultural activities after undergoing the treatment processes in the treatment plant. The treatment plant consists of different chambers like bar and screen chamber, oil and grease tank, collection tank, sedimentation tank, filtration tank and control panel for different processes. Nearly 279 cum water is supplied per day for various purposes and average 9.29 cum/hr wastes are generated. The plant has been designed with a maximum flow capacity of 27.89 cum/hr by considering the peak value as 3. The total estimated cost comes to Rs. 25,75,539 for the construction of the plant.

**Table 2** Design of WWTP

| Sl. | Chambers | Reinforcement details |
|---|---|---|
| 1 | Bar and screen chamber | Long walls:<br>Earth pressure in the absence of water pressure and from outside:<br>Provide bars of 10 mm Ø bars at 100 mm centre-to-centre<br>Water pressure in the absence of earth pressure and from outside:<br>Provide bars of 10 mm Ø bars at 100 mm centre-to-centre<br>Provide bars of 8 mm Ø bars at 170 mm centre-to-centre<br>Short walls:<br>Water pressure in the absence of earth pressure and from outside:<br>Provide bars of 10 mm Ø bars at 100 mm centre-to-centre<br>Design of bottom 1 m:<br>Provide bars of 10 mm Ø bars at 265 mm centre-to-centre<br>Provide bars of 8 mm Ø bars at 170 mm centre-to-centre<br>Saturated soil pressure in the absence of water pressure from inside:<br>Provide bars of 10 mm Ø bars at 100 mm centre-to-centre<br>Bottom one meter:<br>Provide bars of 10 mm Ø bars at 265 mm centre-to-centre<br>Provide bars of 8 mm Ø bars at 170 mm centre-to-centre |
| 2 | Oil and grease tank | Water pressure in the absence of earth pressure and from inside:<br>Provide bars of 10 mm Ø bars at 170 mm centre-to-centre<br>Short wall:<br>Provide bars of 10 mm Ø bars at 100 mm centre-to-centre<br>Design of bottom 1 m:<br>Provide bars of 10 mm Ø bars at 265 mm centre-to-centre<br>Provide bars of 8 mm Ø bars at 170 mm centre-to-centre<br>Saturated soil pressure in the absence of water pressure:<br>Provide bars of 10 mm Ø bars at 115 mm centre-to-centre<br>Bottom one meter:<br>Provide bars of 10 mm Ø bars at 265 mm centre-to-centre<br>Provide bars of 8 mm Ø bars at 170 mm centre-to-centre |
| 3 | Sedimentation and collection tank | Water pressure in the absence of earth pressure and from inside:<br>Provide bars of 10 mm Ø bars at 170 mm centre-to-centre<br>Saturated soil pressure in the absence of water pressure:<br>Provide bars of 10 mm Ø bars at 100 mm centre-to-centre<br>Design of bottom 1 m:<br>Provide bars of 10 mm Ø bars at 260 mm centre-to-centre<br>Provide bars of 8 mm Ø bars at 170 mm centre-to-centre<br>Saturated soil pressure in the absence of water pressure:<br>Provide bars of 10 mm Ø bars at 125 mm centre-to-centre<br>Bottom one meter:<br>Provide bars of 10 mm Ø bars at 265 mm centre-to-centre<br>Provide bars of 8 mm Ø bars at 170 mm centre-to-centre |
| 4 | Filtration tank | Long Walls:<br>Earth pressure in the absence of water pressure and from outside:<br>Provide bars of 10 mm Ø bars at 100 mm centre-to-centre<br>Water pressure in the absence of earth pressure and from inside:<br>Provide bars of 10 mm Ø bars at 170 mm centre-to-centre<br>Short Walls:<br>Water pressure in the absence of earth pressure and from inside:<br>Provide bars of 10 mm Ø bars at 100 mm centre-to-centre<br>Design of bottom 1 m:<br>Provide bars of 8 mm Ø bars at 170 mm centre-to-centre<br>Earth pressure in the absence of water pressure and from outside:<br>Provide bars of 10 mm Ø bars at 100 mm centre-to-centre<br>Bottom one meter:<br>Provide bars of 8 mm Ø bars at 170 mm centre-to-centre |

(continued)

**Table 2** (continued)

| Sl. | Chambers | Reinforcement details |
|---|---|---|
| 5 | Control panel | Water pressure in the absence of earth pressure and from inside:<br>Provide bars of 10 mm Ø bars at 100 mm centre-to-centre<br>Design of bottom 1 m:<br>Provide bars of 10 mm Ø bars at 220 mm centre-to-centre<br>Provide bars of 10 mm Ø bars at 125 mm centre-to-centre<br>Saturated soil pressure in the absence of water pressure:<br>Provide bars of 10 mm Ø bars at 100 mm centre-to-centre<br>Bottom one meter:<br>Provide main bars of 10 mm Ø bars at 260 mm centre-to-centre<br>Provide distribution bars of 8 mm Ø bars at 170 mm centre-to-centre |
| 6 | Base slab | Provide main bars of 10mm Ø bars at 100mm centre-to-centre<br>Provide distribution bars of 8 mm Ø bars at 100 mm centre-to-centre |

**Table 3** Cost estimation as per CPWD and Odisha PWD works manual (CPWD Specifications 2009; Schedule of Rates 2022)

| Sl. no. | Description of item | Quantity | Rate/unit | Amount (in Rs.) |
|---|---|---|---|---|
| 1 | Earthwork in excavation in $m^3$ | 1527.34 | 55.36 | 84569.34 |
| 2 | 1:2:4 CC with granite chips in 40 mm of foundation in $m^3$ | 31.80 | 3475.70 | 110561.70 |
| 3 | 1:2:4 RCC excluding steel bars but including shuttering for side walls in $m^3$ | 76.825 | 7558.78 | 580697.24 |
| 4 | Mild steel reinforcement work in side walls in quintal | 20.80 | 4100 | 85403.00 |
| 5 | Brickwork done in 1:4 mortar in $m^3$ | 50.22 | 3844.02 | 193009.24 |
| 6 | 20 mm thick cement mortar plastering in $m^2$ | 217.80 | 112.40 | 24498.10 |
| 7 | 1:2:4 RCC excluding steel bars for base slab in $m^3$ | 86.60 | 7558.80 | 654743.24 |
| 8 | Mild steel reinforcement work in base slab in quintal | 55.25 | 4100 | 22664 |
| 9 | 1:2:4 RCC excluding steel bars for top slab in $m^3$ | 68.10 | 7558.80 | 514754.28 |
| 10 | Mild steel reinforcement work in top slab in quintal | 24.55 | 4100 | 100655.00 |
| | | | Total | 25,75,539.20 |

**Table 4** Result of the effluent

| Parameters | Values |
|---|---|
| 5Days B.O.D (mg/l) | 7.67 |
| C.O.D (ppm) | 89 |
| Oil and Grease (ppm) | 4.32 |
| Chlorides | 598.86 |
| pH | 7.85 |
| T.D.S (ppm) | 2801.23 |
| T.S.S (ppm) | 192 |

**Fig. 11** Treatment plant

**Fig. 12** Raw water towards treatment plant

**Fig. 13** Gardening

**Fig. 14** Field crop

**Fig. 15** Lawn

**Fig. 16** Gardening

**Fig. 17** Horticulture nursery

The treated water maintains the standards as per the requirement for the use in various agricultural activities, gardening purposes. Ten hectares of land, including field crops, horticultural crops, plantation crops, a herbal garden, a lawn, and a playground, are supplied with treated water.

## References

A book, "Waste Water engineering by S.K. Garg"

A book, "Advanced concrete Design by N. Krishna Raju"

A code book for water tank design "IS-3370-2-2009"

Agrawal RK, Pandey PK (2005) Productive recycling of basic oxygen furnace sludge in integrated steel plant. J Sci Indus Res 64:702–706

Asiwal RS, Sar SK, Singh S, Sahu M (2016) Wastewater treatment by effluent treatment plants. SSRG Int J Civil Eng (SSRG–IJCE) 3:29–35

CPWD Specifications (2009) Central public works department. Government of India

Das B, Prakash S, Reddy PSR, Misra VN (2007) An overview of utilization of slag and sludge from steel industries. Resourc Conserv Recycl 50:40–57

Dhote J, Ingole S, Chavhan A (2012) Review on wastewater treatment technologies. Int J Eng Res Technol 1:1–10

Drolka M, Plazl I, Koloini T (2001) The results of mathematical model and pilot plant research of wastewater treatment. Chem Biochem Eng Q 15:71–74

The Environment (Protection) Rules (1986) Ministry of environment and forests (department of environment, forest and wildlife) notification, New Delhi. https://parivesh.nic.in/writereaddata/ENV/THE%20ENVIRONMENT.pdf

Hoak RD (1961) Steel mill sludge recovery. Water Pollut Control Federation 33:1050–1055

Ibeje AO, Okoro BC (2013) Mathematical modeling of cassava wastewater treatment using anaerobic Baffled Reactor. Am J Eng Sci 2:128–134

IS 4733 (1972) Indian standard methods of sampling and test for sewage effluents. Indian Standard Institution, New Delhi
Kaddoura K, Zayed T, Chen Z, Qasem A, Atabay S (2020) Optimized maintenance model for wastewater treatment plants. Can J Civil Eng 47:929–38
Meneses M, Concepcion H, Vrecko D, Vilanova R (2015) Life cycle assessment as an environmental evaluation tool for control strategies in wastewater treatment plants. J Clean Prod 107:653–661
Meneses M, Concepcion H, Vilanova R (2016) Joint environmental and economical analysis of wastewater treatment plants control strategies: a benchmark scenario analysis. Sustainability 8:360
Qasem A, Zayed T, Chen Z (2009) Condition rating framework for wastewater treatment plants. In: Proceedings of annual general conference for the Canadian society for civil engineering (CSCE), Canada, pp 335–344
Rai JPN (2005) Mathematical model for phytoremediation of pulp and paper industries wastewater. JSIR 64:717–721
Rangwala SC, Rangwala KS, Rangwala PS (2005) Water supply and sanitary engineering. Charotar Publishing House, Gujarat, India
Revollar S, Vilanova R, Vega P, Francisco M, Meneses M (2020) Wastewater treatment plant operation: simple control schemes with a holistic perspective. Sustainability 12:768
Schedule of Rates (2022) Works department. Government of Odisha
Sonaje NP (2015) Modeling of wastewater treatment plant design for pulp and paper industry: a review. Int J Civil Struct Environ Infrastruct Eng Res Develop (IJCSEIERD) 5:59–68

MIX
Papier aus verantwortungsvollen Quellen
Paper from responsible sources
FSC® C105338

If you have any concerns about our products,
you can contact us on
**ProductSafety@springernature.com**

In case Publisher is established outside the EU,
the EU authorized representative is:
**Springer Nature Customer Service Center GmbH**
**Europaplatz 3, 69115 Heidelberg, Germany**

Printed by Libri Plureos GmbH
in Hamburg, Germany